Flexible Piezoelectric Energy Harvesters and Sensors

Flexible Piezoelectric Energy Harvesters and Sensors

Bin Yang
Zhiran Yi
Chengkuo Lee

WILEY-VCH

Authors

Prof. Bin Yang
Shanghai Jiao Tong University
Department of Micro/Nano Electronics
No. 800 Dongchuan Road
200240 Shanghai
China

Dr. Zhiran Yi
Shanghai Jiao Tong University
School of Mechanical Engineering
No. 800 Dongchuan Road
200240 Shanghai
China

Prof. Chengkuo Lee
National University of Singapore
Electrical and Computer Engineering
4 Engineering Drive 3
117583 Singapore
Singapore

Cover Image: © Yuichiro Chino/Getty Images

Library of Congress Card No.: applied for

British Library Cataloguing-in-Publication Data
A catalogue record for this book is available from the British Library.

Bibliographic information published by the Deutsche Nationalbibliothek
The Deutsche Nationalbibliothek lists this publication in the Deutsche Nationalbibliografie; detailed bibliographic data are available on the Internet at <http://dnb.d-nb.de>.

Print ISBN: 978-3-527-34934-0
ePDF ISBN: 978-3-527-83307-8
ePub ISBN: 978-3-527-83301-6
oBook ISBN: 978-3-527-83311-5

Typesetting Straive, Chennai, India
Printing and Binding CPI Group (UK) Ltd, Croydon, CR0 4YY

C114668_090922

Contents

Preface

With the rapid development of Artificial Intelligent Internet of Things (AIoT) technologies based on a large number of wireless sensing nodes, battery is becoming the main limitation of sensor node power supply due to its limited service life, difficult maintenance, environmental pollution, and potential security risks. Piezoelectric energy-harvesting technology, an old but promising technology, is being developed as a replacement of the battery in many areas so that one can obtain a sustainable power supply. However, the frequency of most ambient vibration sources is very low (less than 200 Hz) compared to most traditional silicon-based microelectromechanical systems. High structural stiffness of piezoelectric energy harvesters makes it difficult to obtain effective and stable output at low frequency. At the same time, there is an increasing demand for flexible piezoelectric energy-harvesting and energy-sensing technologies with the emergence of flexible electronics technology due to their high compatibility with some extreme or complex objects. At present, flexible piezoelectric energy-harvesting and energy-sensing technologies are being used in new ones for powering wearable or implantable electronics. This book addresses the design methods, flexible piezoelectric materials, different working mechanisms, micromachining processes, and the application of flexible energy harvesters for wearable and implantable devices.

In particular, cutaneous activities mainly derived from motion in limbs, pulse waves in arteries, or disturbances associated with breathing are studied. The subtle changes caused by cutaneous activities can be measured via the sensors attached to the skin, which is valuable for in situ monitoring of vital signs in healthcare and recognizing different gestures in man–machine interaction. Conventional sensors face some critical demerits, such as rigid structures and high power consumption, which cause inconvenience in daily life for medical applications. These sensors detect physiological signals produced by human activities, but the power supply is critical for their operation, which limits their wearable or implantable application. Therefore, some flexible sensors should be reviewed again and summarized in detail for wearable and implantable applications.

To address researchers, scientists, and application engineers from different backgrounds, this book systematically and comprehensively demonstrates the global progress and achievements in the field of flexible piezoelectric energy-harvesting and energy-sensing technologies. This book covers the fundamentals, developments, discussions in the design and fabrication of energy-harvesting, energy-sensing, and their corresponding applications. The working principle of piezoelectric devices is introduced in Chapter 1. The requirement of flexible energy harvesters and sensors is demonstrated. Chapters 2 and 3 report the design methods and fabrication processes of flexible energy-harvesting techniques. Low-frequency cantilever-based energy harvesters, and free- and forced-vibration nonlinear energy harvesters are introduced in Chapters 4–6. Chapter 7 demonstrates water vortex shedding-induced and aeroacoustics-driven jet-stream wind-energy harvesters. Some examples of wearable and implantable applications are described in Chapters 8 and 9. The working principle and its application for body sensor network are covered in Chapter 10. In Chapter 11, we introduce the approaches for sensing data from the flexible sensor to the fullest with the assistance of machine-learning algorithms. The existing challenges of current intelligent systems and future trends are provided, which give a glimpse of further developments in the 5G/internet of things (IoT) era.

The book can be an introduction for undergraduates, graduates, and developers who are working in this area. This book will bring some new ideas for some researchers and engineers in the flexible energy-harvesting and energy-sensing fields.

Shanghai Jiao Tong University
Minhang Campus
01 October 2021 *Bin Yang*

Foreword

Piezoelectric energy-harvesting and energy-sensing technologies are not new. The dilemma of rapidly developing piezoelectric energy harvesters or sensors is that the performance is contradictory to the structural flexibility. To make a contribution that would obtain a trade-off for this contradiction, we provide a framework for the recent advances and also show a blueprint from fundamental structural design to the practical fabrication and from standalone devices to the integrated systems for promoting their practical applications. To achieve this goal, we needed to cooperate closely with every editor, and we also needed to convince leading scientists to take time from their busy schedules to provide comments for every chapter. Fortunately, nearly all those we approached were very friendly. Many constructive suggestions have been accepted for improving this book. What you hold in your hand is the culmination of our wisdom, although not comprehensive. We are extremely grateful to all of our contributors.

One striking feature of this book is that more than one-third of the chapters focus on discussing the practical applications, and even some challenge issues resulting from battery have been tried to solve based on some ingenious batteryless strategies. It has been alleged that the best science is done in academia. This book may provide an uncommon insight for scientific exploration. The main reason may be that many of the piezoelectric energy-harvesting or energy-sensing techniques involved require expensive equipment and infrastructure as well as large collaborations between scientists from disparate disciplines, especially further study in practical applications.

Although tremendous progress has now pervaded laboratories across the world, the ultimate success of any energy-harvesting or energy-sensing technology is measured in the quantity and quality of energy conversion efficiency or sensing sensitivity that it produces. Flexible piezoelectric energy harvesting or sensing has only been practical for the past decade, too soon to expect it to produce marketed electronic devices, but we believe these will come in time. Moreover, many of the techniques and concepts described in this book will improve the flexible piezoelectric energy-harvesting or energy-sensing technologies in subtle, tangential ways. Ideally, readers will be endowed to improve the methods described here, or even to develop fundamentally new methods for flexible piezoelectric energy harvesters or sensors.

Acknowledgments

Finally, we would like to thank all colleagues and members who greatly contributed in writing several chapters: Lijun Lu, Yang Huang, and Guosheng Hu at Shanghai Jiao Tong University, Zixuan Zhang at the National University of Singapore. Also, we thank the Wiley-VCH editorial team for giving us a great opportunity to edit this book.

1

Introduction

1.1 Background

With the rapid development of information technology, microelectronics technology has been widely used in people's daily life, such as healthcare monitoring, intelligent communications, cloud computing, big data, and Internet of Things, and has made considerable progress in intelligent information technology. However, as the current microelectronic devices mainly rely on the wired power supply or battery power supply, wired power supply greatly limits the portability of microelectronic devices and the freedom of space applications, whereas the battery power supply is unsustainable and there are certain safety and environmental pollution problems. In particular, artificial intelligence, the Internet of Things, 5G high-velocity communications, wearable and implantable electronic devices or microsystems, and other new generations of information technology are emerging and rapidly developing. A large number of distributed microsensors composed of wireless sensor networks (WSNs) is essential for these emerging technologies, which makes the problem increasingly prominent. Traditional battery power supply strategies have become a major factor limiting their progress toward miniaturization and flexibility [1]. Therefore, it is observed that the opportunities for batteryless technologies by harvesting surrounding energy of the powered objects or passive sensors are coming, and they are urgent requirements for maintaining the sustainable development of the emerging technologies.

Energy harvesting is the conversion of surrounding waste or useless kinetic, radiant, thermal, and biological energy into electrical energy (as shown in Figure 1.1), and is new energy enabling technology for wireless sensor network node applications. Kinetic energy is the energy that an object has due to its movement, such as mechanical vibrations, human movement, structure rotation, water flow, and so forth, which is ubiquitous in the environment. In contrast to any other energies in surroundings, kinetic energy is widely distributed, clean and safe, easy to miniaturize and integrate, and has a low environmental impact, making it the main source of energy for research into energy-harvesting technologies. With the rapid development of microelectronics technology, more and more low-power consumption microelectronic devices have emerged, while the rapid progress of

Flexible Piezoelectric Energy Harvesters and Sensors, First Edition. Bin Yang, Zhiran Yi, and Chengkuo Lee.

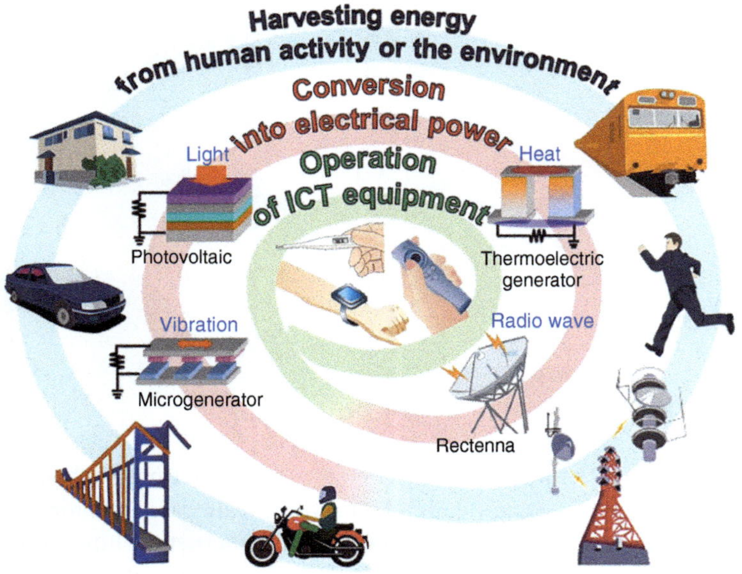

Figure 1.1 Schematic illustration of energy-harvesting strategies based on the ambient energies [2]. Source: Reprinted with permission from Fujitsu Laboratories Ltd. [2]; © 2016 MEMS consulting.

kinetic-energy-harvesting technology in terms of material preparation, conversion methods, device structures, and energy-management circuits has obtained higher and higher conversion efficiency, and the generated electricity has gradually been able to meet the power supply needs of many microelectronic devices, gradually ushering in the development opportunities of batteryless microelectronic systems. Therefore, the research of kinetic-energy-harvesting technology is expected to completely get rid of the shackles of wired power supply and battery supply, and realize batteryless power supply in complex scenarios as well as in situ self-supply of wireless sensor network nodes in the Internet of Things, so as to achieve wireless batteryless sustainable power supply, thus solving the problems of system non-portability, high maintenance costs, environmental pollution, limited life span, and related safety hazards caused by power supply wires or batteries. It is also free from the constraints of battery power supply on the development of flexible and miniaturized electronic devices.

Flexible piezoelectric sensor is one of self-powered sensors, generally be delivered into two types – one is passive sensors, such as piezoelectric sensors, electromagnetic sensors, and thermoelectric sensors; the other one is active sensors with an energy harvester to provide power supply for themself. In this book, it is to avoid the misunderstanding that we described the former as self-power devices and the latter as batteryless devices. Self-powered sensors as passive devices can work almost without power consumption for sensing, which is significant for long-term working and energy saving. In this book, we focus on flexible piezoelectric energy harvesting and expand some corresponding technologies to discuss the development of flexible piezoelectric tactile sensors and their applications in Internet of Things, healthcare monitoring, intelligent recognitions, and some other interesting and possible applications combing with the artificial intelligent technologies.

1.2 Working Principle of Piezoelectric Devices

Piezoelectric energy harvesting or sensing technique is mainly based on the positive piezoelectric effect of materials, which was discovered by the Curie brothers in 1880 in alpha-quartz crystals and reflects the coupling between the elastic and dielectric properties of crystals, meaning that there is no electric field but only the strain or stress generated by the action of an external force, resulting in the generation of electrodes in the crystal, leading to the appearance of a different macroscopic charge on certain surfaces or inside the medium. These macroscopic charges are called polarized charges and cannot leave the dielectric or move freely within the dielectric, so they are also called bound charges [3]. Researches into piezoelectric energy harvesting or sensing technique began at the end of the last century and the main transformation methods include the free vibration method and the forced vibration method, which all rely on mechanical surrounding excitations, such as mechanical vibration and dynamic force. Therefore, piezoelectric effect has been widely investigated for force or stress sensing and energy harvesting.

In terms of mechanical energy harvesting, piezoelectric energy harvesting is one of major components, mainly including piezoelectric, electromagnetic, electrostatic, triboelectric, flexoelectric, and electrochemical mechanisms.

Electromagnetic-energy-harvesting technology is based on Faraday's law of electromagnetic induction, using a part of the conductor of a closed circuit to make a motion of cutting magnetic induction lines in a magnetic field, causing a change in magnetic flux, forming an induced electric potential, and generating an induced current in the conductor. The main transformation methods are cantilever beam, suspension bridge, single-track-reciprocating vibration and rotary motion [4, 5]. Electrostatic-energy-harvesting techniques are used to create a circuit current by forcing a change in charge or voltage between the two plates of a capacitor through mechanical motion [6]. The main structure of an electrostatic energy harvester is a capacitive structure consisting of two conductive electrodes separated by air or dielectric and an elastic structure connected by the electrodes. The basic working principle of triboelectric-energy-harvesting technology consists mainly of contact initiation and electrostatic induction [7]. According to the charge-separation mechanism, the harvesting forms of triboelectric electrical energy harvesters can be divided into contact based on vertical charge polarization and sliding based on in-plane charge polarization. Electrochemically driven method [8] is a new type of kinetic-energy-harvesting method. Electrochemically driven energy harvesters are mainly based on the migration of ions in the composition of electrochemically active materials (e.g. alloys of Li and Si or Ge) subjected to external stresses, resulting in an electrical potential difference. The flexoelectric effect is similar to the piezoelectric effect and refers to the polarization of a dielectric material when it is bent by an

Table 1.1 Advantages and disadvantages of the different kinetic energy harvesters.

Working principles	Advantages	Disadvantages
Piezoelectric	No requirement for external energy at starting, high energy density, simple structure, and easy integration.	Easy self-discharge at low frequencies, high dependence on materials, high output impedance, and difficult flexibility.
Electromagnetic	No mechanical contact is required, low mechanical damping, high reliability, high output current, and low output impedance.	Difficult to miniaturize, low efficiency at low frequencies, low output voltage, and difficult integration.
Electrostatic	MEMS process compatibility, suitable for low frequencies, easy integration, and high electrical damping.	Dielectric breakdown, requirement of external energy at starting, complex fabrication process, high output impedance, low current, and easy to fail due to pull in instability.
Triboelectric	Good flexibility, simple structure, and high output voltage.	Low output current, dependence on friction surfaces, low reliability, susceptible to environmental humidity, difficult to miniaturize, high impedance.
Electrochemical	Flexibility and low frequency.	Low output voltage.
Flexoelectric	High-quality factor and easy integration.	Weak effect for bulk material and low output energy.

external force, resulting in structural deformation when an external electric field is applied. It is a weak effect for bulk materials [9], and therefore studies of flexural electricity are usually focused on the nanoscale.

Table 1.1 lists the advantages and disadvantages of the different mechanisms of kinetic-energy-harvesting methods. From the table, it can be found that electromagnetic and triboelectric methods are difficult to miniaturize due to their mechanism of action, and electromagnetic-energy-harvesting methods are difficult to be compatible with flexible electronic devices; triboelectric-energy-harvesting methods are not yet able to break through the limitation on material triboelectric surface loss; electrostatic methods are highly dependent on the level of process integration and manufacturing, and are prone to microscale adsorption failure problems; electrochemical drive methods and flexoelectric methods are too inefficient in terms of output. It is still difficult to meet the power supply needs of low-power microelectronic devices. In contrast, piezoelectric-energy-harvesting methods have been widely studied for their high energy density, simple structure, easy integration, and no need for external energy to start, but they also suffer from high-frequency dependence, limited piezoelectric material properties, flexibility difficulties, and high output impedance. The flexible piezoelectric-energy-harvesting technology based on mechanical thinning process is expected to solve the problems faced in the development of flexible piezoelectric energy harvesters.

1.3 Requirement for Flexible PEH

The application conditions of piezoelectric-energy-harvesting technology are very demanding. On the one hand, it requires sufficient mechanical energy available in the working environment, and on the other hand, it requires the device to provide sufficient electrical energy under the premise of ensuring applicability and safety, so the working environment of the device and the power consumption of the equipment to be powered become the main limiting factors for the expansion of the application field of piezoelectric-energy-harvesting technology, which is also the main hot spot of the current applied research.

In terms of common microelectronic devices, the application of energy-harvesting technologies is mainly to consider some specific environments where wired power supply is inconvenient and battery power supply is a safety hazard due to the lifetime problem, and the realization of batteryless devices can effectively get rid of these problems. At present, the batteryless research of low-power electronic devices mainly includes some implantable medical devices with low power consumption (as shown in Figure 1.2) and pressure monitoring systems (as shown in Figure 1.3).

As for rapidly developing Internet of Things, it refers to various devices and technologies, such as various information sensors, radio-frequency identification technology, global positioning system, infrared sensors, laser scanners, to collect any object or process that needs to be monitored, connected, and interacted with in real time, collecting its sound, light, heat, electricity, mechanics, chemistry, biology, location, and various other needed information, and through various possible network access, realize the ubiquitous connection between things and things, things and people, and realize the intelligent sensing, identification and management of objects and processes. The Internet of Things is an information carrier based on

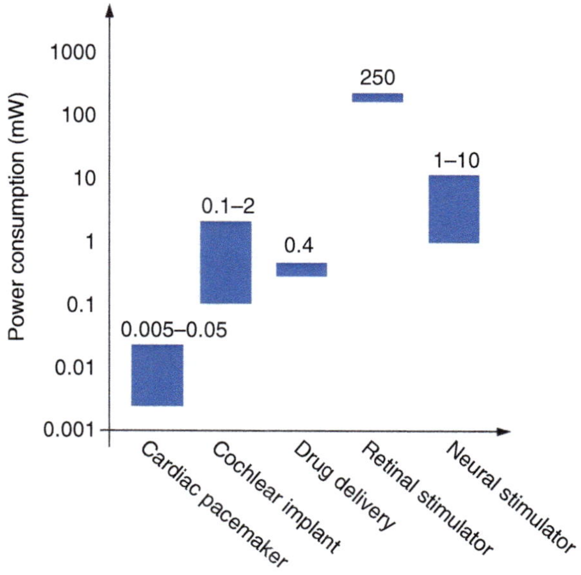

Figure 1.2 Power consumption of different medical and implantable devices [10, 11]. Source: Data from Yang et al. [10].

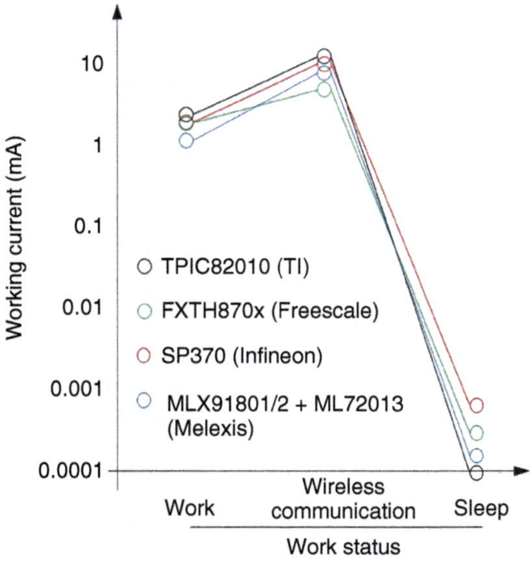

Figure 1.3 The main IC solutions of the tire pressure monitoring system on Market [10]. Source: Data from Yang et al. [10].

the Internet and traditional telecommunication networks, which allows all common physical objects that can be independently addressed to form an interconnected network. Combined with Artificial Intelligent (AI) technology and the new generation of high-speed communication technology (5G or 6G) to achieve high-speed processing of sensory information, high-speed interaction, to achieve timely sensing and timely decision-making of the artificial intelligence & Internet of Things (AIoT) system.

The current power consumption of most IoT devices is shown in Figure 1.4. When using these devices to achieve the interconnection and interoperability of everything, such a huge power supply network is also needed behind the huge connectivity network. Given that wired power supply will greatly limit the scope of its spatial application, the current reliance on battery-powered wireless sensing network technology, however, will certainly cause a large amount of battery consumption and high maintenance costs due to battery life, so the development of the Internet of Things urgently needs a convenient and sustainable power supply network.

With the development of energy-harvesting technology and low-power IoT devices, the environmental energy that can be converted by energy harvesters also gradually meets the power supply needs of some IoT devices, and in the future, with the further development of both, it is very promising to achieve sustainable power supply for IoT. Figure 1.5 shows the power density and output voltage range statistics of common energy-harvesting methods and lithium-ion battery technology. In terms of mechanical kinetic energy harvesting, piezoelectric-energy-harvesting technology can achieve greater energy density and output voltage compared to electromagnetic-energy-harvesting technology and has the potential to be a greater alternative to lithium batteries.

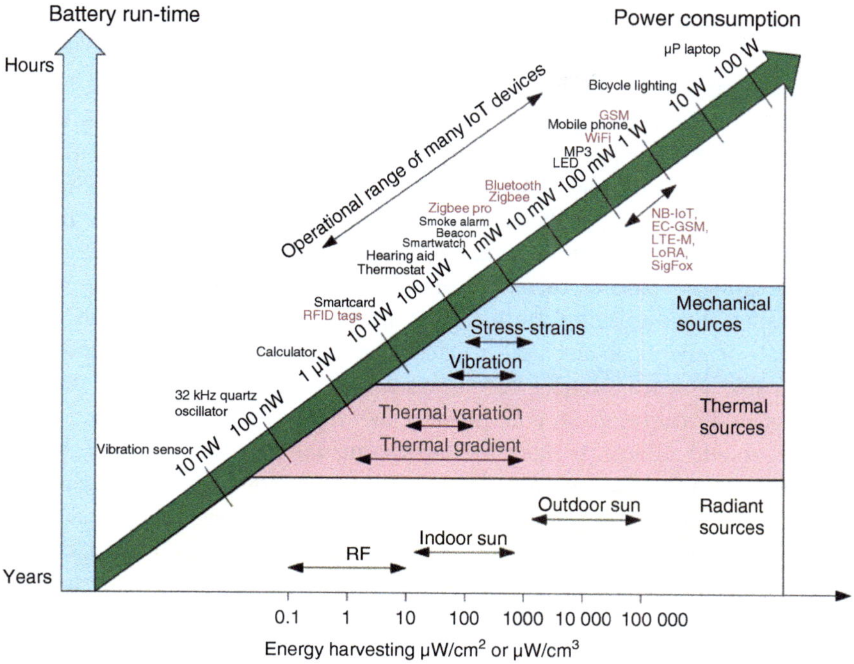

Figure 1.4 Power consumption for various applications of IoT and power densities for various energy sources. Source: Shirvanimoghaddam et al. [12] / CC BY-SA 4.0.

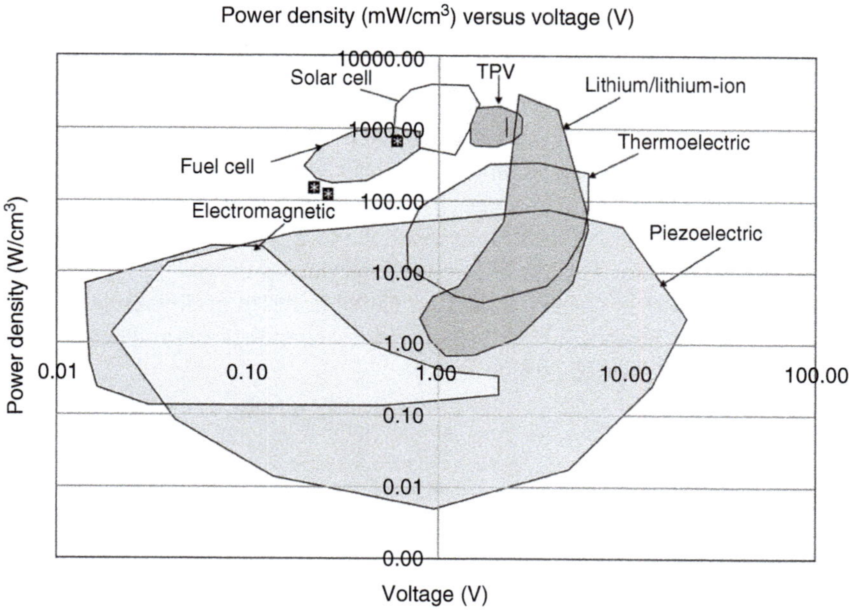

Figure 1.5 Power density versus voltage comparison of common regenerative and lithium/lithium-ion power supply strategies. Source: Cook-Chennault et al. [13] / with permission of IOP Publishing.

1.4 Requirement for Flexible Piezoelectric Sensors

For decades, miniaturization of electronics and advances in materials are paving the way for prosperous development of wearable and implantable devices. Human is experiencing the great conveniences that are brought by diverse electronics, including smartwatch, glasses, shoes, glove, pacemaker, and necklace. With the aid of 5G communication, artificial intelligence, and edge computing, the highly digitized human will undergo a seamless involvement in the whole network or internet of things. Some of them possess the functions of the mobile console for processing and displaying the data. Meanwhile, various sensors with physical, optical, or chemical mechanisms, form another major group in wearable or implantable systems. Hence, our physical activities, vital sign, and physio-biochemical indicators can be continuously monitored for health inspection or performing manipulations, besides, the environmental variations are also able to be detected for further utilization. However, the current commercialized wearable and implantable devices usually rely on the battery-based power supply, such as piezoresistive, optical, and capacitive sensors, and various actuators, which are inevitably encountered with some constraints, including the long-term sustainability and the further shrinkage of device size. As a solution to this urgent power issue, a flexible piezo-microelectromechanical system (piezo-MEMS)-integration strategy was presented on base of piezoelectric thick film with good piezoelectric properties. The sensors based on piezoelectric effect attract broad attention due to their intrinsic high dynamic response and high fidelity [14–19], especially ultra-low power consumption.

In this book, we intend to discuss piezoelectric energy harvesters (PEHs) and piezoelectric sensors on account of the piezo-MEMS thick-film process. The discussion starts with fabrication of piezo-MEMS thick film, design of piezoelectric devices, and some methods of theoretical analysis. Then, we firstly introduce the usual investigations, including common linear-mechanism cantilever piezoelectric energy harvester based on the piezoelectric thick film and nonlinear-mechanism PEHs, to discuss the design and fabrication of the PEHs. Thereafter, we introduce the rotation-driven flexible PEH, the aeroacoustics-driven jet-stream wind energy harvester, and the wearable and implantable PEHs in different chapters, respectively. Finally, we focused on the piezoelectric tactile sensors and advances in artificial-intelligence-assisted flexible sensors.

1.5 Summary

In this book, we focused on introducing the recent advances and important development in flexible piezoelectric energy harvesting and sensing techniques, mainly including the design, fabrication, theoretical analysis, nonlinear mechanism induced bandwidth properties, and some applications in self-powered systems of flexible piezoelectric energy harvesters, and we also introduced the key role of flexible piezoelectric devices in the field of wearable electronics and tactile sensing.

We also provided some introduction about the emerging research directions due to disciplines-crossing, such as artificial intelligence technology-assisted design optimization, data processing, or sensing recognition.

References

1 Gao, W., Emaminejad, S., Nyein, H. et al. (2016). Fully integrated wearable sensor arrays for multiplexed in situ perspiration analysis. *Nature* 529: 509–514.

2 Fujitsu Laboratories Ltd. (2010). Fujitsu Develops Hybrid Energy Harvesting Device for Generating Electricity from Heat and Light. https://www.fujitsu.com/global/about/resources/news/press-releases/2010/1209-01.html.

3 Toprak, A. and Tigli, O. (2014). Piezoelectric energy harvesting: state-of-the-art and challenges. *Applied Physics Reviews* 1: 031104.

4 Williams, C.B., Shearwood, C., Harradine, M.A. et al. (2001). Development of an electromagnetic micro-generator. *IEEE Proceedings-Circuits Devices and Systems* 148: 337–342.

5 Tan, Y., Dong, Y., and Wang, X. (2017). Review of MEMS electromagnetic vibration energy harvester. *Journal of Microelectromechanical Systems* 26: 1–16.

6 Khan, F.U. and Qadir, M.U. (2016). State-of-the-art in vibration-based electrostatic energy harvesting. *Journal of Micromechanics and Microengineering* 26: 103001.

7 Niu, S., Liu, Y., Wang, S.H. et al. (2014). Theoretical investigation and structural optimization of single-electrode triboelectric nanogenerators. *Advanced Functional Materials* 24: 3332–3340.

8 Kim, S., Choi, S.J., Zhao, K.J. et al. (2016). Electrochemically driven mechanical energy harvesting. *Nature Communications* 7: 10146.

9 Bhaskar, U.K., Banerjee, N., Abdollahi, A. et al. (2016). A flexoelectric microelectromechanical system on silicon. *Nature Nanotechnology* 11: 263–266.

10 Yang, Z., Zhou, S., Zu, J., and Inman, D. (2018). High-performance piezoelectric energy harvesters and their applications. *Joule* 2: 642–697.

11 Ohm, O.J. and Danilovic, D. (1997). Improvements in pacemaker energy consumption and functional capability: four decades of progress. *PACE* 20: 2–9.

12 Shirvanimoghaddam, M., Shirvanimoghaddam, K., Abolhasani, M.M. et al. (2019). Towards a green and self-powered internet of things using piezoelectric energy harvesting. *IEEE Access* 7: 94533–94556.

13 Cook-Chennault, K.A., Thambi, N., and Sastry, A.M. (2008). Powering MEMS portable devices – a review of non-regenerative and regenerative power supply systems with special emphasis on piezoelectric energy harvesting systems. *Smart Materials and Structures* 17: 043001.

14 Yi, Z., Huang, J., Liu, Z. et al. (2020). Portable, wireless wearable piezoelectric arterial pulse monitoring system based on near-field communication approach. *IEEE Electron Device Letters* 41: 183–186.

15 Yi, Z., Yang, H.J., Tian, Y.W. et al. (2018). Self-powered force sensor based on thinned bulk PZT for real-time cutaneous activities monitoring. *IEEE Electron Device Letters* 39: 1226–1229.

16 Dagdeviren, C., Su, Y.W., Joe, P. et al. (2014). Conformable amplified lead zirconate titanate sensors with enhanced piezoelectric response for cutaneous pressure monitoring. *Nature Communications* 5: 4496.

17 Ershad, F., Thukral, A., Yue, J.P. et al. (2020). Ultra-conformal drawn-on-skin electronics for multifunctional motion artifact-free sensing and point-of-care treatment. *Nature Communications* 11: 3823.

18 Wang, C., Li, X.S., Hu, H.J. et al. (2018). Monitoring of the central blood pressure waveform via a conformal ultrasonic device. *Nature Biomedical Engineering* 2: 687–695.

19 Wang, H., Wang, L., Sun, N.N. et al. (2020). Quantitative comparison of the performance of piezoresistive, piezoelectric, acceleration, and optical pulse wave sensors. *Frontiers in Physiology* 10: 1563.

2

Design of Flexible Piezoelectric Energy Harvesters

2.1 Introduction

Design of piezoelectric energy harvesters is to generate highly effective strain in piezoelectric layer. The piezoelectric energy harvesters mainly include the free-vibration type based on the inertia principle and the forced-vibration type based on the action of periodic external forces. For a piezoelectric material with defined piezoelectric properties, how to achieve the most effective strain generation by piezoelectric crystals through the structural design of piezoelectric energy harvesters becomes the key to achieving high-conversion efficiency of piezoelectric energy harvesters.

The cantilever beam structure can produce higher average strain for a given force condition and has greater conversion efficiency for application in energy harvesters. It is also simple in structure and thus has received a lot of attention from researchers. At present, most of the studies on cantilever beam vibration energy harvesters focus on the uniform cross-sectional beam structure composed of piezoelectric and nonpiezoelectric layers with the same length, and the output power of the cantilever beam vibration energy harvester is greatly affected by different piezoelectric layer arrangement lengths, but there are relatively few relevant research reports, and there is a lack of systematic theoretical analysis for this nonuniformly distributed cantilever beam study, especially for the commonly used. The performance of the cantilever beam vibration energy harvester with mass bulk is not clear.

In this chapter, we discuss some challenges of flexible piezoelectric energy harvesters in structural design, types of generally used structural designs, methods of theoretical analysis, methods of simulation, and the corresponding experimental verification. Finally, we introduce how to optimize the structural designs by using the data-driven strategy.

2.2 Challenges of Structural Design

2.2.1 Low Frequency

It is important that the resonant frequency of the piezoelectric energy harvester can meet the requirement of matching the frequency of the vibration source in the

Flexible Piezoelectric Energy Harvesters and Sensors, First Edition. Bin Yang, Zhiran Yi, and Chengkuo Lee.

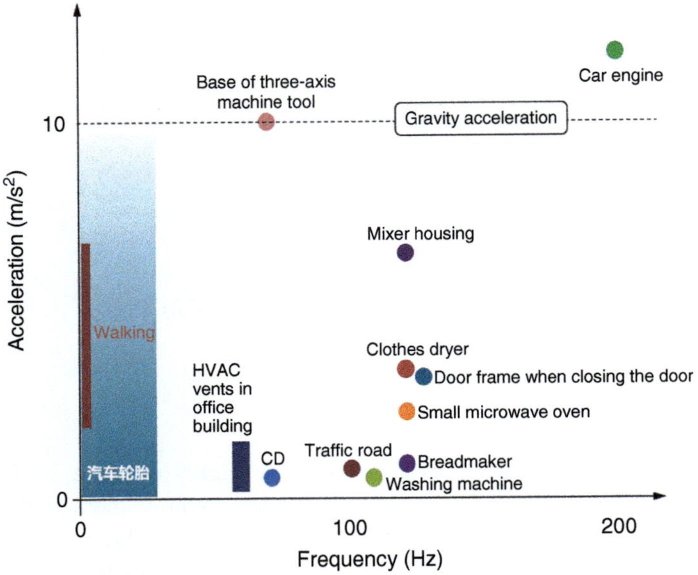

Figure 2.1 Frequency and acceleration levels of various ambient vibration resources. Source: Adapted from [1, 2].

operating environment to achieve the best operating condition of the device. Given that the frequency of common environmental vibration sources is usually less than 200 Hz (Figure 2.1), the development of high-performance piezoelectric energy harvesters with low frequencies has attracted extensive attention and research from scientists.

The research of vibration energy harvesters is based on the environmental state of the harvesting device, and the optimal operating frequency required for the device varies according to its working environment. To achieve high-efficiency energy harvesting, it is necessary to control the resonant frequency of the device in line with the environmental vibration frequency to achieve resonance, and microelectromechanical systems (MEMS) vibration energy harvesters are difficult to control the resonant frequency in the low-frequency range below 200 Hz due to their inherent small size and process-compatible materials, which make most MEMS vibration energy harvesters have poor output performance under actual operating conditions. Therefore, the development of low-frequency MEMS vibration energy harvesters is particularly important to advance their applications in real-world environmental conditions.

The basic composition of the cantilever beam free-vibration energy harvester mainly consists of a piezoelectric layer, a substrate layer, and a mass bulk, whose intrinsic frequency can be expressed as [3]:

$$f \propto \sqrt{\frac{\mathrm{EI}}{mL^4}} \tag{2.1}$$

where EI is the flexural stiffness of the cantilever beam, m is the equivalent mass of the cantilever beam, and L is the total length of the cantilever beam. It can be seen that the inherent frequency is mainly controlled by the flexural stiffness of the cantilever beam structure, the cantilever beam mass distribution, and the total length of the cantilever beam. For the MEMS cantilever beam energy harvester, due to its small-size limitation, its effective length cannot be too large, and adding end-mass bulks to adjust the mass distribution of the cantilever beam structure is an effective way to reduce the intrinsic frequency. However, the conventional piezoelectric energy harvester with high-modulus silicon material as substrate is not only difficult to achieve the design requirements of low frequency, but also easy to be damaged under higher excitation acceleration conditions, and the device is difficult to ensure normal operation under the effect of complex environment. At the same time, the size of the mass bulk is also limited by considering the limitation of process and structure size, so the frequency regulation by adjusting the mass bulk alone has great limitation. The magnitude of flexural stiffness, besides being limited by the cross-sectional dimensions, also depends mainly on Young's modulus of the material used in the cantilever beam itself. The conventional MEMS cantilever beam energy harvester is mainly based on silicon as the substrate material, which has a high Young's modulus and is prone to brittle cracking. Therefore, the current research on low-frequency piezoelectric free-vibration energy harvesters is mainly focused on mass bulk bonding and low-stiffness beams.

On the other hand, the study of the energy-harvesting method of forced vibration based on the ambient vibration method and the consideration of the device size allow the combination of the design of the boosting mechanism to raise the ultralow-frequency-forced vibration to a higher frequency range, thus improving the energy conversion efficiency.

2.2.2 Bandwidth

Usually, the cantilever beam vibration energy harvester only shows high-output performance in a small range near the resonant frequency point, and its output performance decreases very significantly when it deviates from its resonant frequency. Therefore, to solve the problem of single-frequency response of vibration energy harvester, there are two main solutions – array-type cantilever beam topology and flexural bridge-type nonlinear vibration topology.

Array cantilever beam topology is the result of integrating multiple cantilever beam vibration energy harvesters with different resonant frequencies to achieve multifrequency response (Figure 2.2a), and connecting these devices in series or parallel to achieve high-electromechanical-conversion efficiency at multiple frequency points or certain frequency bands (as shown in Figure 2.2b) to meet the needs of different vibration environments.

Bridge-type structures are less frequently used in the study of energy harvesters because of their large resonant frequencies and low strain. However, the resonant frequency and output performance of energy harvesters based on flexible substrates can be improved, and nonlinear vibration effects can be achieved by applying certain bending stress in advance or by using the bending curvature of the structure itself in

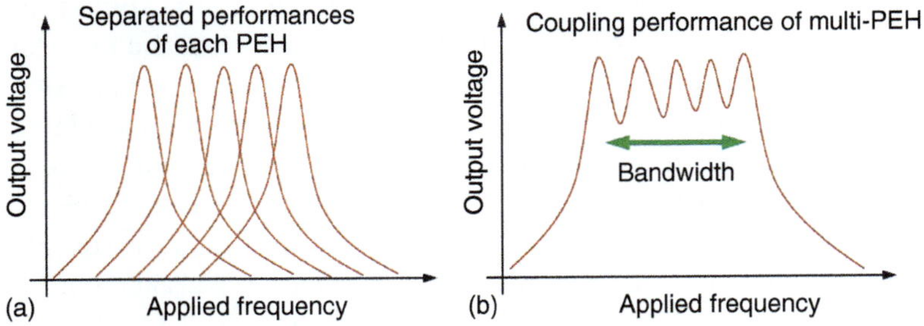

Figure 2.2 Design principle of bandwidth piezoelectric energy harvester based on multifrequencies responses of multistructures. (a) Output performance of each harvester. (b) Output performance in parallel or in series.

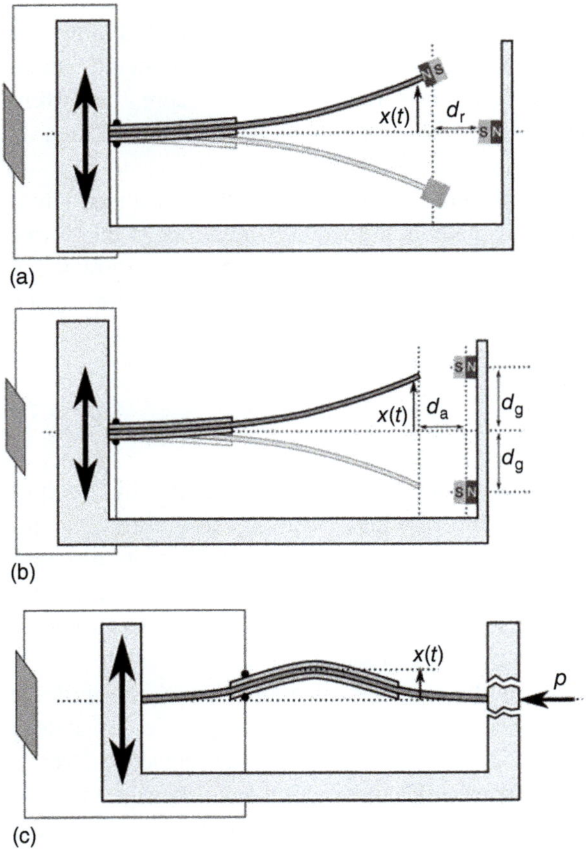

Figure 2.3 Nonlinear vibration energy harvester. Source: Harne and Wang [4] / with permission of IOP Publishing. (a) bistable magnetic repulsion harvester. (b) Magnetic attraction harvester. (c) Buckled beam harvester.

the process to achieve a flexural bridge structure (Figure 2.3). In addition, according to the theoretical analysis, it can be found that the nonlinear effect of softening or hardening can be achieved by controlling the bending angle, and the nonlinear vibration effect of broadening to low frequency or high frequency can be achieved, respectively.

2.2.3 Flexibility

Flexible piezoelectric energy harvesters are proposed mainly because the structural stiffness of the devices may bring inconvenience or some unavoidable side effects to the practical applications of the devices (e.g. wearable or implantable electronic devices), and the electromechanical coupling coefficient of polymer-based flexible piezoelectric materials such as poly (vinylidene fluoride) (PVDF) is low, and the output of the energy harvesters can hardly meet the power supply requirements of the current low-power electronic devices. Therefore, the research on flexible piezoelectric energy-harvesting technology mainly focuses on the development of flexible piezoelectric materials with high electromechanical coupling coefficients and the design and preparation of flexible strain-type piezoelectric energy harvesters.

The research on flexible piezoelectric materials mainly includes the study of high-performance flexible piezoelectric polymers (Figure 2.4a), flexible piezoelectric

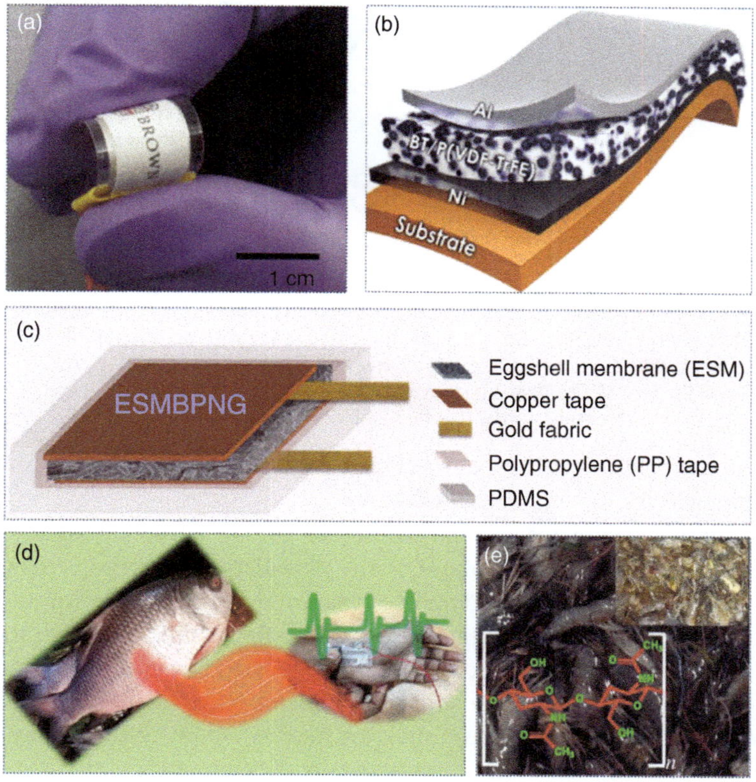

Figure 2.4 Piezoelectric energy harvester based on flexible piezoelectric materials, (a) paper substrate-based PVDF piezoelectric energy harvester. Source: From Won et al. [5] / with permission of AIP Publishing. (b) flexible nanocomposite-based piezoelectric energy harvester. Source: From Siddiqui et al. [6] / with permission of Elsevier. (c) eggshell membrane-based flexible piezoelectric energy harvester. Source: From Karan et al. [7] / with permission of Elsevier. (d) and (e) novel biomaterials, such as fish scale. Source: From Ghosh and Mandal [8] / with permission of American Chemical Society. shrimp shell, Source: From Ghosh and Mandal [9] / with permission of AIP Publishing or other gelatin-rich materials [10], based flexible piezoelectric energy harvesters.

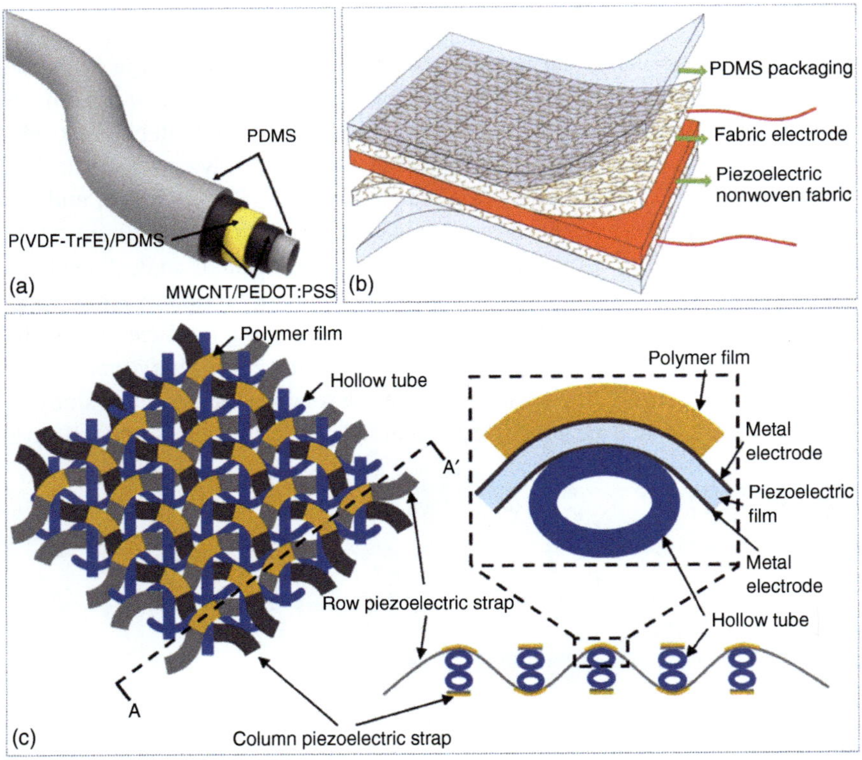

Figure 2.5 Flexible energy harvester based on the piezoelectric fibers, (a) core–shell-type piezoelectric fabric energy harvester. Source: From Ryu et al. [11] / with permission of Elsevier. (b) nonwoven fabric piezoelectric energy harvester. Source: Zeng et al. [12] / with permission of Royal Society of Chemistry. (c) woven fabric piezoelectric energy harvester. Source: Ahn et al. [13] / with permission of IOP Publishing.

composites based on piezoelectric ceramics (Figure 2.4b), and the development of new flexible piezoelectric materials (Figure 2.4c–e).

In terms of device structure, the piezoelectric fiber (Figure 2.5) structure has also become one of the hot spots in the research of flexible piezoelectric energy harvesters due to its low-dimensional size and compliable characteristics that make it have low bending stiffness characteristics. In terms of device preparation, considering rigid piezoelectric ceramics or piezoelectric monocrystals in the process of thin film will show a certain degree of flexibility at low thickness, so based on piezoelectric ceramic, thin film has also become one of the ways to carry out high-performance flexible piezoelectric energy harvesters, but the amount of deformation it can withstand is very limited, deformation or concentrated stress is too large, easily leading to device failure, which is the main challenge of this method.

In terms of device design, the island-connected design (Figure 2.6), which miniaturizes the rigid piezoelectric cell and realizes a soft and rigid device structure through flexible materials and flexible electrical interconnection technology, provides a proven method for rigid piezoelectric materials in achieving flexible energy harvesting.

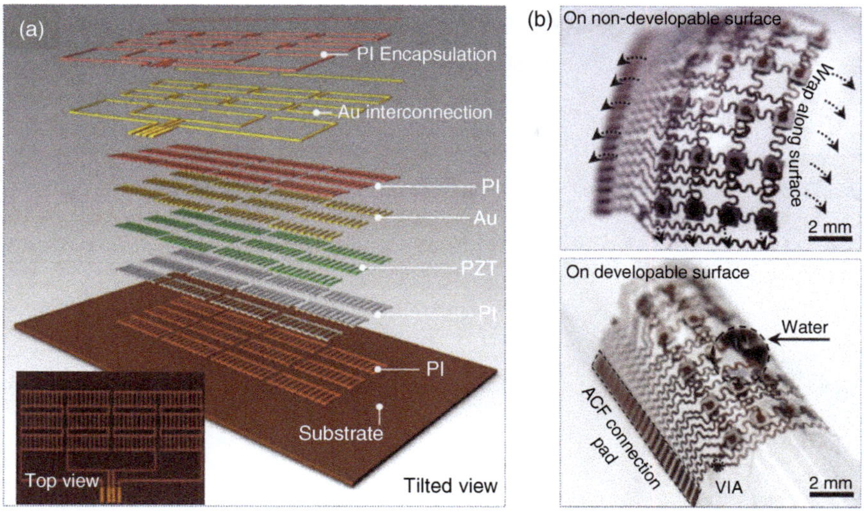

Figure 2.6 Flexible piezoelectric energy harvester based on piezoelectric ceramics, (a) flexible piezoelectric energy harvester based on PZT ribbon array. Source: From Dagdeviren et al. [14] / with permission of National Academy of Sciences. (b) flexible piezoelectric sensor based on 1–3 type piezoelectric composites. Source: Wang et al. [15] / with permission of Springer Nature.

2.3 Types of Structural Design

Piezoelectricity is a coupling property between the mechanical and electrical behavior of piezoelectric materials, and the piezoelectric equation is obtained by analyzing the relationship between each mechanical and electrical physical quantity by considering the piezoelectric crystal as a thermodynamic system. The piezoelectric energy harvester converts the mechanical kinetic energy in the environment into stress–strain in the piezoelectric material and then generates surface polarization potential under the piezoelectric effect to achieve the conversion from mechanical kinetic energy to electrical energy. Among them, the research on obtaining the most effective stress–strain by designing the structure of piezoelectric energy harvesters has become the main content of the research on the structural design of piezoelectric energy harvesters. Currently, piezoelectric energy harvester structures include cantilever beam, bridge, flexure, diaphragm, cymbal, and piezoelectric stack.

As can be seen from Figure 2.7, cantilever beam, bridge, flexure, and diaphragm have a substrate structure, while cymbal and piezoelectric stack generally require only an electrode layer on the upper and lower surfaces of the piezoelectric layer (which has almost no effect on the structural stiffness). The former mainly relies on the deformation of the piezoelectric material with the substrate structure to generate strain to achieve the purpose of energy conversion, while the latter mainly relies on the internal stress of the piezoelectric material to achieve the energy conversion.

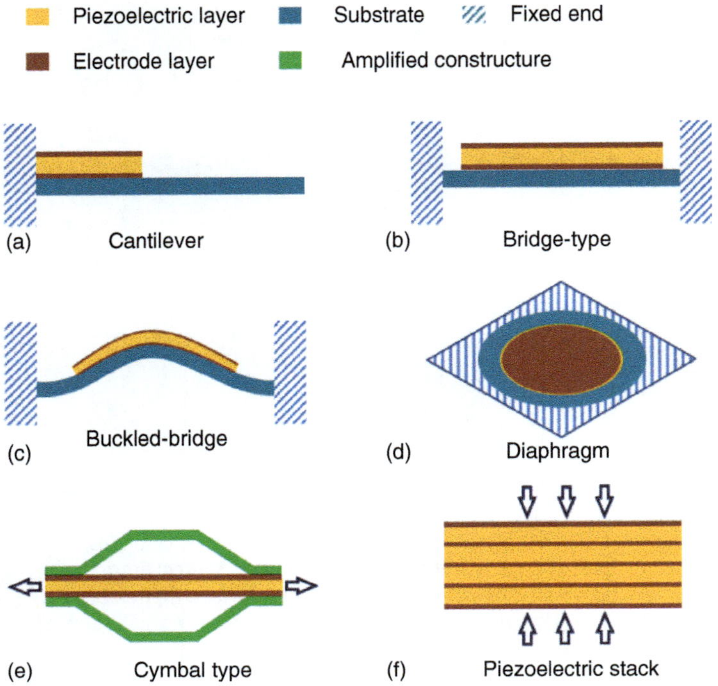

Figure 2.7 Schematic diagram of piezoelectric energy harvesters, (a) cantilever, (b) bridge, (c) buckled-bridge, (d) diaphragm, (e) cymbal, and (f) stack types.

Generally, polarized piezoelectric ceramics are macroscopically isotropic materials with isotropic surfaces defined as 12- or *xy*-planes, so that the piezoelectric ceramic material is symmetric about the 3-axis or *z*-axis (polarization axis). The standard form of the piezoelectric intrinsic equation can have four expressions based on any two of the four field variables (stress T_x, strain S_x, potential shift D_z, and electric field tensor E_z) as independent variables. For the strain-potential shift form of the piezoelectric instantonal equation based on the stress and electric field tensor as independent variables is [3]:

$$\begin{cases} S_x = c_{xx}^E T_x + d_{zx}^T E_z \\ D_z = d_{zx} T_x + \varepsilon_{zz}^S E_z \end{cases} \tag{2.2}$$

This type of piezoelectric energy harvester mainly induces polarized charge by internal stress within the dielectric through the action of external pressure/tension, and generally, the structural strain of the energy harvester is very small, such as cymbal type and piezoelectric stack. They are usually used in dynamic environments with high pressure or tension, such as the pressure on the bottom of the human foot and the relative motion between mechanical structures, for the collection of small excitation kinetic energy in the environment, and generally have low-energy-conversion efficiency. Based on the strain and electric field tensor as the independent variables of the stress-potential shift form of the piezoelectric constitutive equation can be expressed as [3]:

$$\begin{cases} T_x = c_{xx}^E S_x - e_{zx}^T E_z \\ D_z = e_{zx} S_x + \varepsilon_{zz}^S E_z \end{cases} \tag{2.3}$$

Piezoelectric energy harvesters are mainly available in cantilever beam type, bridge type, piezoelectric stack, circular diaphragm type, cymbal type, flexure type, etc. according to device structure.

2.3.1 Cantilever Beam Structure

The cantilever beam structure is used as the main piezoelectric vibration energy harvester structure because it can produce the highest average strain under a certain load [16]. At the same time, its structure is simple, easy to prepare, and compatible with MEMS process. Table 2.1 shows the main cantilever-beam-type piezoelectric energy harvesters studied in recent years, the mode of action is mainly d_{31} type, and d_{33} type is relatively less studied due to the limitation of process accuracy of forked finger electrodes.

2.3.2 Diaphragm Type

An important advantage of the diaphragm structure compared to the cantilever beam is its compatibility with pressure-mode operation [22]. The mode of action is mainly of type d_{31}. The main studies of diaphragm-type piezoelectric energy harvesters in recent years are shown in Table 2.2.

2.3.3 Bridge Type

The main advantage of bridge-type structure of piezoelectric vibration energy harvester is its high structural stability, which has greater advantages in the field of

Table 2.1 High-power density piezoelectric vibration energy harvesters based on the cantilever construction.

Piezoelectric materials	Substrate	Effective volume (mm³)	Averaging output power (μW)	Acceleration (*g*)	Frequency (Hz)	Normalized power density (μW/mm³ g² Hz)
PZT thick film [16]	Steel	494.472	375	0.255	120	0.097
PZT thin film [17]	Silicon	0.2	2.16	1.0	609	0.018
PZT thick film [18]	Copper	613	4000	0.25	100	1.044
PZT thin film [19]	Silicon	0.15	13.0	1.0	2300	0.038
PZT thick film [20]	Silicon	0.4	11.56	1	514.1	0.056
AlN thin film [21]	Silicon	5	1.78	0.06	210	0.471

Table 2.2 Main researches on the diaphragm-type piezoelectric energy harvesters in recent years.

Piezoelectric materials	Area (mm^2)	Performance
PZT [23, 24]	2025.8	Open circuit voltage at 10 kPa pressure: 9.26 V
PZT [25]	27.56	Power density at 149 dB sound pressure intensity source excitation: 0.34 μW/cm^2
ZnO [26]	9.39	Output power at the applied frequency of 1.4 kHz: 1 μW
PZT [27, 28]	5645	Output voltage of 36 V for 1 Hz applied frequency at 6 kPa pressure

Table 2.3 Main researches on the bridge-type piezoelectric energy harvesters in recent years.

Piezoelectric materials	Average output power/voltage	Acceleration (g)	Frequency (Hz)
MFC [29]	100 μW	0.5	68
PZT [30]	7.03 μW	1	10
PZT [31]	57 μW	8	511
PVDF [32]	2.9 μW	—	143
PZT [33]	8.3 V	1	10
	1.85 V		88.1

extreme vibration environments, such as ultrahigh acceleration, but its conversion efficiency and output power density are low compared with cantilever beam structure, so the research of bridge-type vibration energy harvester based on normal ambient vibration conditions is relatively small. The main studies of bridge-type piezoelectric energy harvesters in recent years are shown in Table 2.3, and the mode of action is mainly d_{31} type.

2.3.4 Buckling Type

Flexural vibration energy harvesters mainly use the pre-flexural bending characteristics of the structure to produce a nonlinear large deformation effect under the excitation of external forces, and then enhance the energy-harvesting efficiency and achieve high-power density energy output. The current research mainly focuses on flexural bridge-type structures, flexural flexible piezoelectric energy harvesters, and so on. The main research on flexural piezoelectric energy harvesters in recent years is shown in Table 2.4, and the mode of action is mainly d_{31} type.

Table 2.4 Main researches on the buckled-type piezoelectric energy harvesters in recent years.

Piezoelectric materials	Average output power/voltage	Acceleration (g)	Frequency (Hz)
PZT [34]	30 V	2	30
PZT [35]	5 V	0.2	23
PZT [36]	65.44 μW	—	14
PZT [37]	10.6 μW	0.25	23
PZT [38]	30 pA		
AlN [39]	1.3 nW	2	64
PZT [40]	34 mW	2	8.6

Table 2.5 Main researches on the stack- or cymbal-type piezoelectric energy harvesters in recent years.

Types	Piezoelectric materials	Effective volume (mm^3)	Resonant frequency (Hz)	Loading force (N)	Average output power (μW)
Cymbal type	PMN-PT[41]	15	500	0.55	14
Piezo stack	PZT[42]	400	2.75	220	400
	PZT[43]	1587.6	1204	22.15	347 000

2.3.5 Piezoelectric Stack- and Cymbal-Type Piezoelectric Energy Harvester

The piezoelectric stack-type structure is designed by interlocking multiple layers of piezoelectric structures to enhance the efficiency of longitudinal force utilization, which is a typical d_{33} mode of action. The cymbal-type structure uses structural mechanics amplification mechanism to amplify the excitation force or displacement to enhance the energy-conversion efficiency, which is a d_{31} mode of action. The main researches on piezoelectric energy harvesters, such as stack and cymbal type in recent years, are shown in Table 2.5.

2.4 Theoretical Analysis

The piezoelectric composite-layer-type structure mainly consists of a piezoelectric layer and a substrate layer. The mechanically neutral layer of the piezoelectric composite layer structure (Figure 2.8) is in the substrate layer so that the piezoelectric layer can have the same polarization direction when the structure is deformed, i.e. produce an effective strain. The multilayer composite structure of the

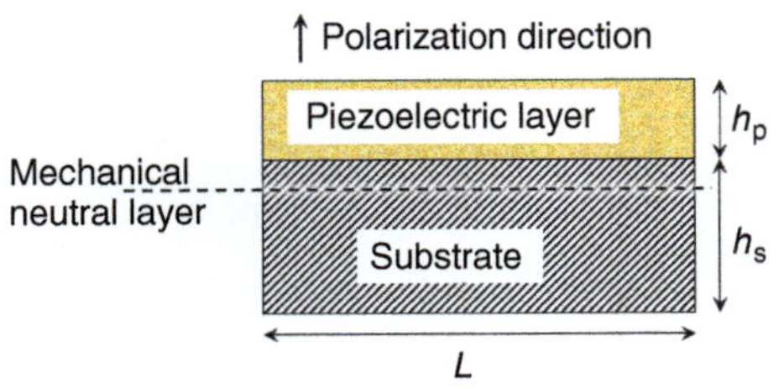

Figure 2.8 Schematic diagram of multilayer piezoelectric composite structure.

piezoelectric cantilever beam energy harvester has a simple structure that facilitates the optimization of structural design parameters and can also be prepared by a mechanically thinning piezoelectric thin-film process. Therefore, the study of the main factors affecting the performance of piezoelectric energy harvesters with the cantilever beam structure is an important guide to explore the design method of high-performance energy harvesters with piezoelectric composite layers.

As shown in Figure 2.9a, the energy harvester mainly consists of a piezoelectric layer with electromechanical-conversion characteristics and a metal layer that takes on the role of support. Meanwhile, it is assumed in the simulation analysis that the two layers are perfectly bonded to each other, the top and bottom electrodes are fully conductive, and a uniform electric field is generated on the piezoelectric layer, and it is also assumed that the arbitrary cross-section of the beam is always constant during the vibration of the cantilever beam. The cantilever beam root is rigidly fixed and the vibration excitation is along the direction of $w(x)$ in Figure 2.9b, i.e. perpendicular to the cantilever beam. The material properties and some parameter settings of the cantilever beam used in the simulation analysis are shown in Tables 2.6 and 2.7, respectively, where the structural parameters are not fixed. To study the effect of multiple factors on the conversion efficiency by the control variable method for multiple groups of cantilever beams with different structures, the simulation analysis is performed using the multi-physics field simulation software COMSOL Multiphysics.

Energy harvesting from ambient energy for activating low-powered electronics has emerged during the last decades as an enabling technology for

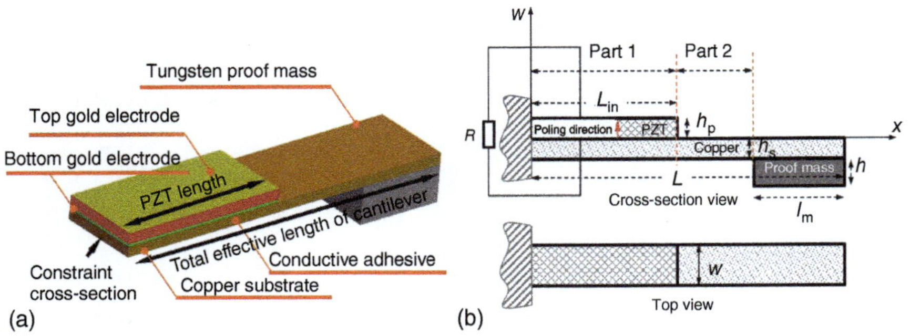

Figure 2.9 Overview of the proposed piezoelectric energy harvester (PEH). (a) Three-dimensional model of the proposed PEH. (b) Structural annotation of the proposed PEH for theoretical analysis.

Table 2.6 Materials properties.

Terms	PZT-5H	Beryllium	Tungsten
Young's modulus (GPa)	—	128	411
Density (kg/m^3)	7500	8250	19350
Poisson ratio	—	0.3	0.28
Elastic modulus matrix (GPa)	$\begin{bmatrix} 127.21 & 80.21 & 84.67 & 0 & 0 & 0 \\ 0 & 127.21 & 84.67 & 0 & 0 & 0 \\ 0 & 0 & 117.44 & 0 & 0 & 0 \\ 0 & 0 & 0 & 22.99 & 0 & 0 \\ 0 & 0 & 0 & 0 & 22.99 & 0 \\ 0 & 0 & 0 & 0 & 0 & 23.47 \end{bmatrix}$		
Piezoelectric coupling matrix (C/m^2)	$\begin{bmatrix} 0 & 0 & 0 & 0 & 17.03 & 0 \\ 0 & 0 & 0 & 17.03 & 0 & 0 \\ -6.62 & -6.62 & 23.24 & 0 & 0 & 0 \end{bmatrix}$		

Table 2.7 Structural dimensions and set parameters.

Terms	Value	Unit
Total length of the cantilever L	15	mm
Width of the cantilever w	2.5	mm
Volume of proof mass	$5 \times 2.5 \times 0.2$	mm^3
Thickness of the substrate h_s	0.05	mm
Thickness of the piezoelectric layer h_p	0.05	mm
Acceleration g	0.5	g

self-powered devices [44–48]. The piezoelectric effect is commonly used for energy harvesters or sensors due to its advantages of simple configuration and high-energy-conversion efficiency [49–51]. Currently, most investigations for power generation of vibration-based piezoelectric energy harvesters (PEHs) focus on high-output power, low resonant frequency, and good reliability [52–55]. Among these studies, PEHs are commonly based on a cantilever structure with a large proof mass at the free end, which is contributed to two reasons [16, 22]. First, for a given force input, the cantilever configuration can result in the highest average strain, which is beneficial to obtain high-output power. Second, the cantilever configuration has a low resonant frequency for a given size, which is easy to meet the low-frequency ambient vibrations.

Most cantilever-based PEHs consist of the piezoelectric and supporting layers with the same or custom length. However, the subtle changes in cantilever configuration can generate great influence on the output power, especially on microscale [56]. To learn the effect of the piezoelectric layer length on the output performance, Gao et al. [57] recently explored the relationship between the length ratio of the nonpiezoelectric layer to the piezoelectric layer and the output voltage under a concentrated static force at the cantilever tip. The results indicated that the optimal voltage was obtained when the piezoelectric layer and the nonpiezoelectric layer have the same length under constant tip displacement. However, in view of most of the practical application scenarios, PEHs are expected to capture the ambient vibration energy, not the static force or displacement. Therefore, a further study for such cantilever PEHs under the excitation of harmonic vibration was carried out by Gao et al. [58]. The results indicated that the maximum power under the optimal resistance occurred at the same length of the piezoelectric layer and the nonpiezoelectric layer. However, the additional mechanical impedance from the defect of the theoretical model influenced their calculation results, leading to that the output power and beam displacement being underestimated. Moreover, the nonpiezoelectric layer thickness has been studied as the only influence factor of the optimal nonpiezoelectric/piezoelectric layer length ratio. Thus, it cannot be applied to the PEHs with a concentrated proof mass at the tip of cantilever. Patel et al. [59] focused on the optimization of the microscale cantilever PEHs under the constant frequency through investigating the influence of geometrical parameters. The theoretical analysis is based on the combination of a distributed parameter electromechanical model and finite element analysis (FEA). Yu et al. [60] presented a vibration-based MEMS piezoelectric energy harvester, where the influence of piezoelectric patch length, width, and thickness on the output voltage was also discussed. Sunithamani et al. [61] presented enhancing the performance of energy harvester by optimizing the lead zirconate titanate (PZT) length. Although these works indicated that performance was improved by optimizing the piezoelectric layer geometry, the optimization schemes are still not systematic. Furthermore, the effect of other key design parameters on the optimal piezoelectric layer design and its underlying mechanism is still not clear. In addition, the natural frequency of 244.4 Hz is higher than the frequency of most natural vibrational sources for practical applications. Therefore, it is significant to optimize the piezoelectric layer length of the low-frequency PEHs with a proof mass for high performance. To determine the optimal PZT layout length, it is necessary to investigate the effects of geometric and material parameters on this length and reveal the foundational optimization mechanism for the piezoelectric cantilever harvester with a tip-proof mass, which contributes to high-conversion efficiency at low-frequency harmonic vibrations.

2.4.1 Electromechanical Coupling Analysis

The nonuniform cantilever piezoelectric energy harvester consists of a piezoelectric layer, a metal layer, and a proof mass (Figure 2.9a). The piezoelectric layer partly covers the clamped end of substrate layer, and the proof mass is attached to the free end. The piezoelectric layer is bonded to the substrate layer. Top and bottom electrodes are perfectively conductive so that a uniform electric field can be defined across the piezoelectric layer. Under the persistent excitation of the translational base acceleration, the piezoelectric cantilever undergoes continual vibration and generates electrical outputs. The arbitrary cross-sections of the beam are assumed to remain plane constantly in this process. The thickness of the stepped beam is designed to be small compared to its overall length such that the effects of shear deformation and rotary inertia are neglected. Therefore, the Euler–Bernoulli beam theory can be used to establish a mathematical model for the piezoelectric energy harvester. Owing to the nonuniform cross-section of the piezoelectric cantilever, for theoretical analysis, the piezoelectric cantilever is divided into two parts, namely, the ceramic–metal composite section (Part I) and metal single section (Part II), as shown in Figure 2.9b. The coordinate systems and geometric dimensions are depicted in Figure 1b.

A continuous system theoretical model is established to obtain the optimal output power with varied L_{in}. The paraphrase and properties of the parameters used in the following theoretical derivation and simulation calculations are shown in Tables 2.6–2.8.

Given that the unique input for this system is from the base periodic vibration, the absolute displacement $w(x, t)$ of the cantilever at position x and time t can be expressed as [62] $w(x, t) = w_{\text{b}}(x, t) + w_{\text{rel}}(x, t)$. If the base movement is resolved into a translation component and a small rotation component, then the base motion in the transverse direction can be expressed as follows [63]: $w_{\text{b}}(x, t) = g(t) + xh(t)$. The transverse relative displacement can be expressed using the method of separation of variables as [64]

$$w_{\text{rel}}(x,t) = \sum_{n=1}^{\infty} \phi_n(x)\eta_n(t) = \begin{cases} w_{\text{rel}}^{n1}(x,t) = \sum_{n=1}^{\infty} \phi_{n1}(x)\eta_n(t), & 0 < x \le L_{\text{in}} \\ w_{\text{rel}}^{n2}(x,t) = \sum_{n=1}^{\infty} \phi_{n2}(x)\eta_n(t), & L_{\text{in}} < x \le L \end{cases}. \tag{2.4}$$

The partial differential motion equation governing the forced vibration of the piezoelectric cantilever, including the effect of the internal damping (strain rate damping c_{s}) and external damping (viscous air damping c_{a}), can be written as [65, 66]:

$$\frac{\partial^2}{\partial x^2}\left[\text{EI}(x)\frac{\partial^2 w_{\text{rel}}(x,t)}{\partial^2 x} + c_{\text{s}} I(x)\frac{\partial^3 w_{\text{rel}}(x,t)}{\partial^2 x \partial t}\right] + c_{\text{a}}\frac{\partial w_{\text{rel}}(x,t)}{\partial t} + m(x)\frac{\partial^2 w_{\text{rel}}(x,t)}{\partial t^2}$$
$$= -[m(x) + M_t \delta(x - L)]\frac{\partial^2 w_{\text{b}}(x,t)}{\partial t^2} \tag{2.5}$$

Both of the damping instances in this case are assumed proportional.

Table 2.8 Nomenclature.

Parameters	Paraphrase
L_{in}	The length of the composite section
L	The overall length of the cantilever
t_{p}	The thickness of the piezoelectric layer
t_{m}	The thickness of the metal layer
w	The width of the cantilever
h	The thickness of the proof mass
l_{m}	The length of the proof mass
ρ_{p}	The *density* of the piezoelectric layer
ρ_{m}	The *density* of the metal layer
c_{s}	The equivalent strain rate damping coefficient
c_{a}	The equivalent viscous air damping coefficient
M_{t}	The mass of the proof mass
$w_{\text{rel}}(x, t)$	The transverse *relative displacement* of the cantilever relative to the clamped base
$w_{\text{b}}(x, t)$	The base movement
$g(t)$	The translation in the transverse direction
$h(t)$	Small rotation in the transverse direction
$\phi_n(x)$	The mass-normalized mode shape function of the nth vibration mode under the condition of the corresponding undamped free vibration
$\eta_n(t)$	The generalized coordinate function in the nth vibration mode
$\phi_{r1}(x)$	The rth mode shape functions for the Part I of the beam
$\phi_{s1}(x)$	The sth mode shape functions for the Part I of the beam
$\phi_{r2}(L)$	The rth mode shape functions for the Part II of the beam
$\phi_{s2}(L)$	The sth mode shape functions for the Part II of the beam
$M(x, t)$	The internal bending moment excluding the internal damping in the *cross-section* of the beam
$S_1^{p_{n1}}(x, y, t)$	The internal axial strains of the piezoelectric layer on the cross-sections of Part I of the beam
$S_1^{S_{n1}}(x, y, t)$	The internal axial strains of the metal layer on the cross-sections of Part I of the beam
$S_1^{S_{n2}}(x, y, t)$	The internal axial strain of the metal layer on the cross-sections of Part II of the beam
d_{31}	The piezoelectric constant
$v(t)$	The *potential difference* (i.e. voltage) across the piezoelectric layer in three-direction

Table 2.8 (Continued)

Parameters	Paraphrase
D_3	the electric displacement of the piezoelectric layer
$C_{n11} - C_{n24}$	The modal amplitude coefficients
β_{n1}, β_{n2}	The wavenumbers for the two parts of the beam of the nth vibration mode
χ_n	$\chi_n = \vartheta \left.\frac{\mathrm{d}\phi_{n1}(x)}{\mathrm{d}x}\right\|_{x=L_{\mathrm{in}}}$ The backward piezoelectric coupling coefficient in modal coordinates
ζ_n	$\zeta_n = \frac{c_{\mathrm{s}}\omega_n}{2E_{n1}} + \frac{c_{\mathrm{a}}}{2m_{n1}\omega_n}$ or $\zeta_n = \frac{c_{\mathrm{s}}\omega_n}{2E_{n2}} + \frac{c_{\mathrm{a}}}{2m_{n2}\omega_n}$ The nth vibration mode mechanical damping ratio
m	$m_{n1} = \rho_{\mathrm{p}} w t_{\mathrm{p}} + \rho_{\mathrm{m}} w t_{\mathrm{m}}, \quad 0 < x \leq L_{\mathrm{in}}$ $m_{n2} = \rho_{\mathrm{m}} w t_{\mathrm{m}}, \quad L_{\mathrm{in}} < x \leq L$ The mass per unit length
I_{t0}	$I_{\mathrm{t0}} = \frac{1}{12} M_{\mathrm{t}} \left(h^2 + l_{\mathrm{m}}^2\right)$ The rotational inertia of tip mass bulk
I_{t}	$I_{\mathrm{t}} = I_{\mathrm{t0}} + M_{\mathrm{t}} \left(\frac{l_{\mathrm{m}}^2}{2}\right)^2$ The rotational inertia about the right end of the beam
ϖ_{rs}	$\varpi_{rs} = \begin{cases} 0, & r \neq s \\ 1, & r = s \end{cases}$ The Kronecker delta
$E_3(t)$	$E_3(t) = -v(t)/t_{\mathrm{p}}$ The electric field in three-direction (i.e. y-direction)
$\delta(x)$	$\int_{-\infty}^{\infty} \frac{\mathrm{d}\delta(x - x_0)}{\mathrm{d}x^{(n)}} f(x)dx = -\frac{df(x_0)}{\mathrm{d}x}$ The Dirac delta function
I	$I_{n1} = \frac{Awt_{\mathrm{m}}^3}{12} + \frac{wt_{\mathrm{p}}^3}{12} + \frac{Awt_{\mathrm{m}}t_{\mathrm{p}}}{4}\frac{(t_{\mathrm{m}} + t_{\mathrm{p}})^2}{At_{\mathrm{m}} + t_{\mathrm{p}}}, \quad 0 < x \leq L_{\mathrm{in}}$ $I_{n2} = \frac{wt_{\mathrm{m}}^3}{12}, \quad L_{\mathrm{in}} < x \leq L$ $A = E_{\mathrm{m}}/E_{\mathrm{p}}$ The moment of inertia of cross-section
EI	$\mathrm{EI}_{n1} = \frac{E_{\mathrm{p}}wt_{\mathrm{p}}^3}{12}\left(\frac{A^2B^4 + 4AB^3 + 6AB^2 + 4AB + 1}{AB + 1}\right), \quad 0 < x \leq L_{\mathrm{in}}$ $\mathrm{EI}_{n2} = \frac{E_{\mathrm{m}}wt_{\mathrm{m}}^3}{12}, \quad L_{\mathrm{in}} < x \leq L$ $B = t_{\mathrm{m}}/t_{\mathrm{p}}$ The flexural rigidity

(Continued)

Table 2.8 (Continued)

Parameters	Paraphrase
$M_{n1}(x,t)M_{n2}(x,t)$	$M_{n1}(x,t) = -\left(\int_{t_a}^{t_b} T_1^{S_{n1}}(x,y,t)yw\ \mathrm{d}y + \int_{t_b}^{t_c} T_1^{p_{n1}}(x,y,t)yw\ \mathrm{d}y\right)$, $0 < x \le L_{\mathrm{in}}$ $M_{n2}(x,t) = -\int_{-t_s/2}^{t_s/2} T_1^{S_{n2}}(x,y,t)yw\,\mathrm{d}y$, $L_{\mathrm{in}} < x \le L$ The internal bending moment
$T_1^{p_{n1}}$	$T_1^{p_{n1}}(x,y,t) = E_p\left[S_1^{p_{n1}}(x,y,t) - d_{31}E_3(t)\right]$ The piezoelectric layer on the cross-sections of Part I of the beam
$T_1^{S_{n1}}$	$T_1^{S_{n1}}(x,y,t) = E_s S_1^{S_{n1}}(x,y,t)$ The metal layer on the cross-sections of Part I of the beam
$T_1^{S_{n2}}$	$T_1^{S_{n2}}(x,y,t) = E_s S_1^{S_{n2}}(x,y,t)$ The metal layer on the cross-sections of Part II of the beam
t_{pa}	$t_{\mathrm{pa}} = \dfrac{t_{\mathrm{p}}^2 + 2At_{\mathrm{p}}t_{\mathrm{s}} + At_{\mathrm{s}}^2}{2(t_{\mathrm{p}} + At_{\mathrm{s}})}$ $t_a = -t_{\mathrm{sa}}$, $t_b = t_{\mathrm{pa}} - t_{\mathrm{p}}$ and $t_c = t_{\mathrm{pa}}$ The distance between the top surface of the piezoelectric layer and the neutral layer of Part II of the beam
t_{sa}	$t_{\mathrm{sa}} = \dfrac{t_{\mathrm{p}}^2 + 2t_{\mathrm{p}}t_{\mathrm{s}} + At_{\mathrm{s}}^2}{2(t_{\mathrm{p}} + At_{\mathrm{s}})}$ The distance between the bottom surface of the metal layer and the neutral layer of Part II of the beam
ϑ	$\vartheta = -\dfrac{wE_{\mathrm{p}}d_{31}}{2t_{\mathrm{p}}}\left(t_c^2 - t_b^2\right)$ The backward piezoelectric coupling coefficient
e_{31}	$e_{31} = d_{31}E_{\mathrm{p}}$ The piezoelectric stress constant
$\boldsymbol{C}_{n1}, \boldsymbol{C}_{n2}$	$\boldsymbol{C}_{n1} = \left\{C_{n11}\ \ C_{n12}\ \ C_{n13}\ \ C_{n14}\right\}^T$ $\boldsymbol{C}_{n2} = \left\{C_{n21}\ \ C_{n22}\ \ C_{n23}\ \ C_{n24}\right\}^T$ The modal amplitude coefficients vectors
ε_{33}^S	$\varepsilon_{33}^S = \varepsilon_{33}^T - d_{31}^2E_{\mathrm{p}}$ The permittivity at constant strain
$\boldsymbol{Z}_{n11}$	$\boldsymbol{Z}_{n11} = \begin{bmatrix} 0 & 1 & 0 & 1 \\ \beta_{n1} & 0 & \beta_{n1} & 0 \\ 0 & 0 & 0 & 0 \\ 0 & 0 & 0 & 0 \end{bmatrix}$ The parameters vectors

Table 2.8 (Continued)

Parameters	Paraphrase
$\boldsymbol{C}_{n12}$	$\boldsymbol{C}_{n12} = \begin{bmatrix} 0 & 0 \\ 0 & 0 \\ -\beta_{n2}^2 \sin(\beta_{n2}L) - \theta_I\beta_{n2}\cos(\beta_{n2}L) & -\beta_{n2}^2\cos(\beta_{n2}L) + \theta_I\beta_{n2}\sin(\beta_{n2}L) \\ -\beta_{n2}^3\cos(\beta_{n2}L) - \theta_M\sin(\beta_{n2}L) & \beta_{n2}^3\sin(\beta_{n2}L) - \theta_M\cos(\beta_{n2}L) \\ \\ 0 & \\ 0 & \\ \beta_{n2}^2\sinh(\beta_{n2}L) - \theta_I\beta_{n2}\cos h(\beta_{n2}L) & \\ \beta_{n2}^3\cosh(\beta_{n2}L) - \theta_M\sinh(\beta_{n2}L) & \end{bmatrix}$ $\theta_I = \beta_{n1}^4 \dfrac{\mathrm{EI}_{n1}I_t}{\mathrm{EI}_{n2}m_{n1}}, \theta_M = -\beta_{n1}^4 \dfrac{\mathrm{EI}_{n1}M_t}{\mathrm{EI}_{n2}m_{n1}}$
$\boldsymbol{C}_{n21}$	$\boldsymbol{C}_{n21} = \begin{bmatrix} \sin(\beta_{n1}L_{\text{in}}) & \cos(\beta_{n1}L_{\text{in}}) \\ \beta_{n1}\cos(\beta_{n1}L_{\text{in}}) & -\beta_{n1}\sin(\beta_{n1}L_{\text{in}}) \\ -\mathrm{EI}_{n1}\beta_{n1}^2\sin(\beta_{n1}L_{\text{in}}) & -\mathrm{EI}_{n1}\beta_{n1}^2\cos(\beta_{n1}L_{\text{in}}) \\ -\mathrm{EI}_{n1}\beta_{n1}^3\cos(\beta_{n1}L_{\text{in}}) & \mathrm{EI}_{n1}\beta_{n1}^3\sin(\beta_{n1}L_{\text{in}}) \\ \\ \sinh(\beta_{n1}L_{\text{in}}) & \cosh(\beta_{n1}L_{\text{in}}) \\ \beta_{n1}\cosh(\beta_{n1}L_{\text{in}}) & \beta_{n1}\sinh(\beta_{n1}L_{\text{in}}) \\ \mathrm{EI}_{n1}\beta_{n1}^2\sinh(\beta_{n1}L_{\text{in}}) & \mathrm{EI}_{n1}\beta_{n1}^2\cosh(\beta_{n1}L_{\text{in}}) \\ \mathrm{EI}_{n1}\beta_{n1}^3\cosh(\beta_{n1}L_{\text{in}}) & \mathrm{EI}_{n1}\beta_{n1}^3\sinh(\beta_{n1}L_{\text{in}}) \end{bmatrix}$
$\boldsymbol{C}_{n22}$	$\boldsymbol{C}_{n22} = \begin{bmatrix} -\sin(\beta_{n2}L_{\text{in}}) & -\cos(\beta_{n2}L_{\text{in}}) \\ -\beta_{n2}\cos(\beta_{n2}L_{\text{in}}) & \beta_{n2}\sin(\beta_{n2}L_{\text{in}}) \\ \mathrm{EI}_{n2}\beta_{n2}^2\sin(\beta_{n2}L_{\text{in}}) & \mathrm{EI}_{n2}\beta_{n2}^2\cos(\beta_{n2}L_{\text{in}}) \\ \mathrm{EI}_{n2}\beta_{n2}^3\cos(\beta_{n2}L_{\text{in}}) & -\mathrm{EI}_{n2}\beta_{n2}^3\sin(\beta_{n2}L_{\text{in}}) \\ \\ -\sinh(\beta_{n2}L_{\text{in}}) & -\cosh(\beta_{n2}L_{\text{in}}) \\ -\beta_{n2}\cosh(\beta_{n2}L_{\text{in}}) & -\beta_{n2}\sinh(\beta_{n2}L_{\text{in}}) \\ -\mathrm{EI}_{n2}\beta_{n2}^2\sinh(\beta_{n2}L_{\text{in}}) & -\mathrm{EI}_{n2}\beta_{n2}^2\cosh(\beta_{n2}L_{\text{in}}) \\ -\mathrm{EI}_{n2}\beta_{n2}^3\cosh(\beta_{n2}L_{\text{in}}) & -\mathrm{EI}_{n2}\beta_{n2}^3\sinh(\beta_{n2}L_{\text{in}}) \end{bmatrix}$

The mass-normalized mode shape function $\phi_{n1}(x)$ and $\phi_{n2}(x)$ can be, respectively, expressed as:

$$\begin{cases} \phi_{n1}(x) = C_{n11} \sin \beta_{n1} x + C_{n12} \cos \beta_{n1} x + C_{n13} \sinh \beta_{n1} x & 0 < x \le L_{\text{in}} \\ \quad + C_{n14} \cosh \beta_{n1} x & \\ \phi_{n2}(x) = C_{n21} \sin \beta_{n2} x + C_{n22} \cos \beta_{n2} x + C_{n23} \sinh \beta_{n2} x & L_{\text{in}} < x \le L \\ \quad + C_{n24} \cosh \beta_{n2} x, & \end{cases} \tag{2.6}$$

Note that the unknown modal amplitude coefficients (C_{n11}–C_{n24}) and the values of two wavenumbers (β_{n1}, β_{n2}) in Eq. (2.3) can be obtained by applying the following conditions:

(a) Boundary conditions
The clamped end of the cantilever:

$$\phi_{n1}(0) = 0, \left.\frac{\mathrm{d}\phi_{n1}(0)}{\mathrm{d}x}\right|_{x=0} = 0$$

The free end of the cantilever:

$$\left.\frac{\mathrm{d}^2\phi_{n2}(x)}{\mathrm{d}x^2}\right|_{x=L} = \frac{\omega^2 I_t}{\mathrm{EI}_{n2}} \left.\frac{\mathrm{d}\phi_{n2}(x)}{\mathrm{d}x}\right|_{x=L}, \left.\frac{\mathrm{d}^3\phi_{n2}(x)}{\mathrm{d}x^3}\right|_{x=L} = -\frac{\omega^2 M_t}{\mathrm{EI}_{n2}} \phi_{n2}(L)$$

(b) Continuity conditions
At the connection:

$$\phi_{n1}(L_{\text{in}}) = \phi_{n2}(L_{\text{in}}), \left.\frac{\mathrm{d}\phi_{n1}(x)}{\mathrm{d}x}\right|_{x=L_{\text{in}}} = \left.\frac{\mathrm{d}\phi_{n2}(x)}{\mathrm{d}x}\right|_{x=L_{\text{in}}}$$

$$\mathrm{EI}_{n1} \left.\frac{\mathrm{d}^2\phi_{n1}(x)}{\mathrm{d}x^2}\right|_{x=L_{\text{in}}} = \mathrm{EI}_{n2} \left.\frac{\mathrm{d}^2\phi_{n2}(x)}{\mathrm{d}x^2}\right|_{x=L_{\text{in}}}, \mathrm{EI}_{n1} \left.\frac{\mathrm{d}^3\phi_{n1}(x)}{\mathrm{d}x^3}\right|_{x=L_{\text{in}}}$$
$$= \mathrm{EI}_{n2} \left.\frac{\mathrm{d}^3\phi_{n2}(x)}{\mathrm{d}x^3}\right|_{x=L_{\text{in}}}$$

(c) Orthogonality conditions

$$\int_0^{L_{\text{in}}} m_{n1}\phi_{r1}(x)\phi_{s1}(x)\,\mathrm{d}x + \int_{L_{\text{in}}}^{L} m_{n2}\phi_{r2}(x)\phi_{s2}(x)\,\mathrm{d}x + \phi_{r2}(L)M_t\phi_{s2}(L)$$
$$+ \left.\frac{\mathrm{d}\phi_{r2}(x)}{\mathrm{d}x}\right|_{x=L} I_t \left.\frac{\mathrm{d}\phi_{s2}(x)}{\mathrm{d}x}\right|_{x=L} = \varpi_{rs}$$

(d) The same natural frequency for two parts
The natural frequency of the nth vibration mode can be expressed as $\omega_n = (\beta_{n1} L_{\text{in}})^2 \sqrt{\frac{\mathrm{EI}_{n1}}{m_{n1} L_{\text{in}}^4}}$ or $\omega_n = (\beta_{n2} L)^2 \sqrt{\frac{\mathrm{EI}_{n2}}{m_{n2} L^4}}$. Therefore, the wavenumber β_{n2} can be represented by the following expression: $\beta_{n2} = \beta_{n1}(\mathrm{EI}_{n1} m_{n2}/\mathrm{EI}_{n2} m_{n1})^{\frac{1}{4}}$.

2.4.2 Electrical Output Analysis of the PEHs

The internal bending moment excluding the internal damping in the cross-section of the cantilever can be expressed as

$$M(x,t) = \mathrm{EI}(x)\frac{\partial^2 w_{\mathrm{rel}}(x,t)}{\partial^2 x} \tag{2.7}$$

The internal bending moments of the two parts of the beam $M_{n1}(x, t)$ and $M_{n2}(x, t)$ are caused by the internal axial stresses on the cross-sections of each area and are given by

$$\begin{cases} M_{n1}(x,t) = -\left(\int_{t_a}^{t_b} T_1^{S_{n1}}(x,y,t)yw\,\mathrm{d}y + \int_{t_b}^{t_c} T_1^{P_{n1}}(x,y,t)yw\,\mathrm{d}y\right), & 0 < x \le L_{\mathrm{in}} \\ M_{n2}(x,t) = -\int_{-t_s/2}^{t_s/2} T_1^{S_{n2}}(x,y,t)yw\,\mathrm{d}y, \quad L_{\mathrm{in}} < x \le L \end{cases} \tag{2.8}$$

Furthermore, $T_1^{S_{n1}}$, $T_1^{S_{n2}}$, and $T_1^{P_{n1}}$ are provided by the metal material and the piezoelectric material constitutive relation, respectively:

$$\begin{aligned} T_1^{S_{n1}}(x,y,t) &= E_s S_1^{S_{n1}}(x,y,t) \\ T_1^{S_{n2}}(x,y,t) &= E_s S_1^{S_{n2}}(x,y,t) \\ T_1^{P_{n1}}(x,y,t) &= E_p\left[S_1^{P_{n1}}(x,y,t) - d_{31}E_3(t)\right] \end{aligned} \tag{2.9}$$

The electric field in three directions $E_3(t)$ (i.e. y-direction) is uniform across the piezoelectric layer, which can be expressed as

$$E_3(t) = -v(t)/t_{\mathrm{p}} \tag{2.10}$$

Based on the assumptions of Euler–Bernoulli beam theory, the strains are caused by pure bending only. Therefore, they indicate a directly proportional relationship with the curvature of the beam, as follows:

$$S_1(x,y,t) = -y\frac{\partial^2 w_{\mathrm{rel}}(x,t)}{\partial x^2} \tag{2.11}$$

After substituting Eqs. (2.6)–(2.9) into (2.5) and arranging these items, the internal bending moment of the two parts of the beam can be obtained as follows:

$$\begin{cases} M_{n1}(x,t) = \mathrm{EI}_{n1}\frac{\partial^2 w_{\mathrm{rel}}(x,t)}{\partial x^2} + \vartheta v(t), & 0 < x \le L_{\mathrm{in}} \\ M_{n2}(x,t) = \mathrm{EI}_{n2}\frac{\partial^2 w_{\mathrm{rel}}(x,t)}{\partial x^2}, & L_{\mathrm{in}} < x \le L \end{cases}$$

where ϑ is the backward piezoelectric coupling coefficient, which can be expressed as

$$\vartheta = -\frac{wE_{\mathrm{p}}d_{31}}{2t_{\mathrm{p}}}\left(t_c^2 - t_b^2\right)$$

To ensure the existence of $v(t)$ after the differential motion equation, the coupling term should be multiplied by $[H(x) - H(x - L_{\mathrm{in}})]$ in this case, where $H(x)$ is the Heaviside function. Hence, the internal bending moment of the two parts of the beam can be described as

$$\begin{cases} M_{n1}(x,t) = \mathrm{EI}_{n1}\frac{\partial^2 w_{\mathrm{rel}}(x,t)}{\partial x^2} + \vartheta v(t)[H(x) - H(x - L_{\mathrm{in}})], & 0 < x \leq L_{\mathrm{in}} \\ M_{n2}(x,t) = \mathrm{EI}_{n2}\frac{\partial^2 w_{\mathrm{rel}}(x,t)}{\partial x^2}, & L_{\mathrm{in}} < x \leq L \end{cases}$$

Thus, the coupled mechanical equation of motion in modal coordinates is obtained:

$$\frac{\mathrm{d}^2\eta_n(t)}{\mathrm{d}t^2} + 2\zeta_n\omega_n\frac{\mathrm{d}\eta_n(t)}{\mathrm{d}t} + \omega_n^2\eta_n(t) + \chi_n v(t) = f_n(t)$$

$$\begin{aligned} f_n(t) = &- m_{n1}\left[\frac{\partial^2 g(t)}{\partial t^2}\int_0^{L_{\mathrm{in}}}\phi_{n1}(x)\mathrm{d}x + \frac{\partial^2 h(t)}{\partial t^2}\int_0^{L_{\mathrm{in}}} x\phi_{n1}(x)\,\mathrm{d}x\right] \\ &- m_{n2}\left[\frac{\partial^2 g(t)}{\partial t^2}\int_{L_{\mathrm{in}}}^{L}\phi_{n2}(x)\mathrm{d}x + \frac{\partial^2 h(t)}{\partial t^2}\int_{L_{\mathrm{in}}}^{L} x\phi_{n2}(x)\,\mathrm{d}x\right] \\ &- M_t\phi_{n2}(L)\left[\frac{\partial^2 g(t)}{\partial t^2} + L\frac{\partial^2 h(t)}{\partial t^2}\right] \end{aligned}$$

Given that the only input of the electrical circuit coupling is the mechanical strain (in one-direction in the piezoelectric layer) caused by pure bending, the relevant piezoelectric constitutive equation can be reduced to the following expression [66]: $D_3 = e_{31}S_1^{p_{n1}} + \varepsilon_{33}^s E_3$. As the electrical load is the pure resistance R, the electric current generated from the piezoelectric layer can be obtained on the basis of the Gauss law as

$$\frac{\mathrm{d}}{\mathrm{d}t}\left(\int_A \boldsymbol{D}\cdot\boldsymbol{n}\mathrm{d}A\right) = \frac{v(t)}{R}$$

where $\boldsymbol{D}$ is the electric displacement vector of the piezoelectric layer, $\boldsymbol{n}$ is the outward unit normal vector, and A is the area of the piezoelectric layer covered by the electrodes.

$$\varepsilon_{33}^s\frac{wL_{\mathrm{in}}}{t_{\mathrm{p}}}\frac{\mathrm{d}v(t)}{\mathrm{d}t} + \frac{1}{R}v(t) = -e_{31}t_{\mathrm{pc}}w\int_0^{L_{\mathrm{in}}}\frac{\partial^3 w_{\mathrm{rel}}(x,t)}{\partial x^2\partial t}\mathrm{d}x$$

where t_{pc} is the distance from the center of the piezoelectric layer to the neutral layer.

$$C_{\mathrm{p}}\frac{\mathrm{d}v(t)}{\mathrm{d}t} + \frac{1}{R}v(t) = i_{\mathrm{p}}(t) \tag{2.12}$$

where C_{p} is the equivalent internal capacitance of the piezoelectric layer; $i_{\mathrm{p}}(t)$ is the effective internal current source of the piezoelectric layer. These two parameters are obtained as

$$C_{\mathrm{p}} = \varepsilon_{33}^s\frac{wL_{\mathrm{in}}}{t_{\mathrm{p}}}, i_{\mathrm{p}}(t) = \sum_{n=1}^{\infty}\kappa_n\frac{\mathrm{d}\eta_n(t)}{\mathrm{d}t}$$

where κ_n is the forward piezoelectric coupling coefficient which can be expressed as

$$\kappa_n = -e_{31}t_{\mathrm{pc}}w\left.\frac{\mathrm{d}\phi_{n1}(x)}{\mathrm{d}x}\right|_{x=L_{\mathrm{in}}}$$

In this case, the base motion is assumed persistent harmonic translation for convenience and simplicity. Thus, the translation $g(t)$ and small-rotation $h(t)$ in the transverse direction are also assumed to be the following forms: $g(t) = Y_0 e^{j\omega t}$, $h(t) = \theta_0 e^{j\omega t}$, where Y_0 and θ_0 are the amplitudes of the translation and rotation with the initial phase information, ω is the excitation frequency, and j is the imaginary number. Equations (2.11) and (2.12) indicate that the proposed cantilever piezoelectric energy harvester is a linear system. That is, the output of the system should be harmonic if the input indicates a harmonic form. Therefore, the steady-state output voltage response, the generalized coordinate function, and the generalized force can be written as $v(t) = V_o e^{j\omega t}$, $\eta_n(t) = H_n e^{j\omega t}$, $f_n(t) = F_n e^{j\omega t}$, respectively, where V_o is the amplitude of the output voltage response, H_n is the amplitude of the generalized coordinate function of the nth vibration mode, F_n is the generalized force of the nth vibration mode, and it can be arranged into the following form:

$$F_n = \omega^2 \left\{ m_{n1} \left[Y_0 \int_0^{L_{\text{in}}} \phi_{n1}(x)\mathrm{d}x + \theta_0 \int_0^{L_{\text{in}}} x\phi_{n1}(x)\,\mathrm{d}x \right] + m_{n2} \left[Y_0 \int_{L_{\text{in}}}^{L} \phi_{n2}(x)\mathrm{d}x + \theta_0 \int_{L_{\text{in}}}^{L} x\phi_{n2}(x)\mathrm{d}x \right] + M_t \phi_{n2}(L)(Y_0 + L\theta_0) \right\}$$

Although $v(t)$, $\eta_n(t)$, and $f_n(t)$ share similar expressions with each other, phase differences among them persist. Thus, setting Y_0 and θ_0 as real numbers and V_0 and H_n as complex numbers is reasonable.

After substituting $v(t) = V_0 e^{j\omega t}$, $\eta_n(t) = H_n e^{j\omega t}$, and $f_n(t) = F_n e^{j\omega t}$ into Eqs. (2.11) and (2.12) and then rearranging the terms, the mechanical–electrical coupling equations for the proposed PEH are finally obtained as the following:

$$\left(\omega_n^2 - \omega^2 + j2\zeta_n\omega_n\omega\right) H_n + \chi_n V_0 = F_n$$

$$\left(j\omega C_\text{p} + \frac{1}{R}\right) V_0 = j\omega \sum_{n=1}^{\infty} \kappa_n H_n$$

By solving the above simultaneous equations, the steady-state output voltage response and the generalized coordinate function can be respectively expressed as

$$v(t) = \frac{\sum_{n=1}^{\infty} \frac{j\omega\kappa_n F_n}{\omega_n^2 - \omega^2 + j2\zeta_n\omega_n\omega}}{\frac{1}{R} + j\omega C_\text{p} + \sum_{n=1}^{\infty} \frac{j\omega\kappa_n\chi_n}{\omega_n^2 - \omega^2 + j2\zeta_n\omega_n\omega}} e^{j\omega t} \tag{2.13}$$

Given that the pure resistance R is the electrical load, the output current and power can be calculated as

$$i(t) = \frac{v(t)}{R}, p(t) = \frac{v^2(t)}{R} \tag{2.14}$$

Finally, the relative displacement response for the stepped beam over the entire length can be expressed by substituting Eq. (2.15) into Eq. (2.4) as

$$\eta_n(t) = \left(F_n - \chi_n \frac{\sum_{n=1}^{\infty} \frac{j\omega\kappa_n F_n}{\omega_n^2 - \omega^2 + j2\zeta_n\omega_n\omega}}{\frac{1}{R} + j\omega C_\text{p} + \sum_{n=1}^{\infty} \frac{j\omega\kappa_n\chi_n}{\omega_n^2 - \omega^2 + j2\zeta_n\omega_n\omega}} \right) \frac{1}{\omega_n^2 - \omega^2 + j2\zeta_n\omega_n\omega} e^{j\omega t} \tag{2.15}$$

$$w_{\rm rel}^{n1}(x,t)=\sum_{n=1}^{\infty}\left[F_n-\chi_n\left(\frac{\sum_{n=1}^{\infty}\frac{j\omega\kappa_n F_n}{\omega_n^2-\omega^2+j2\zeta_n\omega_n\omega}}{\frac{1}{R}+j\omega C_{\rm p}+\sum_{n=1}^{\infty}\frac{j\omega\kappa_n\chi_n}{\omega_n^2-\omega^2+j2\zeta_n\omega_n\omega}}\right)\right]\times\frac{\phi_{n1}(x)}{\omega_n^2-\omega^2+j2\zeta_n\omega_n\omega}e^{j\omega t},\quad 0<x\leq L_{\rm in} \tag{2.16}$$

$$w_{\rm rel}^{n2}(x,t)=\sum_{n=1}^{\infty}\left[F_n-\chi_n\left(\frac{\sum_{n=1}^{\infty}\frac{j\omega\kappa_n F_n}{\omega_n^2-\omega^2+j2\zeta_n\omega_n\omega}}{\frac{1}{R}+j\omega C_{\rm p}+\sum_{n=1}^{\infty}\frac{j\omega\kappa_n\chi_n}{\omega_n^2-\omega^2+j2\zeta_n\omega_n\omega}}\right)\right]\times\frac{\phi_{n2}(x)}{\omega_n^2-\omega^2+j2\zeta_n\omega_n\omega}e^{j\omega t},\quad L_{\rm in}<x\leq L \tag{2.17}$$

2.4.3 Output Performance of the PEHs

Figure 2.10 shows the effect of PZT layout length on the output power under the varied external resistances based on the results from theoretical analysis.

Meanwhile, the effects of the width of cantilever, thickness of proof mass and PZT layer, and the total length of cantilever on the optimal PZT layout length are analyzed

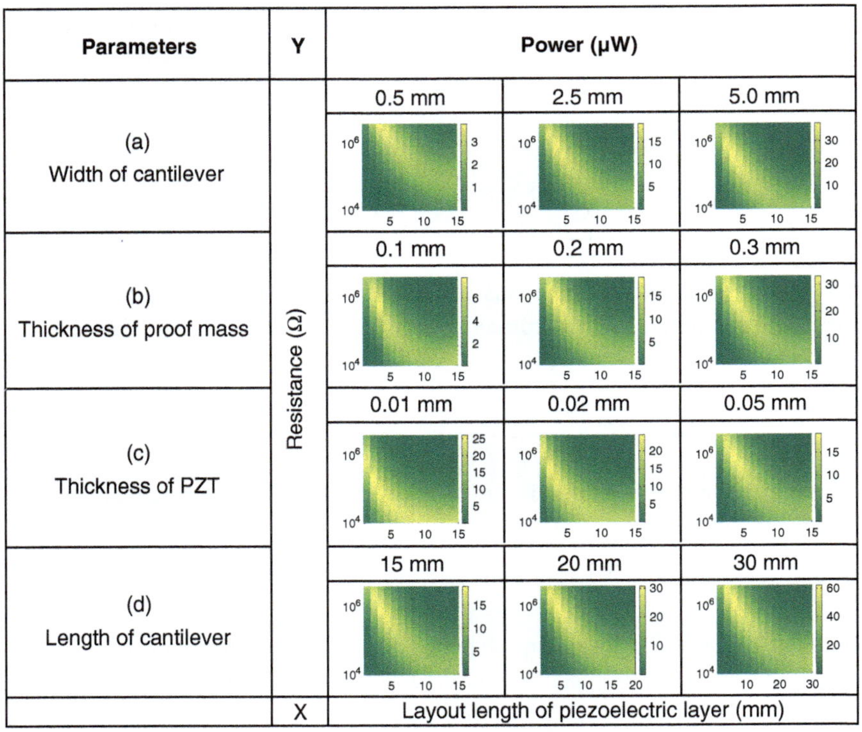

Figure 2.10 Effect of length of piezoelectric layer on output power and the effects of width of cantilever (a), thickness of proof mass (b), thickness of piezoelectric layer (c), and total length of cantilever (d) on the optimal length of piezoelectric layer. The applied acceleration amplitude is 0.5g. The data in (a)–(c) came from theoretical analysis and FEA.

by FEA and theoretical analysis. Given the widths of 0.5, 2.5, and 5 mm, the optimal PZT layout lengths are 3.0, 2.5, and 2.0 mm, thereby indicating that the ratios of PZT layout length and the total length of cantilever are 0.2, 0.17, and 0.13, as shown in Figure 2.10a. Thus, the optimal PZT layout length decreases as cantilever width increases. The output power increases and the optimal external resistance decreases because of the increase of the efficient functional area, which indicates the same change proportion as the width increases.

Figure 2.10b shows the effect of thickness of the proof mass on the optimal PZT layout length. By analyzing the results with thicknesses of 0.1, 0.2, and 0.3 mm, the output power approximately increases with the same increasing proportion of the thickness of the proof mass. However, the thickness of the proof mass indicates a minimal effect on the optimal PZT layout length. The optimal resistance indicates minimal increase as the thickness of the proof mass increases. Figure 2.10c describes the effect of the thickness of the piezoelectric layer on the optimal PZT layout length. As the thickness of the piezoelectric layer decreases from 0.05 to 0.01 mm, the effect becomes increasingly low, although the optimal PZT layout length remains 3 mm. When the thickness is decreased to 0.01 mm, the maximum output power obtained at PZT layout length of 3 mm indicates a minimal distance from other layout lengths. However, the thin piezoelectric layer can provide low optimal external resistance for the maximum output power. Figure 2.10d provides the effect of the total length of the cantilever on the optimal PZT layout length. The harvesters with total lengths of 15, 20, and 30 mm are analyzed. The results show the optimal PZT layout lengths of the harvesters are 2.5, 3.5, and 5.5 mm, thereby indicating that the ratios of PZT layout length and the total length of cantilever are 0.17, 0.175, and 0.18. The increase in the total length of the cantilever only makes the PZT layout length slightly increase, thereby demonstrating that the change in total length of cantilever indicates a minimal little effect on the PZT layout length.

Figure 2.11 shows the effects of the modulus of substrate and applied acceleration on the optimal PZT layout length. When the substrate is substituted by the

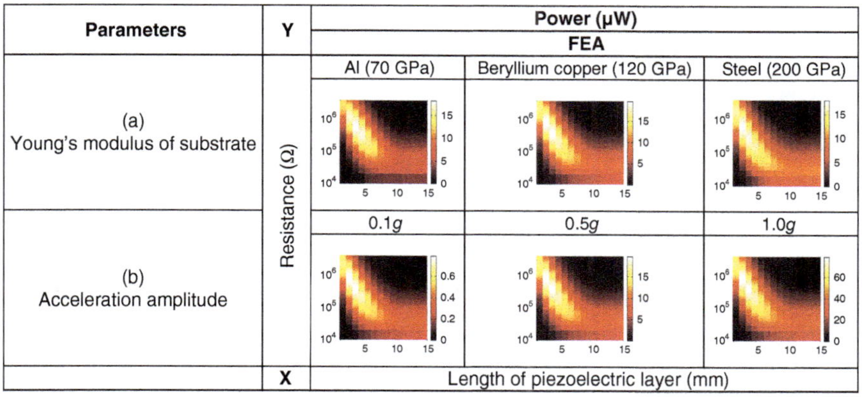

Figure 2.11 Effects of length of piezoelectric layer on output power and the effects of Young's modulus of substrate (a) and the applied acceleration amplitude (b) on the optimal length of piezoelectric layer. The data in (a) and (b) came from FEA.

aluminum, beryllium copper, and steel materials, thereby indicating Young's modulus values of 70, 120, and 200 GPa. The results shown in Figure 2.11a reveal that the optimal PZT layout lengths occur at 3, 2.5, and 2 mm when the harvester is with aluminum, beryllium copper, and steel substrate, respectively. This result demonstrates that the harvester with high modulus of substrate will indicate a short optimal PZT layout length. However, the varied applied accelerations indicate a minimal effect on the optimal PZT layout length, as shown in Figure 2.11b.

2.5 Finite Element Analysis

The piezoelectric energy harvester with proposed shapes was simulated in COMSOL Multiphysics 5.3a using solid mechanical, electrostatic, and circuit interfaces for setting the boundary conditions. The goal is to study the deformation, strain, and generated voltage and power of piezoelectric energy harvesters. The multilayer 3D models were developed based on FEA by applying mesh in models. The cantilever has been fixed at the clamping edge in the solid mechanical interface. For electrostatic boundary conditions, the bottom of the piezoelectric layer has been set as ground or zero potential, and the top of PZT is terminated to the external circuit along with the piezoelectric layer. The charge on the peripheral surface of the beam is considered to be zero. The eigenfrequency analysis is essential to estimate the resonance mode of the models. In addition, the frequency analysis based on harmonic excitation of the base is also carried out for the first resonant frequencies to calculate the maximum average power and output voltage generated by the harvesters under different external resistances. The effect of the FEA mesh is precise on the resonant frequency (Figure 2.12a) and maximum output power (Figure 2.12b). The FEA results only have a little difference in normal, fine, and finer mesh precision. Therefore, the practical simulation adopts the normal mesh precision.

2.5.1 Output Performance of the PEHs

2.5.1.1 Effect of Piezoelectric Layer Length on Cantilever Beam Energy Harvesting

Cantilever beam energy harvesters have been favored by many researchers because of their simple structure, easy processing, and good electromechanical-conversion efficiency. However, most of the current studies on cantilever-beam-type vibration energy harvesters mainly focus on the uniform cross-sectional beam structure composed of piezoelectric and substrate layers of the same length [22, 67]. Different piezoelectric layer arrangement lengths have a great influence on the output power of cantilever beam vibration energy harvesters, but there are relatively few related research reports, and no clear design guidance scheme for piezoelectric layer lengths is given. In this section, the finite element method is used to study the effect of the arrangement length of PZT piezoelectric ceramics (PZT-5H) along with the length of the cantilever beam on the output power of vibration energy harvesting, to optimize the structural design of the cantilever-beam-type energy harvester and

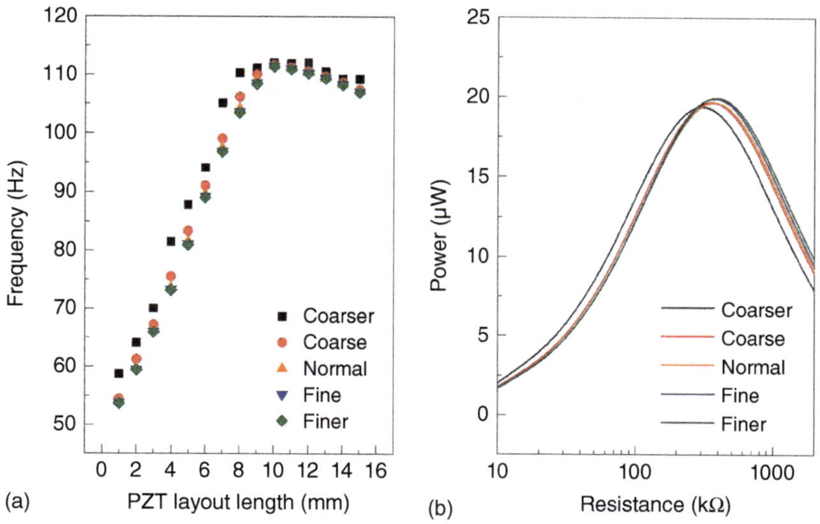

Figure 2.12 The effect of the FEA mesh precise on (a) the resonant frequency and (b) maximum output power.

improve the power density. The effect of different PZT arrangement lengths on the frequency, stress distribution, internal impedance, and maximum achievable power of the energy harvester is mainly investigated, which is an important guideline for the preparation of high-power density cantilever-beam-type vibration energy harvester.

First, the analytical model is modeled in three dimensions, and the first-order inherent frequencies of cantilever beams with different PZT arrangement lengths along with the x-direction from the root of the cantilever beam can be analyzed by parameterizing the length of the PZT arrangement. Figure 2.13a,b show the schematic diagrams of the piezoelectric cantilever beam with no mass bulk bonded to the end of the beam and with a mass bulk bonded to the end of the beam, both of which show an increase and then a decrease in the first-order intrinsic frequency with the increase of the PZT arrangement length (Figure 2.13c,d).

The eigenfrequency of the harvester increases first and then decreases with increasing the PZT layout length. The maximum eigenfrequencies occur at PZT layout lengths of 9 and 10 mm (i.e. the ratio of PZT and cantilever length is 0.6 and 0.67) for the harvester without and with proof mass, respectively. As the PZT layout length increases, the flexural rigidity of cantilever increases because of the increasing equivalent inertia moment, thereby causing an increase in the eigenfrequency. However, the varied PZT layout length changes the mass distribution of the cantilever, thereby also affecting the eigenfrequency. Generally, adding the proof mass is an efficient means to decrease the eigenfrequency of the cantilever harvester. When the PZT layout length contributes more to the mass distribution rather than to the flexural rigidity of the cantilever harvester, the eigenfrequency decreases with the increasing PZT layout length. In terms of the harvester without the proof mass, the increase rate of the eigenfrequency is approximately 17.73 Hz/mm and the

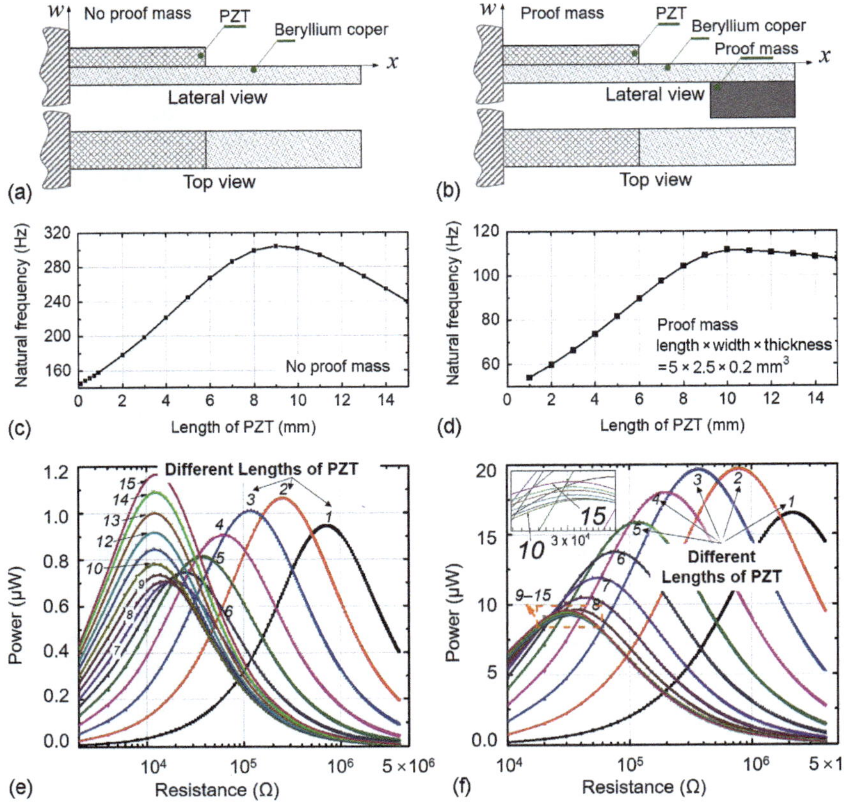

Figure 2.13 Comparison of the cantilever piezoelectric energy harvesters with and without the proof mass under the varied layout lengths of the piezoelectric layer, (a) piezoelectric cantilever without proof mass, (b) piezoelectric cantilever with proof mass, (c) effects of length of PZT on natural frequency of cantilever without proof mass. (d) Effects of length of PZT on natural frequency of cantilever with a proof mass. (e) The output power varied with loading resistance without proof mass. (f) The output power varied with loading resistance with a proof mass.

descending speed is approximately −10.91 Hz/mm. By contrast, the increase rate of the eigenfrequency is only 6.39 Hz/mm and the descending speed is approximately −0.86 Hz/mm for the harvester with the proof mass. Thus, the PZT layout length has less effect on the eigenfrequency of the harvester with the proof mass than the harvester without the proof mass because it indicates less effect on the mass distribution for the harvester with the proof mass.

Figure 2.13e,f show the output powers as a function of the external resistances of the harvester without and with the proof mass under the same excitation condition, respectively. Generally, the harvester with proof mass indicates obvious advantages in the output performance compared with that without proof mass. The maximum output power increases with the PZT layout length increasing first, then decreasing, and finally increasing. Under the condition of no proof mass, the maximum output power occurs at the equal length piezoelectric layer and cantilever, thereby indicating that the end of the cantilever occupies a higher mass proportion. Peak output power is found at the PZT layout length of 2 mm (i.e. PZT layout length/total length of cantilever is 1/7.5). When the harvester is bonded with a proof mass, the maximum output power occurs at the PZT layout length of 3.0 mm (i.e. PZT layout length/total length of cantilever is 1/5). Although the PZT layout length also changes the mass distribution for the harvester with the proof mass, the effect of the changes is negligible because of the existence of the proof mass. In the practical application, obtaining high-output power to meet the needs of electrical energy by adding mass at the end is a usual means of improving the output performance. Therefore, this study mainly explores the effect of PZT layout length on the harvester with a proof mass. Furthermore, the results of Figure 2.13d, f also show that the PEH with the optimal PZT layout length (i.e. PZT layout length/the total length of cantilever is 1/5) and a proof mass is nearly twice the maximum output power and half the first-mode shape natural frequency of the traditional PEHs with full coverage of PZT layer on the substrate.

2.5.1.2 Effect of Piezoelectric Layer Thickness on Cantilever Beam Energy Harvesting

Based on the study of cantilever beam vibration energy harvesters with different piezoelectric layer arrangement lengths with mass bulks, it is found that the optimal output power occurs when the PZT arrangement length is about 2–3 mm (i.e. 1/7.5–1/5 of the total cantilever beam length). Since the cantilever beam energy harvester is greatly affected by structural changes, whether the optimized piezoelectric layer arrangement length ratio is affected by different piezoelectric layer thicknesses, cantilever beam widths, cantilever beam lengths, and mass bulk size changes, as well as its influence law becomes necessary to guide the actual demand design of the optimal device structure, the following is the simulation analysis results for different piezoelectric layer thicknesses.

Figure 2.14 shows the distribution of the output power with the external resistance and the length of the piezoelectric layer arrangement when the thickness of the PZT layer is 0.01, 0.02, and 0.05 mm (other parameters are shown in Tables 2.1 and 2.2), respectively, and the excitation acceleration amplitude is 0.5g. By comparison, it can be seen that the optimal matching resistance still decreases with the

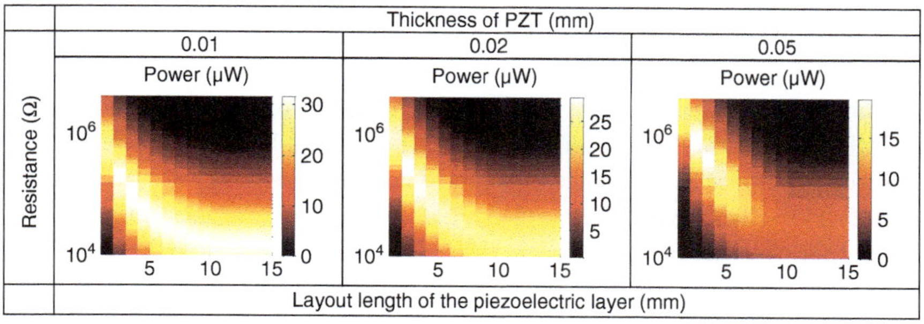

Figure 2.14 Effect of layout length of the piezoelectric layer on output power at the PZT thickness of 0.01, 0.02, and 0.05 mm when the applied acceleration amplitude is 0.5*g*.

increase of the length of the piezoelectric layer arrangement, and both of them level off after the length of 10 mm. This feature is mainly due to the fact that the area of the upper and lower electrodes of the piezoelectric layer increases with the length of the piezoelectric layer, which is equivalent to multiple equivalent capacitors in parallel, and the resonant frequency of the piezoelectric layer increases gradually before the length of the piezoelectric layer is 10 mm. The piezoelectric layer arrangement length after 10 mm due to its increased length for the overall piezoelectric layer arrangement length increase ratio gradually reduced, the matching resistance of the decrease gradually reduced, the piezoelectric layer arrangement length after 10 mm its resonant frequency began to decline, so reflected in the output matching resistance when the change tends to smooth.

When the thickness of piezoelectric layer is 0.01 mm, the maximum power value appears in the piezoelectric layer arrangement length of about 5–6 mm, and after the piezoelectric layer arrangement length is more than 3 mm, the difference in the maximum output power value is relatively small, and the influence of the piezoelectric layer arrangement length on the maximum output power is not significant at this time. When the piezoelectric layer thickness is 0.02 mm, the maximum power value appears in the piezoelectric layer arrangement length of about 3–4 mm, and it is obviously higher than the maximum power output value produced by other piezoelectric layer arrangement lengths. When the thickness of piezoelectric layer is 0.05 mm, the maximum power value appears around 2–3 mm of piezoelectric layer arrangement length, which is also significantly higher than the maximum power output value produced by other piezoelectric layer arrangement lengths. In summary, comparing three different thicknesses of piezoelectric layers shows that the influence of the piezoelectric layer arrangement length on the output power decreases gradually as the piezoelectric layer thickness increases, the piezoelectric layer arrangement length that produces the maximum output power tends to decrease, that is, the ratio of the piezoelectric layer arrangement length decreases and the output power value tends to decrease.

2.5.1.3 Effect of Cantilever Beam Width on Cantilever Beam Energy Harvesting

Based on the previous structures, the simulations were performed for the energy harvesters with cantilever beam widths (the width of the mass bulk also varies with) of 0.5 and 5.0 mm, respectively, and the results obtained are shown in Figure 2.15. It can be seen that the maximum output power appears at the cantilever beam width of 0.5 mm at the piezoelectric layer arrangement length of about 3.0 mm, at the width of 2.5 mm at the piezoelectric layer arrangement length of about 2.5 mm, and at the width of 5.0 mm at the piezoelectric layer arrangement length of about 2.0 mm, that is, the maximum output power appears at the piezoelectric layer arrangement length of about 2.0 mm when the cantilever beam width is 0.5, 2.5 mm; and the maximum output power occurs at 0.2, 0.17, and 0.13 of the total length of the cantilever beam for the piezoelectric layer arrangement at 5.0 mm, respectively, indicating that the optimal piezoelectric layer arrangement length is shortening with the increase of the cantilever beam width in this structure case. Among them, according to the

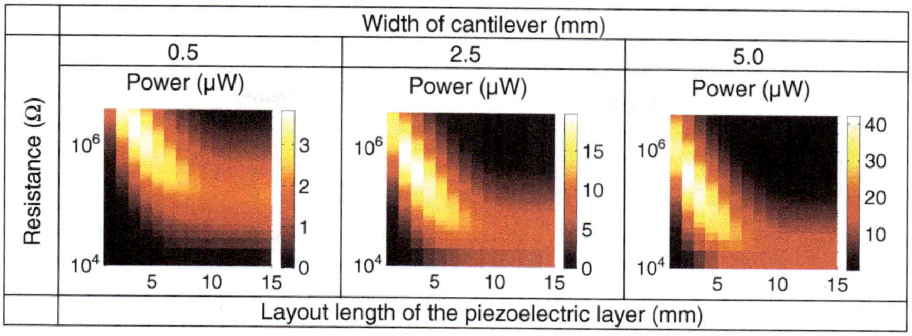

Figure 2.15 Effect of layout length of the piezoelectric layer on output power at the cantilever width of 0.5, 2.5, and 5.0 mm when the applied acceleration amplitude is 0.5*g*.

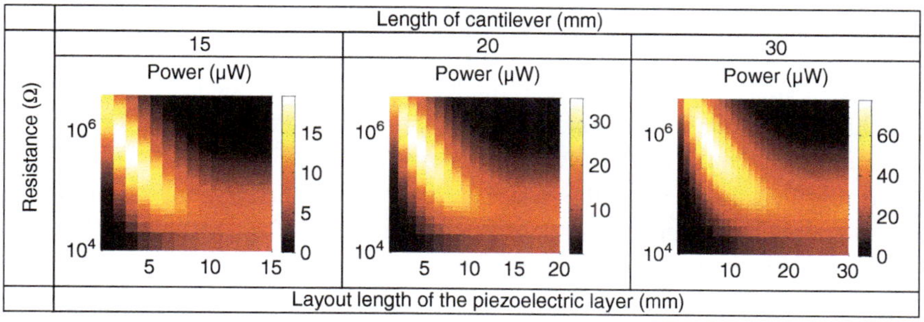

Figure 2.16 Effect of layout length of the piezoelectric layer on output power at the cantilever length of 15, 20, and 30 mm when the applied acceleration amplitude is 0.5*g*.

output power value can be seen that the width increase ratio is basically similar to the output power increase ratio, and as the width increases make the internal resistance decrease, the optimal external resistance is also decreasing.

2.5.1.4 Effect of Cantilever Beam Length on Cantilever Beam Energy Harvesting

As shown in Figure 2.16, the power output of different PZT arrangement lengths for different cantilever beam total lengths, respectively, it can be seen that the optimal PZT arrangement lengths are about 2.5, 3.5, and 5.5 mm for cantilever beam total lengths of 15, 20, and 30 mm (other parameters are shown in Tables 2.6 and 2.7), i.e. the PZT length to cantilever beam total length ratios are about 0.167, 0.175, and 0.183. It can be seen that the increase in the total length of the cantilever beam makes the optimized PZT arrangement length ratio increase slightly, which indicates that the increase of the cantilever beam length has a small effect on the optimized piezoelectric layer arrangement length ratio. However, the impact on the output power is larger, and the maximum output power that can be generated on the basis of this structure increases at a growth rate of approximately 3 μW/mm as the cantilever beam length increases.

2.5.1.5 Effect of Mass Bulk on Energy Harvesting From Cantilever Beam

The increase in mass bulk thickness is equivalent to the increase in weight of the mass bulk unit at the end of the cantilever beam, and the output power distribution for different piezoelectric layer arrangement lengths and different external resistances for mass bulk thicknesses of 0.1, 0.2, and 0.3 mm (other parameters are shown in Tables 2.6 and 2.7) is shown in Figure 2.17. Comparing the results of the analysis, it is found that several groups of results show that the power output is maximum at the piezoelectric layer arrangement length of about 2.5 mm, and the variation of the output power for different piezoelectric layer arrangement lengths remains the same for the three groups of different mass bulk thicknesses. Only the optimal external resistance increases slightly with the thickness of the mass bulk, as well as the power increases proportionally with the thickness (the growth rate is about 80 μW/mm), which is mainly due to the lower resonant frequency of the cantilever beam under the weight of the mass bulk, and the larger deformation generated by the low-frequency resonance, so the power can achieve greater output, and the matching resistance increases slightly due to the lower resonant frequency. Therefore, it shows that the thickness variation of the mass bulk under this structural condition has less influence on the optimal PZT arrangement length.

Taken together, the trend of the intrinsic frequency of the cantilever-beam-type vibration energy harvester (copper substrate – length × width × thickness = 15 mm × 2.5 mm × 0.05 mm, PZT – width × thickness = 2.5 mm × 0.05 mm) with and without the mass bulk increases and then decreases with the increase of the length of the piezoelectric layer arrangement, with the difference that the inflection point occurs at different locations. In the case of no mass bulk, with the changing length of the piezoelectric layer arrangement, the cantilever beam has the maximum intrinsic frequency at the piezoelectric layer arrangement length of 9.0 mm (i.e. the ratio of piezoelectric layer to cantilever beam length is 0.6), while after bonding the tungsten mass bulk (5 mm × 2.5 mm × 0.2 mm), the maximum intrinsic frequency of the cantilever beam appears at the piezoelectric layer arrangement length of 10.0 mm (i.e. the ratio of piezoelectric layer to cantilever beam length is 0.6). Secondly, by analyzing the power output at different piezoelectric layer arrangement lengths, it was

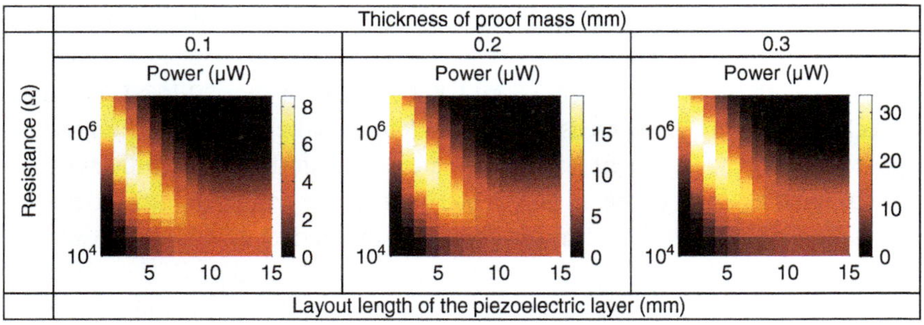

Figure 2.17 Effect of layout length of the piezoelectric layer on output power at the proof mass thickness of 0.1, 0.2, and 0.3 mm when the applied acceleration amplitude is 0.5*g*.

found that the optimal output power of the cantilever beam vibration energy harvester with mass bulk occurs at a piezoelectric layer arrangement length of about 2.5 mm (~1/5 of the total length of the cantilever beam). It is also seen that the cantilever beam energy harvester is affected by the changes of cantilever beam width, length, thickness, and mass bulk thickness as well as its influence law, and it is found that in this structure case, the optimal piezoelectric layer arrangement length is shortening as the cantilever beam width increases, where it can be seen that the ratio of width increase is basically similar to the ratio of output power increase according to the output power value; the thinner the piezoelectric layer will make the ratio of piezoelectric layer arrangement length to the thinner the piezoelectric layer, the smaller the effect of the piezoelectric layer arrangement length on the output power, but only affects the optimal matching resistance of the device; in addition, it is found that the increase of the total length of the cantilever beam and the thickness change of the mass bulk have little effect on the optimized piezoelectric layer arrangement length ratio.

2.5.1.6 Effect of Different Substrate Materials on the Performance of Composite Layer-Based Cantilever Beam Piezoelectric Energy Harvesters

Figure 2.18 shows the output power distribution for different external resistances and different piezoelectric layer arrangement lengths when the substrates are polyethylene terephthalate (PET polymer), aluminum Al, beryllium copper, and

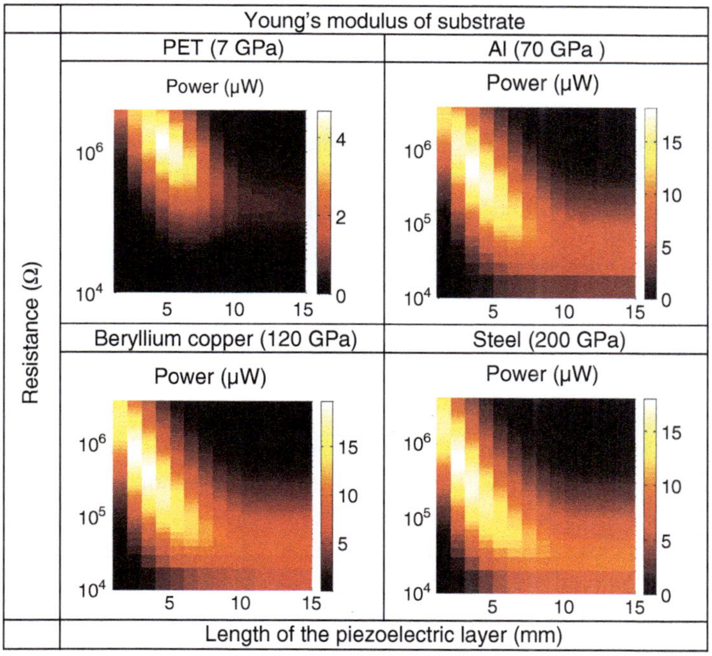

Figure 2.18 Effect of layout length of the piezoelectric layer on output power when the substrate materials are PET, Al, beryllium copper, and steel, respectively, when the applied acceleration amplitude is 0.5g.

stainless steel materials, respectively. In simulation analysis at room temperature 293.15 K, different substrate materials have different Young's modulus (PET 7 GPa, Al 70 GPa, beryllium copper 120 GPa, steel 200 GPa), which makes the cantilever beam structure have different flexural stiffness and different neutral layer positions; other parameters are shown in Tables 2.6 and 2.7.

According to the output power distribution of the simulation analysis, the optimal piezoelectric layer arrangement lengths are 4, 3, 2.5, and 2 mm when the substrate materials are PET, aluminum, beryllium bronze, and stainless steel, respectively, indicating that the shorter piezoelectric layer arrangement lengths can produce greater power output as Young's modulus of the substrate layer increases. However, the peak power output does not change much when the substrate materials are aluminum, beryllium bronze, and stainless steel, respectively, mainly because the modulus of PZT material in the bending direction is 84.6 GPa, and the thickness of piezoelectric layer and substrate layer is the same, the high-modulus substrate makes the resonant frequency of the device increase, and the structural deformation decreases, which is not conducive to the enhancement of power output, but the structural neutral layer is shifted toward the substrate layer, which is conducive to the enhancement of power output. The maximum power output capability is basically the same when the substrate of three materials is used. However, when the substrate material is a flexible PET polymer, the structural neutral layer is in the piezoelectric layer due to its ultralow modulus compared to the piezoelectric material, and the piezoelectric layer is subjected to two opposite strains in the lateral direction during the structural deformation, which makes the polarization charge direction partially cancel, thus resulting in a lower power output performance.

2.5.1.7 Effect of Excitation Intensity on the Performance of a Flexible Substrate Cantilever Beam Piezoelectric Energy Harvester

The maximum generated power [55] of the d_{31}-mode piezoelectric energy harvester with cantilever beam structure is reported to be expressed by the following equation:

$$\text{Power}_{\text{max}} = \frac{1}{8\pi}\left(\frac{e_{31}^2}{\varepsilon_r \varepsilon_0} \cdot \frac{m}{E}\right)\frac{\text{acc}^2}{f}$$

where m is the mass, acc is the acceleration amplitude, f is the excitation frequency, E is Young's modulus, ε_r is the relative dielectric constant, ε_0 is the vacuum dielectric constant, and e_{31} is the piezoelectric constant.

Figure 2.19 shows the output power distribution of the substrate with different external resistances and different piezoelectric layer arrangement lengths for excitation accelerations of 0.1, 0.5, and 1.0g, respectively, and the device composition materials and dimensional parameters are shown in Tables 2.6 and 2.7. It can be seen from the figure that the optimal piezoelectric layer arrangement length remains the same for different excitation accelerations, which are still at 2.5 mm. However, as the acceleration increases, the output power increases significantly, which is mainly due to the growth of the input mechanical energy as a quadratic function of the acceleration, thus indicating that the change of excitation acceleration has no effect on the optimal piezoelectric layer arrangement length.

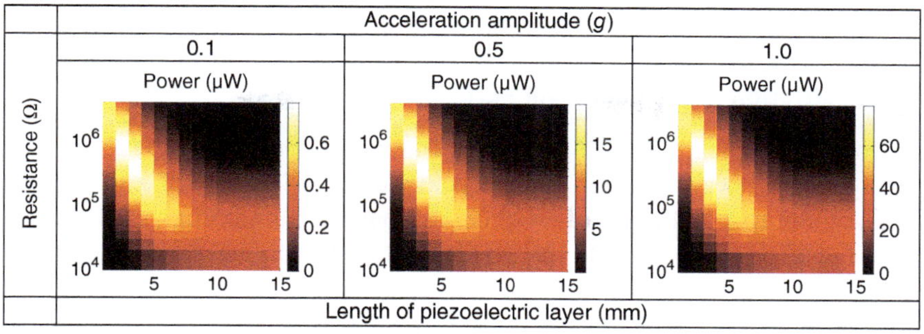

Figure 2.19 Effect of layout length of piezoelectric layer on output power when the applied acceleration amplitudes are 0.1, 0.5, and 1.0*g*, respectively.

2.5.2 Structural Strain Distribution Analysis

Based on the definition of piezoelectric materials, it can be seen that the ability to generate effective stresses or strains in the piezoelectric layer is very important for the structural design to improve the efficiency of piezoelectric kinetic energy conversion and is a major consideration for the structural design. Figure 2.20 shows the distribution of the maximum strain on the piezoelectric layer along the length of the cantilever beam for a cantilever vibration energy harvester with different piezoelectric layer lengths. The strain distribution shows that a small part of the length (less than 0.3 mm) at the root of the cantilever beam structure is limited by the fixed end during vibration, and a peak strain point jumps up near the root of the cantilever beam, and then the strain gradually decreases along with the length. According to the previous analysis, the piezoelectric layer has the maximum power output when the piezoelectric layer length is between 2 and 3 mm, and the piezoelectric layer lengths of 2 and 3 mm produce larger strain peaks than the cantilever beams of other lengths. The strain curves show that the cantilever beams with 2 and 3 mm piezoelectric layer lengths produce higher average strain and therefore higher power output.

Figure 2.20 Simulated strain distributions for the cantilever harvester with the proof mass along the arc length with the varied PZT layout length.

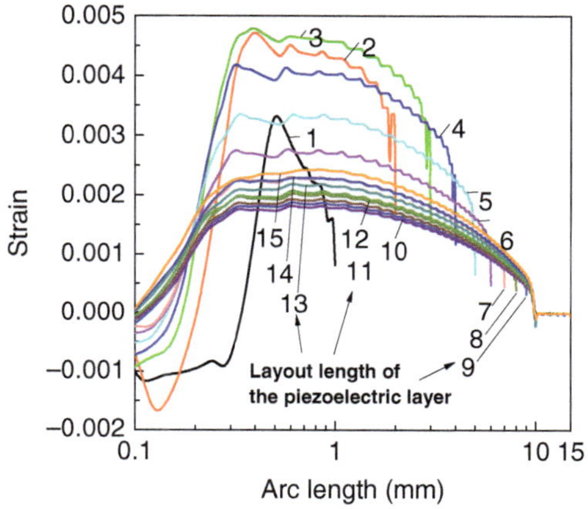

2.6 Experimental Verification

2.6.1 Fabrication of the Cantilever PEH

Firstly, a 400 μm thick bulk PZT was polished, and bottom electrode (20 nm Cr, 180 nm Au) layer was sputtered on the polished side. Secondly, the beryllium bronze was thinned and polished by chemical mechanical polishing (CMP) to a thickness of 50 μm, and this side was bonded with the polished side of bulk PZT using conductive epoxy as the intermediate layer (3 μm) at 175 °C for 180 minutes. To avoid PZT depolarization, the bonding process is attained below 175 °C, which is smaller than the Curie temperature of the PZT (295 °C) that we used. To obtain bulk PZT thick film, the bulk PZT was thinned to 50 μm and polished by CMP, and 200-nm-thick Cr/Au was sputtered on the smooth surface of PZT as the top electrode. The top and bottom electrodes were separated by ion-beam milling. Then, the designed structure with 3 mm wide cantilever beam was patterned by ultraviolet laser method from backside, thereby effectively avoiding ablation of top electrode and PZT. The additional tungsten mass was assembled at the end of cantilever beam to further reduce the resonant frequency of this device. Finally, the cantilever beam was assembled on printed circuit board (PCB), and the top and bottom electrodes were connected to bonding pads using gold wires.

2.6.2 Measurements

The output performance of the PEH was evaluated using the experimental platform that included a vibration control system, an electrodynamic shaker, a power amplifier, an acceleration sensor, and an oscilloscope. The PEH was mounted on the shaker, which can simulate the external vibration sources. The shaker was driven by a sinusoidal signal from the vibration control system and power amplifier and controlled the vibration conditions of the device by the acceleration sensor connected with the shaker. The electrical output from the device was monitored with the oscilloscope, in which the input impedance is 10 MΩ.

2.6.3 Results and Discussions

In order to verify the optimized results for PZT length based on the theoretical model and FEA presented above, the corresponding experiments are conducted. The experimental procedures are similar to those reported in our previous work [68, 69]. Figure 2.21 shows the output power of the harvester as a function of the external resistances with varied PZT layout lengths under 1.0g applied acceleration amplitude. The load resistance is connected in serial with this device and varies from 1 to 1100 kΩ to determine the optimum load resistance. It shows a comparison between the FEA and the corresponding experimental results, which demonstrates that the output power of the PEH indicates a high dependence on the external resistance. The results of FEA and experiments all show that the maximum output power occurs at the PZT layout length of 3 mm. Besides, the optimum load resistance for maximum effective power decreases with increasing covered length of piezoelectric layer, which is mainly attributed to the increasing electrode area.

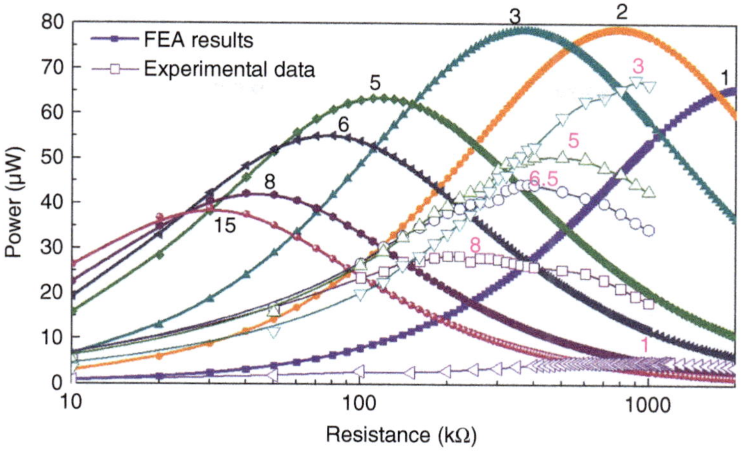

Figure 2.21 Comparison of output power for the presented harvester with the varied PZT layout length under different external resistance between FEA and experimental results. The applied acceleration amplitude is 1.0*g*. Source: Adapted from [68, 69].

At this optimal structure, the maximum output power can be achieved with 67.8 μW in experiments. The difference of power corresponding to resistance between FEA and experimental results mainly comes from the manufacturing and testing errors. Generally, the agreement of the optimal PZT length from the FEA and experimental results proves the validity of the presented optimization scheme and reveals the important guiding significance of the architecture design for practical applications.

Then, a high-performance piezoelectric energy harvester was fabricated by this guide. Figure 2.22a illustrates the response of open-circuit output voltages with the excited frequency under the applied acceleration amplitudes of 0.1, 0.5, 1.0, 1.5, 1.7, and 2.0g. The maximum output peak voltages are 3.5, 14.8, 25.4, 33.6, 36.0, and 39.0 V under the resonant frequency of 46 Hz, respectively. The resonant frequencies of the maximum output voltages apparently shifted to lower frequency with the increase of the applied acceleration amplitude because of the nonlinear response of PZT materials under large stress.

Figure 2.22b shows the output power of the harvester as a function of the external load resistances under varied applied acceleration amplitudes. The load resistance is connected in serial with this fabricated device and varies from 1 to 1100 kΩ to determine the optimum load resistance. The optimum load resistance for maximum effective power (0.965 mW) transfer is reached at the resistance of 400 kΩ. The maximum power density is approximately 30 mW/cm^3 at the applied acceleration amplitude of 2.0g. To investigate the stability of the PEH, high acceleration at resonant frequency is applied to evaluate the PEH's capability to cope with high workload situations. Figure 2.22c describes the measured open-circuit voltage at 72 000 seconds (approximately 3.3 million cycles) at 1.5g acceleration amplitude with the resonant frequency. Obviously, the PEH indicates high stability during this period. Compared with the previously reported PEH, the PEH with the optimal layout length of piezoelectric layer indicates a higher normalized power density, as shown in Figure 2.23 and Table 2.9. The piezoelectric compliant mechanism (PCM)

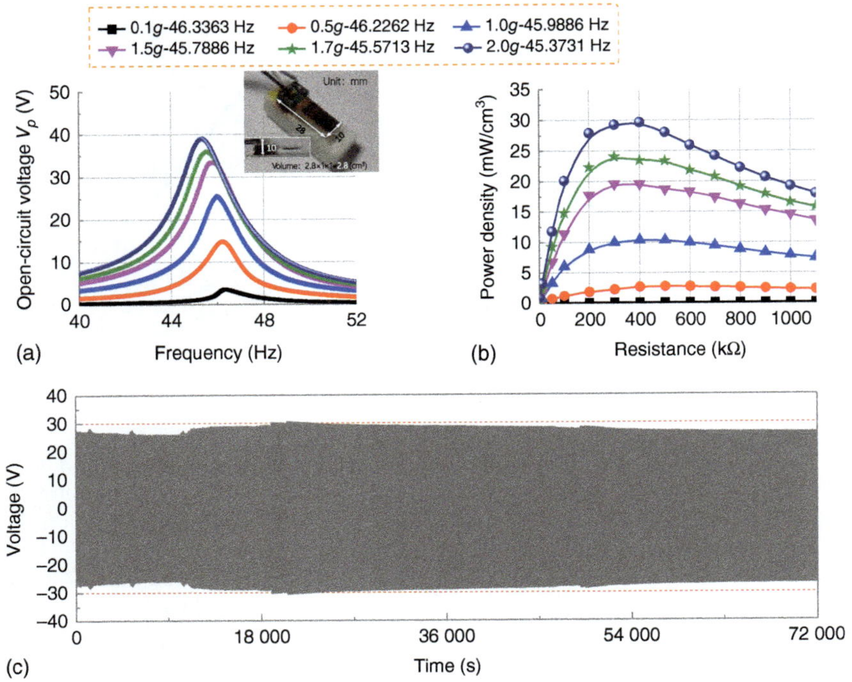

Figure 2.22 (a) Open-circuit voltage as a function of the frequency. (b) Power density as a function of the resistance. (c) The stability of the PEH with the optimal length of piezoelectric layer. The data in (a)–(c) came from experimental results.

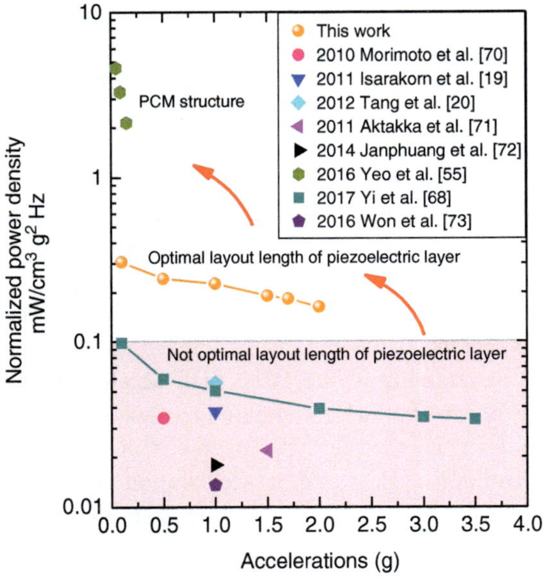

Figure 2.23 Comparison of normalized power density of recently reported PEHs with the optimal PEH in this chapter.

Table 2.9 Comparison of characteristics of recently reported PEHs with the optimal PEH in this chapter.

Authors	Piezoelectric material	Supporting material	Volume (mm^3)	Power (μW)	Acceleration (g)	Frequency (Hz)	Power density ($\mu W/mm^3$)	Normalized power density ($\mu W/mm^3$ g^2 Hz)
Morimoto et al. [70]	Epitaxial PZT films	Stainless steel	4.88	5.3	0.5	126	1.086	0.0345
Isarakorn et al. [19]	Epitaxial PZT films	Silicon	0.15	13.0	1.0	2300	86.67	0.0377
Tang et al. [20]	PZT	Silicon	0.4	11.56	1.0	514.1	28.9	0.0562
Aktakka et al. [71]	Thinned bulk PZT	Silicon	27	205	1.5	154	7.597	0.0219
Janphuang et al. [72]	Thinned bulk PZT	Silicon	47.82	82.4	1.0	96	1.72	0.0179
Yeo et al. [55]	Sputtered PZT	Ni	724.25	210	0.15	6.2	0.29	2.148
Won et al. [73]	Mn-doped KNN	Silicon	2	3.62	1.0	132	1.81	0.0136
Yi et al. [68]	Thinned bulk PZT films	Beryllium bronze	30.6	979	3.5	77.2	31.99	0.0338
Hu et al. [74]	Thinned bulk PZT films	Beryllium bronze	32	965	2.0	45.37	30.15	0.166

was presented to improve the performance of PEH, but this structure was based on a flexible piezoelectric film, which is not a cantilever structure. Meanwhile, the PEH is affected by gravity and cannot adapt to relatively high frequencies and high acceleration.

This session focuses on the structural design optimization for cantilever PEH under mechanical vibration. The optimal PZT layout length of the cantilever structure is studied systematically for achieving high performance of the PEH through theoretical analysis, FEA, and experiments. The results show that the PZT layout length of the PEH for the maximum power is approximately 1/5 length of the cantilever with a proof mass. Meanwhile, the effects of all the key design parameters of the harvester on the optimal length of PZT layer are systematically investigated in detail. It indicates that the width of cantilever, the thickness of PZT layer, and the modulus of substrate have a significant effect on the optimal PZT layout length. The optimal PZT layout length decreases with the cantilever width. As the piezoelectric layer thickness decreases, its effect on the optimal PZT layout length becomes increasingly low. The large substrate modulus will indicate a short optimal PZT layout length. However, the thickness of proof mass, the total length of cantilever, and the applied acceleration have a minimal effect on the optimal PZT layout length. Furthermore, the underlying mechanism for the improvement in power of such a nonuniform cantilever structure is revealed. The maximum output power results from the large average strain distribution in the PZT layer with the optimal length. Using this optimization scheme, a nonuniform cantilever piezoelectric harvester with the optimal piezoelectric layer length based on the thinned PZT thick film was fabricated. It can achieve high-power density (30 mW/cm^3) and good stability (3.3 million cycles) at the low resonant frequency of 46 Hz. The proposed standard distributed parameter theoretical model can be applied for the harvesters with the same nonuniform cantilever structure to predict and design the optimal length of the functional layer. In general, this study provides a significance of this simple structure optimization with such cantilever strain-induced vibration energy harvesters for the practical application of such cantilever strain-induced vibration energy harvesters.

2.7 Data-Driven Further Optimization

Although manually set and test single geometric parameters at a fixed interval each time can reveal the influence of individual parameter on output performance, optimizing the cantilever PEHs is a multiparameter coupling problem. In addition to univariate optimization, multivariate optimization can be implemented using data-driven methods. Taking the same method to study multivariate optimization problems may require researchers to spend more time trying each possible combination of geometric parameters to obtain the optimal output performance. Therefore, several data-driven optimization strategies have been proposed recently. For example, data-driven optimization can solve the optimization problems that are difficult to formulate or solve in that they are non-convex, multi-objective, or multi-modal. It can solve optimization problems based on discretely derivative-free data. Ghoddus and Kordrostami [75] presented an optimization approach based on particle swarm optimization (PSO) algorithm to maximize the output power of PEHs with four different structures, which optimized multiple geometric parameters. In addition, Nabavi and Zhang [76, 77] used genetic algorithm to optimize multiple objectives, including resonant frequency, output power, and device volume. However, expensive objective function evaluations may be required for optimization of the multiparameter coupling problem for PEHs. Therefore, as population-based stochastic optimization techniques, the PSO and genetic algorithm (GA) implemented in the previous works [75–77] may encounter low efficiency when dealing with the multiparameter coupling problem. Moreover, parameter tuning is needed to obtain a better convergence for PSO algorithm and GA [78], such as particle number, accelerate constant, inertia weight, and population size. Although the global search ability of GA and PSO may be better, generalized pattern search (GPS) algorithm is more suitable for the time-consuming multiparameter coupling problem due to its effortless parameter tuning and relatively fewer objective function evaluations [79].

In this session, a data-driven optimization strategy based on GPS algorithm [80] is proposed. Multiple geometric parameters were simultaneously optimized for maximum output performance based on the strategy. It can not only optimize PEHs to achieve high-output performance but also save time for analytical modeling and complex parameter tuning.

2.7.1 Methods

As shown in Figure 2.24a, b, the unimorph PEH is composed of a piezoelectric layer, a structural layer, and a proof mass. Figure 2.25 depicts the overall function of the presented strategy. The data-driven optimization scheme mainly includes a finite element method (FEM) (COMSOL Multiphysics 5.4) solver and a GPS (MATLAB 2020a) optimizer. The material properties used in FEM simulation were concluded

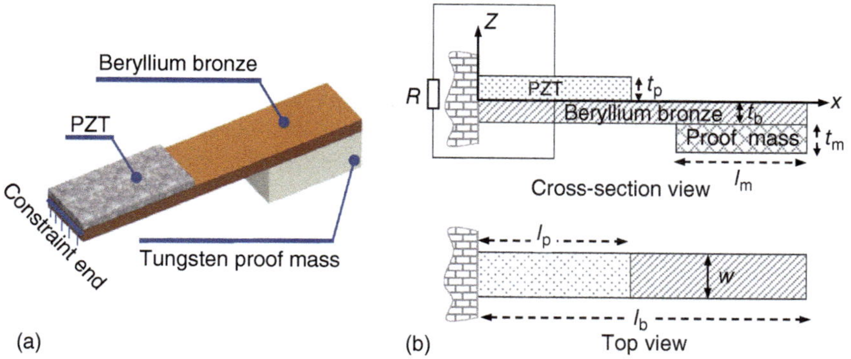

Figure 2.24 Schematic diagram [81] of the piezoelectric energy harvester (PEH) and the proposed strategy. (a) Three-dimensional view. (b) Annotation for geometric parameters of the cantilever PEH. Source: From Huang et al. [81] / MDPI / CC BY-4.0.

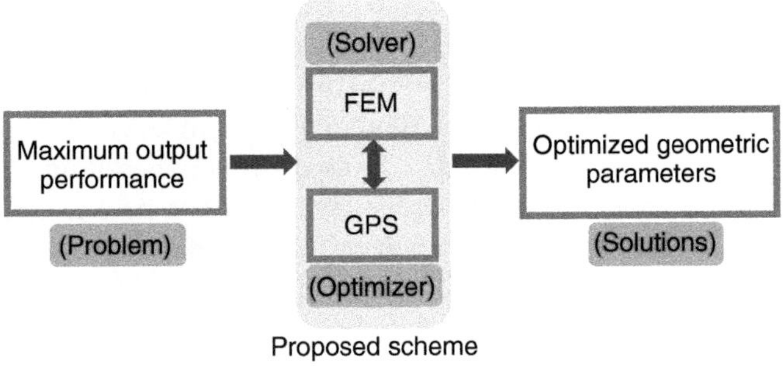

Figure 2.25 Overview [81] of the presented optimization working process. Source: Huang et al. [81] / MDPI / CC BY-4.0.

in Tables 2.6 and 2.7. COMSOL LiveLink™ for MATLAB module was used for the communication between solver and optimizer. Different types of pattern search algorithms have been utilized for solving engineering optimization problems [78, 79, 82], and the convergence analysis of GPS algorithm has been performed by Audet and Dennis [83].

Figure 2.26a further illustrates the execution operation between the solver and the optimizer, and summarizes the data interaction and feedback mechanism of the overall scheme execution. The first step of initialization is mainly used for the parameter settings of the GPS algorithm, including search initial points, search and polling methods, and convergence criteria. Proper settings can speed up optimization. The subsequent update of the structural parameters, the sequential execution of the research steps, the extraction of the results, and the realization of the GPS algorithm are implemented by MATLAB. In addition, the specific workflow of the GPS algorithm is presented in Figure 2.26b, which mainly includes two steps – search and polling. When introducing polling operations, two important concepts need to be introduced: Mesh and pattern. In each iteration, the GPS algorithm searches for a set of points to be tested within a certain range near the current point of the current iteration and tries to find a better point with its objective function value better than the current point's. Then, the better point found in this iteration is set as the current point of the next iteration, and this process is called polling. After continuous polling, GPS will find a series of points close to the optimal value, which will not stop until the convergence criterion is met. Among them, the points within a certain range around the current point in each iteration are called mesh, and the operators used to generate the mesh are called pattern. Specifically, let M_k denote the mesh at kth iteration, and $x_k^{(i)}$ denotes the ith point in the M_k.The mesh is defined as below:

$$\left(M_k \triangleq \left\{x \in \mathbb{R}^n \mid x = x_k' + \Delta_k \cdot \nu_k, k \in \{1,2,\ldots,n\}\right\}\right)$$

where x_k' denotes the current point at the kth iteration while Δ_k denotes mesh size, and ν_k denotes pattern. The polling process can be defined as below [81]:

(1) Suppose $F(x)$ denotes the objective function in the optimization problem. At each iteration, GPS updates the geometrical parameters and communicates with FEM.
(2) $F\left(x_k^{(i)}\right)$ is computed by FEM and GPS finds a better point $x_k^{(j)}$ so that $F\left(x_k^{(j)}\right) < \left(F\left(x_k'\right)\right)$, then set $x_{k+1}' = x_k^{(j)}$ and $\Delta_{k+1} = 2 \cdot \Delta_k$.

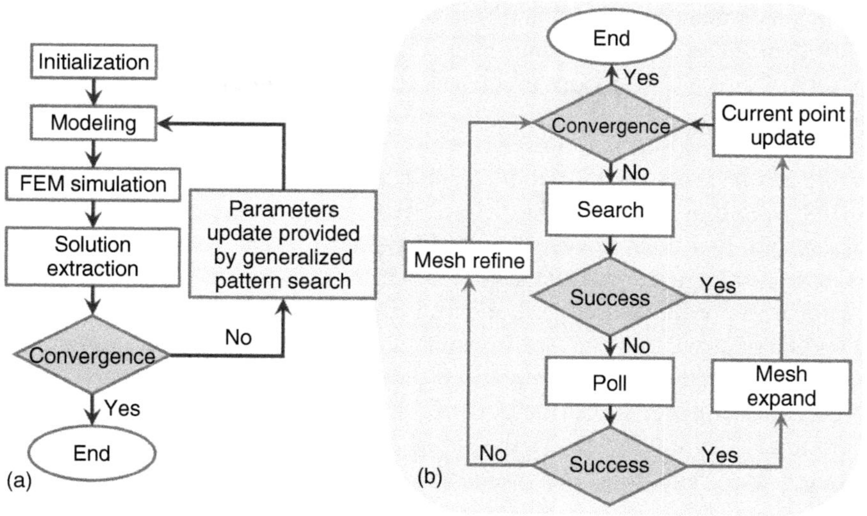

Figure 2.26 Workflow [81] of the proposed data-driven optimization strategy. (a) Overall working mechanism. (b) Detailed workflow of generalized pattern search. Source: Huang et al. [81] / MDPI / CC BY-4.0.

(3) Otherwise, if the polling fails to find a better point at kth iteration, then set $x'_{k+1} = x'_k$ and $\Delta_{k+1} = 1/2 \cdot \Delta_k$.
(4) Then perform convergence judgment, if satisfied, stop; if not, enter the next iteration and repeat step (1).

In addition, the search method runs before polling can select a different current point, which may accelerate the optimization if tuning well.

To improve the output performance, the objective function can be defined as normalized power density or output power. The definition of normalized power density is shown as the following equation [81]:

$$P_\mathrm{n} = \frac{P}{a^2 \cdot f_1 \cdot V_\mathrm{eff}}$$

where P_n denotes the normalized power density, P denotes the output power, f_1 represents the first-modal resonant frequency, V_eff denotes the effective volume of the specific structure, and a represents the excitation acceleration. The maximizing of normalized power density may indicate the trend of maximizing output power and minimizing the first-modal resonant frequency and effective volume [81]. Therefore, the optimization problem is defined as below:

$$\begin{aligned} \text{maximize} \quad & P_\mathrm{n} \text{ or } P \\ \text{subject to} \quad & 0.5\ \text{mm} \le l_\mathrm{p} \le 15\ \text{mm} \\ & 0.5\ \text{mm} \le w \le 3\ \text{mm} \\ & 0.001\ \text{mm} \le t_\mathrm{p} \le 0.05\ \text{mm} \\ & 0.03\ \text{mm} \le t_\mathrm{b} \le 0.05\ \text{mm} \end{aligned}$$

The geometric parameters to be optimized were set as l_p, w, t_p, and t_b. As the thickness of the whole beam gets thinner, there is higher probability of device failure.

From a conservative consideration, the upper and lower bounds of t_b were set as 30–50 mm based on the experimental data in the previous work [74].

2.7.2 Results and Discussion

Figure 2.27a presents the trajectory of optimization, which directly shows how the GPS algorithm optimizes the defined geometrical parameters. Each point represents the best point at every iteration, respectively. The optimal length and thickness of PZT are 6.0 mm and 4.6 μm, respectively. Figure 2.27b demonstrates the comparison of the output performance between the reference structure (Ref) and optimized structure (4OPT-1). It can be observed that the output power and normalized power density were, respectively, about 3.71 and 10.10 times larger and the resonant frequency is 2.54 times lower, showing an improvement in output performance. Figure 2.27c shows the strain of the piezoelectric layer along the arc length $\left(S_1(x, y, t) = -y\frac{\partial^2 z(x,t)}{\partial x^2}\right)$, which demonstrates a higher average strain distribution. However, it is noted that the effects of a reduction in material properties like piezoelectricity with the decrease in the thickness of PZT are neglected theoretically. Although the output performance of 4OPT-1 may be augmented without the consideration of the effects of material properties, the optimal result (4OPT-1) given by the proposed data-driven optimization strategy provides the trend of improving output performance.

Taken together, a data-driven optimization strategy [81] based on FEM simulation (solver) and GPS algorithm (optimizer) was proposed and implemented. The proposed strategy can search the variable space more precisely than manually varying the geometric parameters at fixed intervals. The optimal ratio of the length of PZT and structural layer increases as the thickness of PZT decreases compared with the previous work [74]. The optimization results showed that the optimal length ratio is 0.40 (6.0/15.0 mm) when the thickness of PZT is 4.6 μm, while the optimal length ratio is 0.23 (3.4/15.0 mm) when the thickness of PZT is 50 μm.

2.8 Summary

This chapter introduces the structural design method of a multilayer piezoelectric composite structure energy harvester, and specifically analyzes the influence law of relevant structural design factors through a typical cantilever beam piezoelectric structure, as follows.

A simulation and analysis model of flexible piezoelectric composite layer cantilever-beam-type vibration energy harvester is established, and the device output performance under different piezoelectric layer arrangement lengths is analyzed, and it is found that the maximum output power can be increased nearly one time by optimizing the piezoelectric layer arrangement. At the same time, it is found that the optimized piezoelectric layer arrangement length is shortening as the width of the cantilever beam increases, where the ratio of width increase is basically similar to the output power increase according to the output power value;

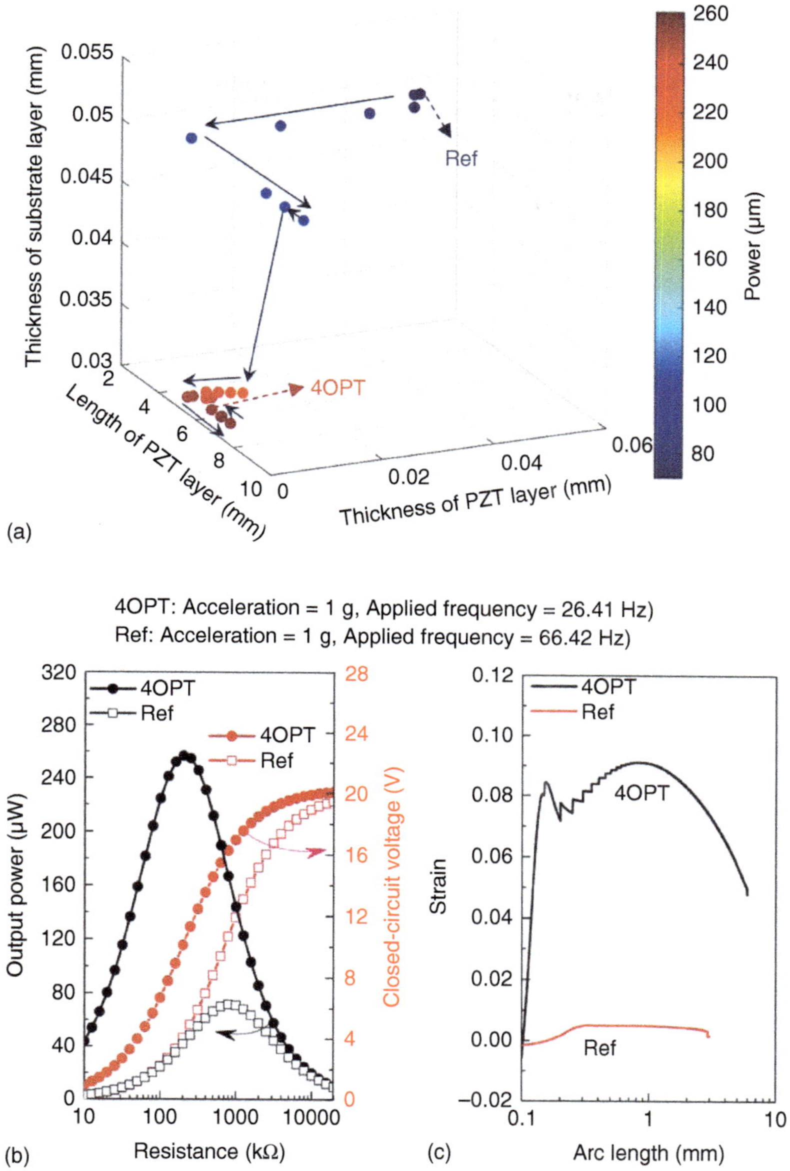

Figure 2.27 Evaluation [81] of the proposed data-driven optimization strategy from various perspectives. (a) Trajectory of the parameters during optimization. (b) Output power versus external resistance for previously reported high-output PEH and optimized structure. (c) Strain distributions along the arc length of previously reported high-output PEH and optimized structure. Source: Huang et al. [81] / MDPI / CC BY-4.0.

the thinner the piezoelectric layer will make the influence of the piezoelectric layer arrangement length ratio on the output power size smaller, but affect the optimal matching resistance of the device; the increase of the total length of the cantilever beam and the thickness change of the mass bulk have little influence on the optimized piezoelectric layer arrangement. The effect of increasing the total length of the cantilever beam and changing the thickness of the mass bulk on the optimal piezoelectric layer arrangement length ratio is not significant.

The output power variation characteristics are studied for different substrate materials, and the changes in the neutral layer, resonant frequency, and piezoelectric layer strain under resonance brought by the change of substrate material are investigated to show that for an energy harvester with a determined piezoelectric layer stiffness there exists an optimal stiffness substrate to make it have the maximum output power.

By studying the influence of the excitation acceleration of the vibration energy harvester on the optimal arrangement length of the piezoelectric layer, it is found that the mechanical energy input increases quadratically as a function of the acceleration, and the output power also increases quadratically as a function of the acceleration, but the optimal arrangement length of the piezoelectric layer remains unchanged.

By analyzing the strain distribution of the cantilever beam vibration energy harvester, it can be seen that the suitable structural design makes the piezoelectric layer produce high average strain which is the main reason for the high-voltage electrical energy conversion efficiency.

Finally, utilizing GPS algorithm can shorten the total optimization time for complex high-dimensional optimization problems in real life. Besides, owing to the extensive application of the FEM simulation, the proposed strategy is not limited to the optimization of PEHs. It can also be implemented for other structures that are challenging to formulate.

References

1 Romero, E., Warrington, R.O., and Neuman, M.R. (2009). Energy scavenging sources for biomedical sensors. *Physiological Measurement* 30: R35–R62.

2 Roundy, S., Wright, P.K., and Rabaey, J. (2003). A study of low level vibrations as a power source for wireless sensor nodes. *Computer Communications* 26: 1131–1144.

3 Erturk, A. and Inman, D.J. (2011). *Piezoelectric Energy Harvesting*, 392. John Wiley & Sons Ltd.

4 Harne, R.L. and Wang, K.W. (2013). A review of the recent research on vibration energy harvesting via bistable systems. *Smart Materials and Structures* 22: 023001.

5 Won, S.S., Sheldon, M., Mostovych, N. et al. (2015). Piezoelectric poly(vinylidene fluoride trifluoroethylene) thin film-based power generators using paper substrates for wearable device applications. *Applied Physics Letters* 107: 202901.

6 Siddiqui, S., Kim, D.I., Duy, L.T. et al. (2015). High-performance flexible lead-free nanocomposite piezoelectric nanogenerator for biomechanical energy harvesting and storage. *Nano Energy* 15: 177–185.

7 Karan, S.K., Maiti, S., Paria, S. et al. (2018). A new insight towards eggshell membrane as high energy conversion efficient bio-piezoelectric energy harvester. *Materials Today Energy* 9: 114–125.

8 Ghosh, S.K. and Mandal, D. (2017). Sustainable energy generation from piezoelectric biomaterial for noninvasive physiological signal monitoring. *ACS Sustainable Chemistry & Engineering* 5: 8836–8843.

9 Ghosh, S.K. and Mandal, D. (2017). Bio-assembled, piezoelectric prawn shell made self-powered wearable sensor for non-invasive physiological signal monitoring. *Applied Physics Letters* 110: 123701.

10 Ghosh, S.K., Adhikary, P., Jana, S. et al. (2017). Electrospun gelatin nanofiber based self-powered bio-*e*-skin for health care monitoring. *Nano Energy* 36: 166–175.

11 Ryu, J., Kim, J., Oh, J. et al. (2019). Intrinsically stretchable multi-functional fiber with energy harvesting and strain sensing capability. *Nano Energy* 55: 348–353.

12 Zeng, W., Tao, X.M., Chen, S. et al. (2013). Highly durable all-fiber nanogenerator for mechanical energy harvesting. *Energy & Environmental Science* 6 (9): 2631–2638.

13 Ahn, Y., Song, S., and Yun, K.-S. (2015). Woven flexible textile structure for wearable power-generating tactile sensor array. *Smart Materials and Structures* 24: 075002.

14 Dagdeviren, C., Yang, B.D., Su, Y.W. et al. (2014). Conformal piezoelectric energy harvesting and storage from motions of the heart, lung, and diaphragm. *Proceedings of the National Academy of Sciences of the United States of America* 111 (5): 1927–1932.

15 Wang, C., Li, X., Hu, H. et al. (2018). Monitoring of the central blood pressure waveform via a conformal ultrasonic device. *Nature Biomedical Engineering* 2: 687–695.

16 Roundy, S. and Wright, P.K. (2004). A piezoelectric vibration based generator for wireless electronics. *Smart Materials and Structures* 13: 1131–1142.

17 Fang, H.B., Liu, J.Q., Xu, Z.Y. et al. (2006). A MEMS-based piezoelectric power generator for low frequency vibration energy harvesting. *Chinese Physics Letters* 23 (3): 732–734.

18 Reilly, E.K., Burghardt, F., Fain, R., and Wright, P. (2011). Powering a wireless sensor node with a vibration-driven piezoelectric energy harvester. *Smart Materials and Structures* 20: 125006.

19 Isarakorn, D., Briand, D., Janphuang, P. et al. (2011). The realization and performance of vibration energy harvesting MEMS devices based on an epitaxial piezoelectric thin film. *Smart Materials and Structures* 20 (2): 025015.

20 Tang, G., Liu, J.Q., Yang, B. et al. (2012). Fabrication and analysis of high-performance piezoelectric MEMS generators. *Journal of Micromechanics and Microengineering* 22 (6): 065017.

21 Jia, Y. and Seshia, A.A. (2016). Power optimization by mass tuning for MEMS piezoelectric cantilever vibration energy harvesting. *Journal of Microelectromechanical Systems* 25: 108–117.

22 Toprak, A. and Tigli, O. (2014). Piezoelectric energy harvesting: State-of-the-art and challenges. *Applied Physics Reviews* 1: 031104.

23 Kim, S., Clark, W.W., and Wang, Q.M. (2005). Piezoelectric energy harvesting with a clamped circular plate: experimental study. *Journal of Intelligent Material Systems and Structures* 16: 855–863.

24 Kim, S., Clark, W.W., and Wang, Q.M. (2005). Piezoelectric energy harvesting with a clamped circular plate: analysis. *Journal of Intelligent Material Systems and Structures* 16: 847–854.

25 Horowitz, S.B., Sheplak, M., Cattafesta, L.N., and Nishida, T. (2006). A MEMS acoustic energy harvester. *Journal of Micromechanics and Microengineering* 16: S174–S181.

26 Kuehne, I., Marinkovic, D., Eckstein, G., and Seidel, H. (2008). A new approach for MEMS power generation based on a piezoelectric diaphragm. *Sensors and Actuators A: Physical* 142: 292–297.

27 Mo, C.K., Radziemski, L.J., and Clark, W.W. (2010). Experimental validation of energy harvesting performance for pressure-loaded piezoelectric circular diaphragms. *Smart Materials and Structures* 19: 075010.

28 Mo, C.K., Radziemski, L.J., and Clark, W.W. (2010). Analysis of piezoelectric circular diaphragm energy harvesters for use in a pressure fluctuating system. *Smart Materials and Structures* 19: 025016.

29 Gafforelli, G., Corigliano, A., Xu, R., and Kim, S.-G. (2014). Experimental verification of a bridge-shaped, nonlinear vibration energy harvester. *Applied Physics Letters* 105: 203901.

30 Lee, B., Jeong, S., Cheon, S. et al. (2015). Generating characteristics of a tension activated pi-shaped piezoelectric harvester. *Ceramics International* 41: S695–S701.

31 Li, G., Yang, B., Hou, C. et al. (2017). A piezoelectric energy harvester utilizing $Pb[Zr_xTi_{1-x}]O_3$ thick film on phosphor bronze. *Sensors and Materials* 29 (12): 1723–1731.

32 Jung, S.-M. and Yun, K.-S. (2010). Energy-harvesting device with mechanical frequency-up conversion mechanism for increased power efficiency and wideband operation. *Applied Physics Letters* 96: 111906.

33 Seo, M.-H., Choi, D.-H., Kim, I.-H. et al. (2012). Multi-resonant energy harvester exploiting high-mode resonances frequency down-shifted by a flexible body beam. *Applied Physics Letters* 101: 123903.

34 Xu, C., Liang, Z., Ren, B. et al. (2013). Bi-stable energy harvesting based on a simply supported piezoelectric buckled beam. *Journal of Applied Physics* 114 (11): 114507.

35 Zhu, Y. and Zu, J.W. (2013). Enhanced buckled-beam piezoelectric energy harvesting using midpoint magnetic force. *Applied Physics Letters* 103: 041905.

36 Xie, Z., Kwuimy, C.A.K., Wang, Z., and Huang, W. (2018). A piezoelectric energy harvester for broadband rotational excitation using buckled beam. *AIP Advances* 8: 015125.

37 Fu, H., Sharif-Khodaei, Z., and Aliabadi, F. (2019). A bio-inspired host-parasite structure for broadband vibration energy harvesting from low-frequency random sources. *Applied Physics Letters* 114: 143901.

38 Qi, Y., Kim, J., Nguyen, T.D. et al. (2011). Enhanced piezoelectricity and stretchability in energy harvesting devices fabricated from buckled PZT ribbons. *Nano Letters* 11 (3): 1331–1336.

39 Massaro, A., De Guido, S., Ingrosso, I. et al. (2011). Freestanding piezoelectric rings for high efficiency energy harvesting at low frequency. *Applied Physics Letters* 98 (5): 053502.

40 Arrieta, A.F., Hagedorn, P., Erturk, A., and Inman, D.J. (2010). A piezoelectric bistable plate for nonlinear broadband energy harvesting. *Applied Physics Letters* 97: 104102.

41 Ren, B., Or, S.W., Zhao, X.Y., and Luo, H.S. (2010). Energy harvesting using a modified rectangular cymbal transducer based on $0.71Pb(Mg_{1/3}Nb_{2/3})O_3$–$0.29PbTiO_3$ single crystal. *Journal of Applied Physics* 107: 034501.

42 Feenstra, J., Granstrom, J., and Sodano, H. (2008). Energy harvesting through a backpack employing a mechanically amplified piezoelectric stack. *Mechanical Systems and Signal Processing* 22: 721–734.

43 Xu, T.B., Siochi, E.J., Kang, J.H. et al. (2013). Energy harvesting using a PZT ceramic multilayer stack. *Smart Materials and Structures* 22 (6): 065015.

44 Jeong, C.K., Lee, J., Han, S. et al. (2015). A hyper-stretchable elastic-composite energy harvester. *Advanced Materials* 27 (18): 2866–2875.

45 Abdelkefi, A. (2016). Aeroelastic energy harvesting: a review. *International Journal of Engineering Science* 100: 112–135.

46 Selvan, K.V. and Ali, M.S.M. (2016). Micro-scale energy harvesting devices: review of methodological performances in the last decade. *Renewable & Sustainable Energy Reviews* 54: 1035–1047.

47 Ma, T., Yang, H.X., Guo, X.D. et al. (2018). Development of inline hydroelectric generation system from municipal water pipelines. *Energy* 144: 535–548.

48 Jasim, A., Yesner, G., Wang, H. et al. (2018). Laboratory testing and numerical simulation of piezoelectric energy harvester for roadway applications. *Applied Energy* 224: 438–447.

49 Lee, H.S., Chung, J., Hwang, G.T. et al. (2014). Flexible inorganic piezoelectric acoustic nanosensors for biomimetic artificial hair cells. *Advanced Functional Materials* 24 (44): 6914–6921.

50 Wu, N., Cheng, X.F., Zhong, Q.Z. et al. (2015). Cellular polypropylene piezoelectret for human body energy harvesting and health monitoring. *Advanced Functional Materials* 25 (30): 4788–4794.

51 Shaikh, F.K. and Zeadally, S. (2016). Energy harvesting in wireless sensor networks: a comprehensive review. *Renewable & Sustainable Energy Reviews* 55: 1041–1054.

52 Kim, D.H., Shin, H.J., Lee, H. et al. (2017). In vivo self-powered wireless transmission using biocompatible flexible energy harvesters. *Advanced Functional Materials* 27 (25): 1700341.

53 Wang, X., Chen, C.S., Wang, N. et al. (2017). A frequency and bandwidth tunable piezoelectric vibration energy harvester using multiple nonlinear techniques. *Applied Energy* 190: 368–375.

54 Wu, J., Shi, H., Zhao, T. et al. (2016). High-temperature $BiScO_3$-$PbTiO_3$ piezoelectric vibration energy harvester. *Advanced Functional Materials* 26: 7186–7194.

55 Yeo, H.G., Ma, X., Rahn, C., and Trolier-McKinstry, S. (2016). Efficient piezoelectric energy harvesters utilizing (001) textured bimorph PZT films on flexible metal foils. *Advanced Functional Materials* 26: 5940–5946.

56 Takei, R., Makimoto, N., Okada, H. et al. (2016). Design of piezoelectric MEMS cantilever for low-frequency vibration energy harvester. *Japanese Journal of Applied Physics* 55: 06GP14.

57 Gao, X., Shih, W.-H., and Shih, W.Y. (2009). Induced voltage of piezoelectric unimorph cantilevers of different nonpiezoelectric/piezoelectric length ratios. *Smart Materials and Structures* 18: 125018.

58 Gao, X., Shih, W.H., and Shih, W.Y. (2010). Vibration energy harvesting using piezoelectric unimorph cantilevers with unequal piezoelectric and nonpiezoelectric lengths. *Applied Physics Letters* 97: 233503.

59 Patel, R., McWilliam, S., and Popov, A.A. (2011). A geometric parameter study of piezoelectric coverage on a rectangular cantilever energy harvester. *Smart Materials and Structures* 20: 085004.

60 Yu, H., Zhou, J., Deng, L., and Wen, Z. (2014). A vibration-based MEMS piezoelectric energy harvester and power conditioning circuit. *Sensors* 14: 3323–3341.

61 Sunithamani, S., Lakshmi, P., and Eba Flora, E. (2013). PZT length optimization of MEMS piezoelectric energy harvester with a non-traditional cross section: simulation study. *Microsystem Technologies* 20: 2165–2171.

62 Timoshenko, S.P., Young, D.H., and Weaver, W. (1974). *Vibration Problems in Engineering*, 538. Chichester: Wiley.

63 Roshko, A. (1954). On the Development of Turbulent Wakes from Vortex StreetsNACA Report 1191.

64 Caughey, T.K. and O'Kelly, M.E.J. (1965). Classical normal modes in damped linear dynamic systems. *Journal of Applied Mechanics* 32: 583–588.

65 Clough, R.W. and Penzien, J. (1975). *Dynamics of Structures*. Wiley.

66 ANSI/IEEE Std 176-1987 (1987). *IEEE Standard on Piezoelectricity*. New York: The Institute of Electrical and Electronics Engineers https://doi.org/10.1109/IEEESTD.1988.79638.

67 Yang, Z., Zhou, S., Zu, J., and Inman, D. (2018). High-performance piezoelectric energy harvesters and their applications. *Joule* 2: 642–697.

68 Yi, Z., Yang, B., Li, G.M. et al. (2017). High performance bimorph piezoelectric MEMS harvester via bulk PZT thick films on thin beryllium-bronze substrate. *Applied Physics Letters* 111 (1): 013902.

69 Tang, G., Yang, B., Hou, C. et al. (2016). A piezoelectric micro generator worked at low frequency and high acceleration based on PZT and phosphor bronze bonding. *Scientific Reports* 6: 38798.

70 Morimoto, K., Kanno, I., Wasa, K., and Kotera, H. (2010). High-efficiency piezoelectric energy harvesters of c-axis-oriented epitaxial PZT films transferred onto stainless steel cantilevers. *Sensors and Actuators A: Physical* 163: 428–432.

71 Aktakka, E.E., Peterson, R.L., and Najafi, K. Thinned-PZT on SOI process and design optimization for piezoelectric inertial energy harvesting. In: *2011 16th International Solid-State Sensors, Actuators and Microsystems Conference*, https://doi.org/10.1109/TRANSDUCERS.2011.5969857, 1649–1652.

72 Janphuang, P., Lockhart, R., Uffer, N. et al. (2014). Vibrational piezoelectric energy harvesters based on thinned bulk PZT sheets fabricated at the wafer level. *Sensors and Actuators A: Physical* 210: 1–9.

73 Won, S.S., Lee, J., Venugopal, V. et al. (2016). Lead-free Mn-doped ($K_{0.5}$, $Na_{0.5}$)NbO_3 piezoelectric thin films for MEMS-based vibrational energy harvester applications. *Applied Physics Letters* 108 (23): 232908.

74 Hu, Y., Yi, Z.R., Dong, X.X. et al. (2019). High power density energy harvester with non-uniform cantilever structure due to high average strain distribution. *Energy* 169: 294–304.

75 Ghoddus, H. and Kordrostami, Z. (2018). Harvesting the uultimate electrical power from MEMS piezoelectric vibration energy harvesters: an optimization approach. *IEEE Sensors Journal* 18: 8667–8675.

76 Nabavi, S. and Zhang, L. (2017). Design and optimization of piezoelectric MEMS vibration energy harvesters blased on genetic algorithm. *IEEE Sensors Journal* 17: 7372–7382.

77 Nabavi, S. and Zhang, L. (2019). Frequency tuning and efficiency improvement of piezoelectric MEMS vibration energy harvesters. *Journal of Microelectromechanical Systems* 28: 77–87.

78 Güneş, F. and Tokan, F. (2010). Pattern search optimization with applications on synthesis of linear antenna arrays. *Expert Systems with Applications* 37: 4698–4705.

79 Zhao, Z., Meza, J.C., and Hove, M.V. (2006). Using pattern search methods for surface structure determination of nanomaterials. *Journal of Physics: Condensed Matter* 18: 8693–8706.

80 Torczon, V. (1997). On the convergence of pattern slearch algorithms. *SIAM Journal on Optimization* 7: 1–25.

81 Huang, Y., Yi, Z., Hu, G., and Yang, B. (2021). Data-driven optimization of piezoelectric energy harvesters via pattern search algorithm. *Micromachines* 12: 561.

82 Abramson, M.A. (2004). Mixed variable optimization of a load-bearing thermal insulation system using a filter pattern search algorithm. *Optimization and Engineering* 5: 157–177.

83 Audet, C. and Dennis, J.E. Jr. (2002). Analysis of generalized pattern searches. *SIAM Journal on Optimization* 13: 889–903.

3

Fabrication of Flexible Piezoelectric Energy Harvesters

3.1 Introduction

Energy harvesting from ambient energy for the purpose of powering low-powered electronics has emerged during the last decade as an enabling technology for self-powered devices [1] This concept has some significant ecological ramifications, such as reducing noxious waste produced by replacing batteries and decreasing maintenance costs [2]. With the rapid development of low-power-consuming electronics and the need to provide wireless solutions for powering numerous sensors, some micro-energy harvesters scavenging vibration [3, 4], wind [5], solar [6], thermal [7], and fluid energy [8] have been proposed. One of the most hotspots is the use of the piezoelectric effect to convert ambient vibration into useful electrical energy due to its advantages of simple configuration and high-energy-conversion efficiency [2]. Currently, most investigations for power generation of vibration-based piezoelectric energy harvesters (PEHs) focus on low resonant frequency, broad bandwidth, high output power, and good reliability [1, 9]. To improve the output power, bulk lead zirconate titanate (PZT) ceramics are a good choice due to their higher electromechanical-coupling efficiency and higher conversion efficiency (the ratio between the output electrical energy and input mechanical energy) than deposited piezoelectric films, such as AlN or sol–gel PZT [10], and metal substrate is preferable compared to silicon for obtaining good reliability of the PEH [11]. Meanwhile, the thin metal foils are deployed to enable lower resonance frequency than traditional silicon substrate and avoid silicon-etching process.

Currently, the mostly above-reported PEHs based on silicon substrate technology can be easily broken down and only work at low-vibration amplitude from ambient sources due to the brittle properties of silicon [12, 13]. In addition, the frequencies of environmental vibration sources are relatively low (normally less than 200 Hz). For example, a small microwave oven works at the resonant frequency of 121 Hz, a base of three-axis machine tool at 70 Hz, a kitchen blender casing at 121 Hz, and a compact disk (CD) on notebook computer at 75 Hz [14]. Morimoto et al. [15] reported a cantilever PEH using epitaxial PZT thin films transferred onto stainless steel, which showed output power of 5.3 μW at 0.51g and 126 Hz

Flexible Piezoelectric Energy Harvesters and Sensors, First Edition. Bin Yang, Zhiran Yi, and Chengkuo Lee.

without a proof mass. Yeo et al. [16] presented a high efficient bimorph PEH with piezoelectric compliant mechanism design via depositing PZT film on Ni foil based on radio frequency (RF) magnetron sputtering. The copper and nickel metals have larger thermal expansion coefficients than many perovskite films, which produce compressive stresses in the tetragonal PZT film on cooling, increasing the volume fraction along the out-of-plane polarization direction. Therefore, the high compressive stresses in PZT films are beneficial to collect vibration energy. However, the high-quality PZT film deposition was realized by expensive sputtering equipment and controlled by complicated processes with seed and buffer layers. Although fewer papers have reported low-frequency PEHs, their output power is very low and often operates at low applied acceleration.

Development of preparation technology is an important factor limiting the power density improvement of PEHs [1, 17], especially the preparation technology of piezoelectric thin films. The method of thin-film preparation of piezoelectric bulk materials based on mechanical-thinning process is considered as one of the key methods to obtain high-performance piezoelectric thin films at present [18]. The thickness of piezoelectric thin films prepared by mechanical-thinning process is usually controlled from 1 to 100 μm. The most researched silicon substrates with high modulus and brittleness have been dominated for so long, making the device inherit high frequency and prone to structural fracture damage, which cannot adapt to low-frequency environmental vibration source energy harvesting and is currently a key barrier to practical applications. However, with the development of flexible electronics, the flexible microelectromechanical system (MEMS) process based on the development of the traditional MEMS process provides process technology support for the replacement of silicon substrates.

This chapter is based on the preparation of piezoelectric thin-film process with bulk-like piezoelectric properties by MEMS mechanical-thinning technology, the study of piezoelectric composite layer structure process preparation technology for low-stiffness-flexible substrate, the development of flexible metal foil and other flexible substrate cantilever beam structure preparation technology, combined with the theoretical analysis results to explore the design and preparation method of high-power-density piezoelectric vibration energy harvester, and verify the reliability of the relevant theoretical analysis results. The design and preparation of low-frequency, high-power-density piezoelectric vibration energy harvester is proposed. Because the density of the composite materials and the dispersion degree of the piezoelectric ceramics highly depends on different processes, the piezoelectric property of flexible PEHs is deeply affected by the preparation process. Nowadays, a large number of studies have shown that the main factors determining the piezoelectric property are crystallinity content, molecular orientation, dipole orientation along the electric field direction, heat treatment, intensity of polarization electric field, polarization temperature, and time. However, the key factor influencing the performance of PEHs lies in their preparation process. Generally, the preparation methods include sputtering and chemical–mechanical polishing methods of piezoelectric ceramics-based PEH (PZT, AlN, and others) [1], spin coating, and electrospinning.

3.2 Piezoelectric Ceramic Thin Film

Piezoelectric film is the foundation of piezoelectric devices [19]. Depositing processes, based on chemical or physical principles, magnetic sputtering, chemical vapor deposition (CVD), or sol–gel deposition, are common piezoelectric thin-film fabrication processes [18, 20]. Generally, piezoelectric thin films obtained by these processes are with lower piezoelectric-coupling coefficient than the piezoelectric bulk materials due to the compactness or crystallinity [18]. Piezoelectric bulk ceramics are fabricated through high-temperature sintering process, which is useful to obtain high compactness and crystallinity [20–22].

Recently, piezoelectric thick-film process is developed by the mechanical-thinning technology of piezoelectric bulk materials [20, 23–26], which keeps the same piezoelectric properties as bulk materials. Before the thinning process, the piezoelectric bulk ceramic will be bonded on some rigid substrates, such as metal, silicon, or glass [21, 22], which restrict its applications in flexible electronics. With the rapid development of piezoelectric thick-film technology, more possible applications are presented [27, 28]. Among them, flexible piezo-MEMS devices are given more attention due to their high sensitivity, good dynamic response, and low power consumption [29].

Piezoelectric effect with high dynamic property is broadly used to develop tactile sensors or pressure sensors for some high-frequency dynamic application scenarios [30, 31]. However, rigid structural devices based on silicon-based MEMS process are not suitable for some measurements, such as large curved surfaces and skin. Flexible piezoelectric MEMS process is not a new technology. Most flexible piezoelectric MEMS devices are based on the materials, such as piezoelectric polymer [32], AlN [33], ZnO nanowires [34], with poor piezoelectric properties, which restrict the development of high-performance devices.

Currently, the tendency of the piezoelectric MEMS technologies mainly focuses on the high-electromechanical-coupling factor, flexibility, low frequency, width bandwidth, high power density, low noise, high sensitivity, and so forth. The detailed discussion will be as follows:

Dielectric materials are substances with a resistivity greater than $10^{10}\ \Omega\,\text{cm}$ that exhibits their electrical properties in an electric field by induction rather than conduction. The piezoelectric effect is a dielectric crystal material that, when subjected to pressure, generates a polarization potential on the upper and lower surfaces, and can only occur in heteropolar crystals where there is no center of symmetry in the structure. In a certain temperature range, pyroelectric materials with spontaneous polarization dipole moments that can change with the direction of the applied electric field become ferroelectric materials, and several concepts are attributed as shown in Figure 3.1. Piezoelectric materials include piezoelectric monocrystals, piezoelectric polycrystals, piezoelectric polymers, and piezoelectric composites. The main performance parameters of piezoelectric materials include piezoelectric constants, electromechanical-coupling coefficients, dielectric constants, dielectric losses, frequency constants, mechanical quality factors, elastic constants, and Curie temperature.

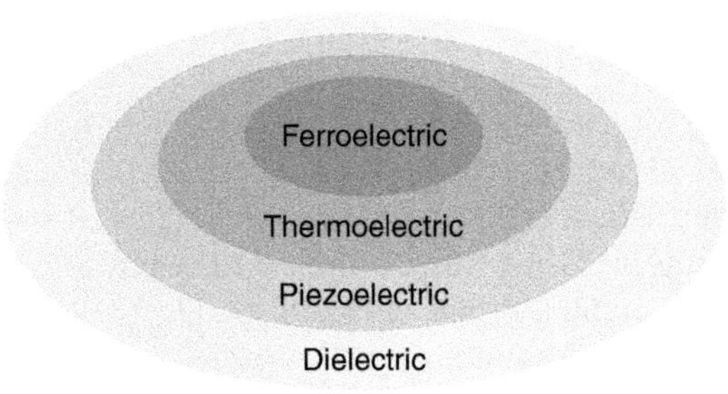

Figure 3.1 Attribution of the piezoelectric crystal materials. Source: Zhang and Wang [35].

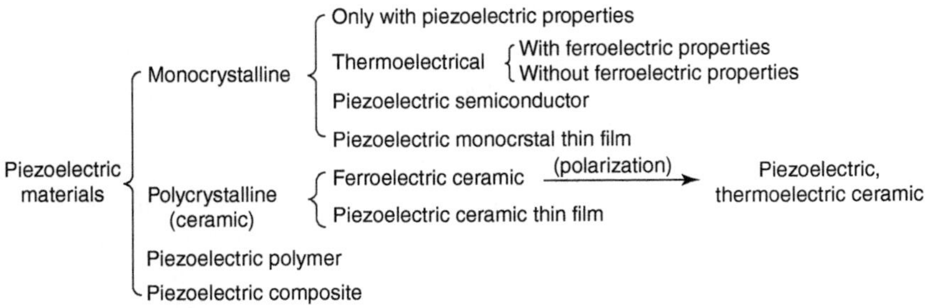

Figure 3.2 Classification of the piezoelectric materials.

The main piezoelectric materials include piezoelectric monocrystals, piezoelectric ceramics, piezoelectric polymers, and piezoelectric composites, as shown in Figure 3.2. For PEHs, the primary objective is to convert mechanical energy into electrical energy more efficiently, and it is generally desirable for piezoelectric materials to have a higher electromechanical-coupling coefficient. The electromechanical-coupling factor k_{31} can be described as [36]:

$$k_{31}^2 = \frac{d_{31}^2}{s_{11}^E \varepsilon_{33}^T} \tag{3.1}$$

Therefore, piezoelectric materials with high piezoelectric constant d_{31}, low elastic flexibility $s_{11,}^E$ and low dielectric constant ε_{33}^T possess higher electromechanical conversion efficiency. Usually, piezoelectric monocrystals possess higher electromechanical-coupling coefficients compared to piezoelectric ceramics, but their phase-transition temperatures are lower and high-temperature processing needs to be avoided in the process. With the development of consumer electronic devices and medical devices, the research on wearable and implantable piezoelectric energy harvesting is also increasing.

Although piezoelectric ceramic materials have large piezoelectric constants (Figure 3.3), their brittleness leads to more serious damage under large deformation, and flexible piezoelectric polymers are then widely studied, but their electromechanical-coupling coefficients are much lower than those of piezoelectric ceramics, which makes it difficult to achieve efficient electromechanical

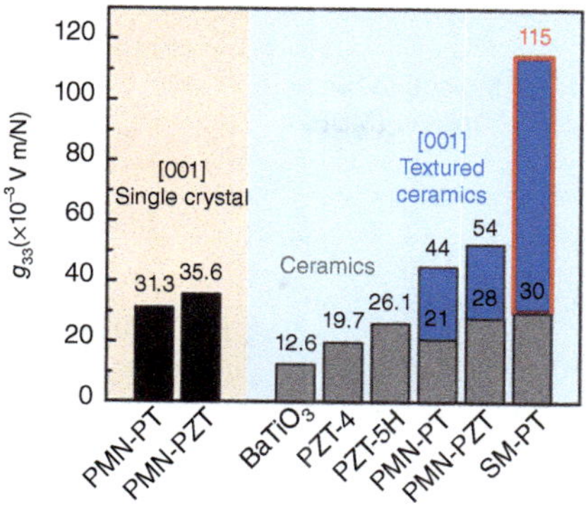

Figure 3.3 Research status of the piezoelectric materials. Source: [41] / Springer Nature / CC BY-4.0.

conversion, and thus promotes the research of piezoelectric complexes. Among them, PZT is lead zirconate titanate piezoelectric ceramics, PMN-PT lead magnesium niobate-lead titanate, PVDF polyvinylidene fluoride, PMN–PZT lead magnesite niobate -lead zirconate titanate solid solution.

The piezoelectric complex is composed of two phases – piezoelectric ceramic and polymer. The piezoelectric ceramic produces the piezoelectric effect, and the polymer reduces the material density and dielectric constant to enhance the elastic flexibility coefficient. This scheme is mainly to enhance the material flexibility by sacrificing part of the piezoelectric properties, which needs to balance the output performance and stiffness according to the specific application scenarios, and is often applied to hydrophones, low-frequency and high-frequency ultrasonic energy harvesters. 1–3-Type piezoelectric composites are materials with piezoelectric effects made by one-dimensional piezoelectric ceramic columns embedded parallelly in a three-dimensional-coupled polymer. New high-performance piezoelectric materials are also emerging, such as piezoelectric materials with certain orientation or piezoelectric solid solutions with piezoelectric effects optimized in the so-called quasi-homotypic phase boundary, but the preparation of these new piezoelectric materials is still to be perfected, and there is a lack of compatibility processes for use in piezoelectric energy-harvesting devices.

The MEMS process developed from semiconductor technology still belongs to two-dimensional planar process technology, and the thin film of piezoelectric materials is a key technology for the development of high-performance piezoelectric sensors and actuators compatible with the MEMS process. At present, the development of piezoelectric material-thinning technology is carried out along two main lines – additive manufacturing methods (such as molecular beam epitaxy [MBE], pulsed laser deposition [PLD], CVD, aerosol deposition (AD), and screen printing (SP), as shown in Figure 3.4) and reduced material technology (bulk thinning).

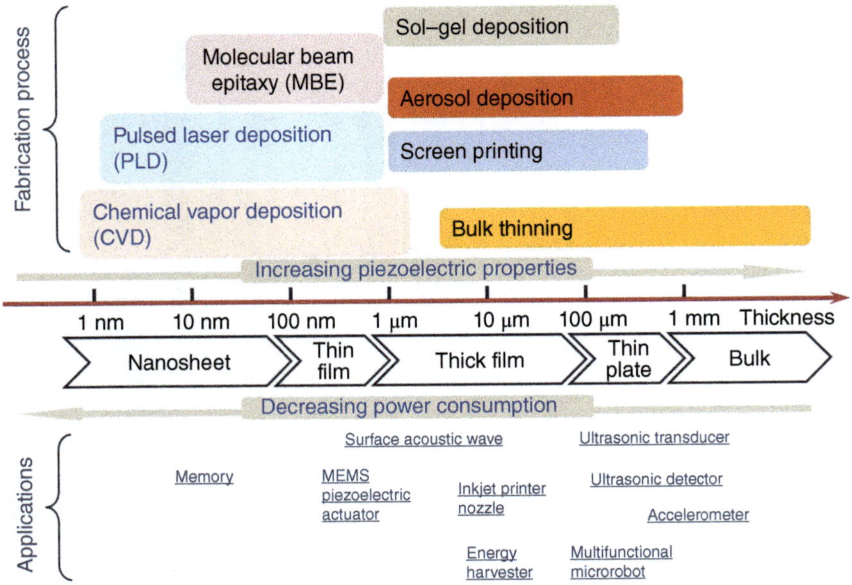

Figure 3.4 Classification of piezoelectric materials based on the key thickness and the corresponding available fields. Source: Adapted from [1, 9, 18, 37].

3.3 Piezoelectric Ceramic Thick Film

What both pathways keep breaking through is the thickness limitation with a view to achieve a wider thickness range of preparation capabilities, greater freedom in device design, and better application prospects. According to the literature, for the aspect of piezoelectric energy-harvesting applications, the piezoelectric film thickness is usually controlled in the thick-film region of 1–100 μm [18]. However, most of the current high-quality additive preparation methods (e.g. PLD, CVD, and MBE) are difficult to achieve the preparation of piezoelectric thick-film zone films (also called piezoelectric thick films), and the continuous growth by layer stacking will introduce a large number of material defects and inconsistencies in orientation, resulting in poor piezoelectric properties of the prepared piezoelectric films. Another class of methods, such as sol–gel method, aerosol method, and screen printing method, have greater advantages in the preparation of piezoelectric thin films, and some of them can realize the preparation of 1–10 μm piezoelectric thin films, but the denseness of the prepared materials is poor compared with the piezoelectric bulk material due to the process limitation, and the piezoelectric properties also have a certain gap with the bulk material (shown in Figure 3.5). Therefore, mechanical polishing technology based on the preparation of reduced material is highly favored by the research of piezoelectric energy-harvesting technology, through mechanical polishing process to thin the piezoelectric body material prepared by high-temperature sintering, to achieve the preparation of body material to piezoelectric thin film, and maintain the good denseness and piezoelectric properties of the body material.

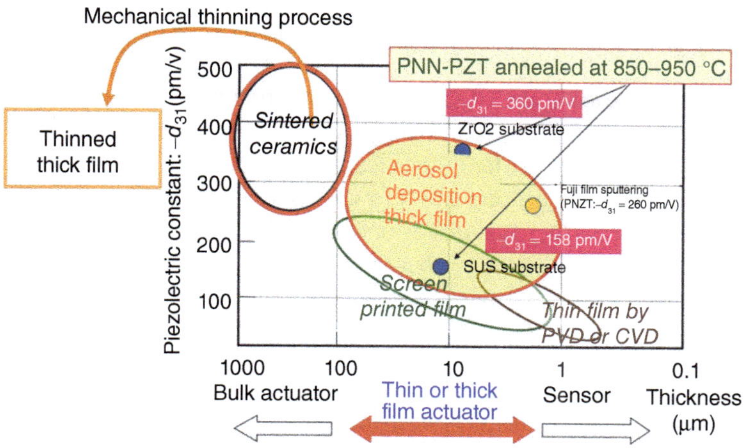

Figure 3.5 Thickness distribution and piezoelectric constant performance of the piezoelectric film obtained by various different fabrication approaches. PNN–PZT, Lead nickel niobate-lead zirconate titanate; PNZT, Niobium doped lead zirconate titanate; SUS, stainless steel; PVD, physical vapor deposition. Source: Akedo et al. [18] / with permission of IOP Publishing.

Given that the flexible layered composite structure of mechanically thinned piezoelectric films is the main microfabrication method used in this chapter, the following is an example of the preparation of a single-crystal cantilever-beam-vibrating energy harvester with a partial arrangement of piezoelectric layers to introduce the specific method. As shown in Figure 3.6, the preparation of a unimorph cantilever beam vibration energy harvester with partial piezoelectric layer arrangement is based on the preparation process of metal substrate piezoelectric thin-film multilayer composite structure, which mainly includes surface polishing of the materials to be bonded, hot-pressing bonding of conductive silver adhesive,

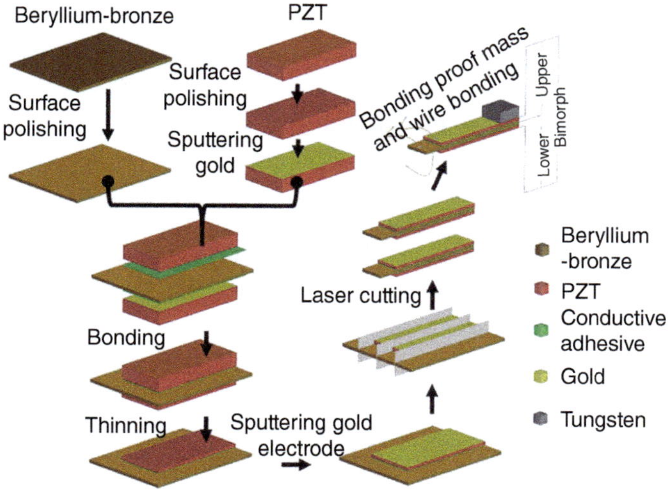

Figure 3.6 Fabrication process of cantilever piezoelectric vibration energy harvester with different layout lengths of the piezoelectric layer. Source: From Yi et al. [20] / AIP Publishing.

mechanical thinning of piezoelectric layer, metal electrode sputtering, laser cutting, cantilever beam fixation, and lead process. The specific processes are as follows:

Metal foil polishing: Firstly, using the UV laser cutting method to cut the metal foil into small units according to the size of a certain multiple of the designed area; fixing the cut metal foil in the center of the clean glass sheet with paraffin wax, setting the temperature of the heating table at 150 °C (the melting point of paraffin wax is about 80 °C, consider the heat diffusion of the heating table), paying attention to the flatness between the piezoelectric bulk and the glass sheet when fixing to prevent the bubbles. After the paraffin wax is completely melted, the glass sheet with the piezoelectric bulk attached is placed on a flat room temperature test bench, and a certain pressure is applied to the upper surface of the piezoelectric bulk with a mass fast to make the attachment tight and uniform; the glass sheet with the metal foil fixed is bonded to the grinding head using double-sided adhesive, and W5 abrasive paste is applied to the surface of the polishing disc to polish the surface of the metal foil, and the speed of the grinding disc is set at 150 *r*/min and the oscillation speed of the grinding head is set at 80 *r*/min. The surface of the metal foil is polished by applying W5 polishing paste on the surface of the polishing disc, with the disc speed of 150 *r*/min and the oscillating speed of the grinding head set at 80 *r*/min for about 10 minutes.

Piezoelectric sheet polishing: The same UV laser cutting method was used to cut the piezoelectric ceramic bulk material into small units at a certain multiple of the size with the area of the piezoelectric layer covered by a single cantilever beam; the same metal-foil-polishing method was used to fix it on the glass sheet and bond the glass sheet to the grinding head; the surface polishing was carried out with W5 abrasive paste to reduce the roughness and remove the surface dirt, which was done at a grinding speed of 150 *r*/min and the oscillating speed of the grinding head is set to 80 *r*/min for about five minutes (Figure 3.7); finally, the piezoelectric sheet is removed on the heating table at 150 °C and left to use.

Sputtering electrode on the polished surface of the piezoelectric sheet: A Cr/Au conductive layer is sputtered on the polished surface of the piezoelectric sheet by magnetron sputtering as the lower electrode layer of the piezoelectric layer.

Piezoelectric and metal foil bonding: Take the polished metal foil and piezoelectric sheet, use acetone, alcohol and deionized water were ultrasonically cleaned for five minutes after 80 °C oven drying; in the polished surface of the metal foil coated with a layer of conductive silver glue, and the piezoelectric sheet with electrodes attached to its side, with fingers gently press (to prevent excessive pressure damage to the piezoelectric material) to make the conductive glue uniform, and use cotton balls to wipe away the surrounding. After fine-tuning the attachment position under the microscope, the wafer is laminated to the cut dust-free cloth with polyimide (PI) tape to play a simple fixing role, and then covered with a piece of dust-free cloth of the same size on the upper layer, placed in the fixture, fixed fixture, apply a certain pressure to the wafer, and put the fixture into a constant temperature (175 °C) vacuum oven for 3.5 hours for hot-pressing bonding. Depending on the specifications of the conductive adhesive used, the bonding strength obtained at 0.1 MPa bonding pressure is above 15 MPa.

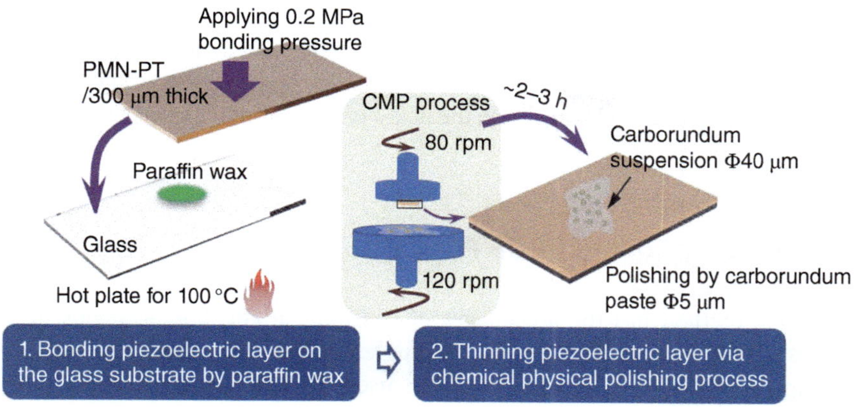

Figure 3.7 Mechanical polishing process of the piezoelectric layer. CMP, chemical-mechanical polishing.

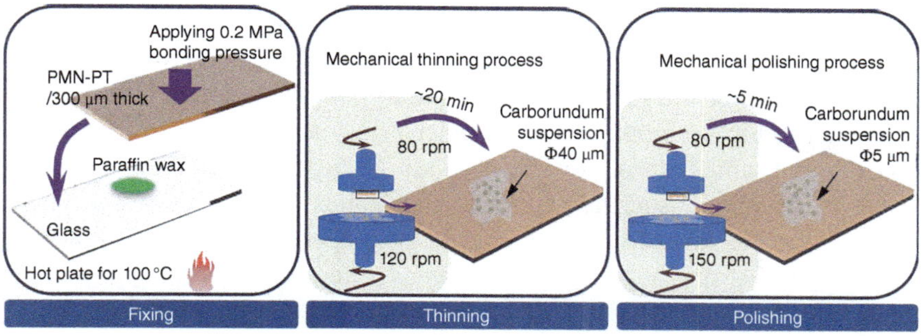

Figure 3.8 Fabrication process of mechanical thinning of the piezoelectric composite unit.

Piezoelectric layer thinning (Figure 3.8): Firstly, the bonded piezoelectric body with metal foil backing is fixed on the glass sheet with paraffin wax. To prevent warping of the extra metal part during the attachment process, a dust-free cloth is placed on the area as a pressure buffer layer before applying pressure so that the metal foil is evenly attached to the glass sheet surface as a whole. Then, mechanical thinning is done by configuring diamond W14 abrasive suspension, and using metal abrasive disc to thin the piezoelectric layer at 120 and 80 *r*/min oscillating speed of abrasive head. After the piezoelectric layer is thinned to the required thickness, the grinding head is cleaned, the diamond abrasive on the surface is removed, the polishing disc is replaced, and the surface of the piezoelectric layer is polished with W5 abrasive paste at a grinding disc speed of 150 *r*/min and the oscillating speed of the grinding head is set at 80 *r*/min.

Piezoelectric layer upper electrode preparation: A Cr/Au conductive layer is sputtered on the polished surface of the piezoelectric sheet as the upper electrode layer of the piezoelectric layer by magnetron sputtering, and polyimide tape is used as a hard mask to attach to the edge of the piezoelectric layer to avoid the conductivity of the sputtered gold electrode with the metal substrate.

Leading wire: The upper and lower film electrodes of the piezoelectric layer are led through 10-μm gold wires using conductive silver glue, and then connected to the copper-plated electrodes on the printed circuit board by electric soldering, and led from the copper-plated electrodes on the board with DuPont wires for subsequent testing.

3.4 Spinning-Coated PVDF Film

Spin-coating method has the advantages of controllable and homogeneous film thickness and high reproducibility. As a simple and cost-effective technique for preparing fine films, spin coating has been used for nearly a century and has become the preferred candidate for preparing most polymer functional films. The basic principle of spin coating is to transfer a drop of polymer solution to the center of the smooth substrate, then rotate the substrate at high speed. Due to centrifugal force at this step, most of the solution is thrown to the edge and finally out of the substrate, leaving a remaining layer of material on the surface of the substrate for the further formation of functional film. Noted that the volatile solvent is usually adopted for the subsequent evaporation. The properties of resulted film can be optimized by controlling the parameters, such as the rotation time, speed, droplet amount, and ratio of solution [38–43]. Currently, despite the existence of some shortcomings, such as without pattern, this strategy possesses many advantages, such as low cost, simple process, easy operation, and low pollution, which provides a strong impetus for the research and application of spin-coating method [44].

The schematic of the preparation of PVDF by spin coating is shown in Figure 3.9a [45]. The crucial procedures are spinning and annealing that directly result in the final thickness and the content of piezoelectric β-phase. Considering the different annealing methods, such as by oven and hot plate, the behaviors of solvent evaporation and crystallization are very different. For example, when annealed in the oven, the deposited PVDF is isotropically heated from surface to inner, inducing a more rapid formation of crystals on film surface. However, solvent evaporation is weakened in the inner region. In other words, this method may promote the content of β-phase with the decline of the electroactive properties through solvent trapping. On the other side, the hot-plate method can produce a unidirectional heat effect from the bottom to the top, which can avoid solvent trapping as possible benefits from the faster crystallization and solvent evaporation from the bottom to the top.

The smooth and flat surface of PVDF is highly desired considering the later electrodes deposition and poling process. Therefore, Cardoso et al. studied the surface morphology of PVDF influenced by thermal annealing in detail [47]. For that, a controlled 20 wt% PVDF/*N*,*N*-dimethylformamide (DMF) solution was adopted for further spin coating, and the rotation speed, acceleration, and time were 1000 rpm, 750 rpm/s, and 30 seconds, respectively. Finally, they were dried and annealed at different temperatures (room temperature 20, 30, 50, 70, and 80 °C).

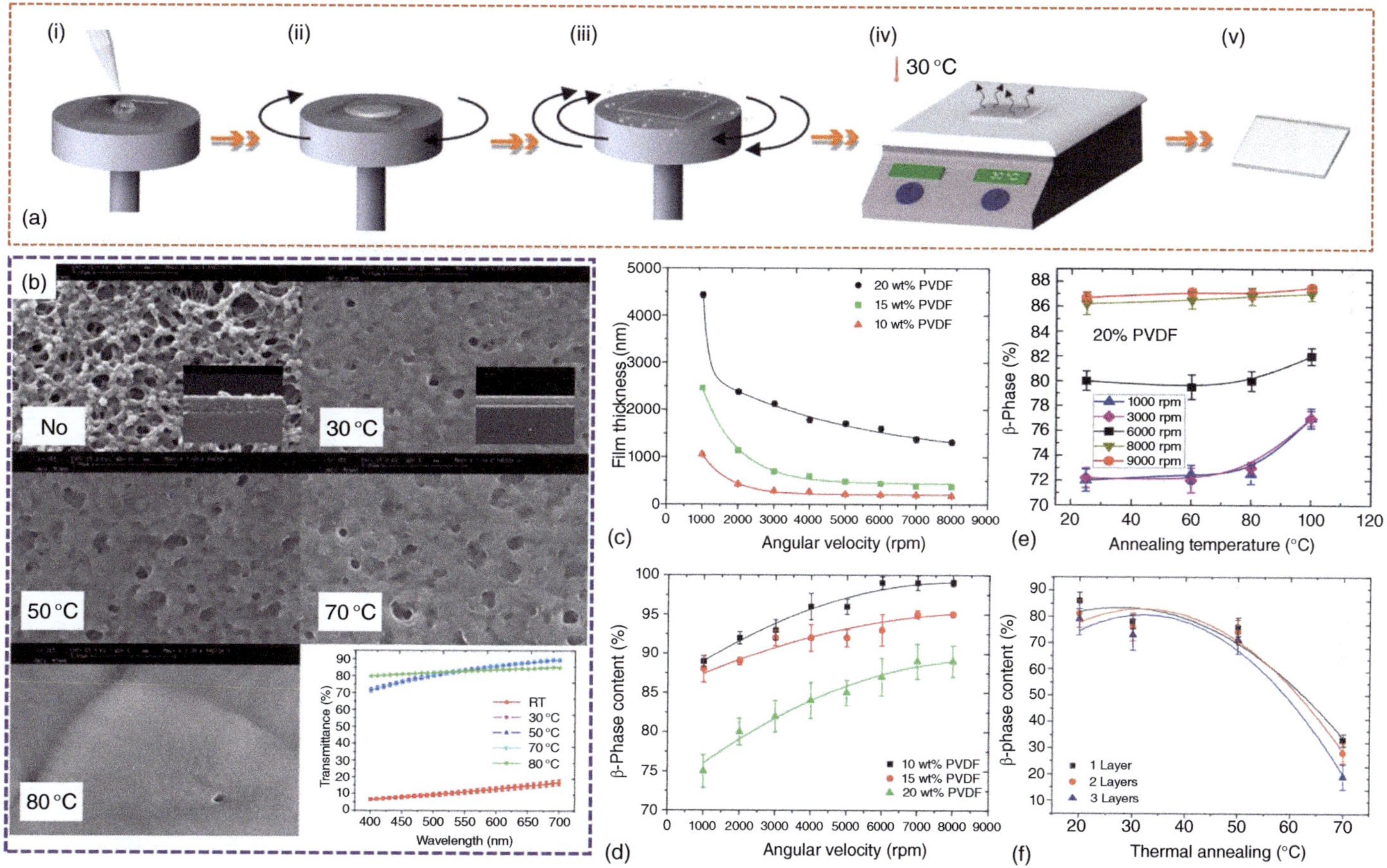

Figure 3.9 (a) Schematic of the preparation of β-PVDF by spin coating. Source: Ribeiro et al. [45] / with permission of Springer Nature. (b) SEM micrographs of the PVDF films without and with post-thermal annealing at: 30, 50, 70, 80 °C and the corresponding transmittance spectra of the films. (c) Film thickness as a function of the angular velocity of the spin-coater for various mass fractions of PVDF/DMF and a thermal annealing of 30 °C and (d) the corresponding β-phase content. (e) Variation of β-phase content of 20% PVDF films at different annealing temperatures. Source: Shaik et al. [46] / with permission of Springer Nature. (f) β-phase content of PVDF thin films with different number of layers. Source: (e) [53] / with permission of Springer Nature.

Clear observation in their scanning electron microscopy (SEM) images revealed that the lower temperature treatment contributed more porosity, which may be originated from the slower evaporation rate of DMF (Figure 3.9b). More intuitively, the transmittance spectra of the films proved the pore structure produced without annealed process while the smooth surface was generated after being annealed at a higher temperature. It is also noted that a certain degree of thermal annealing treatment is necessary for transparency requirements. In addition, the thickness can be adjusted by the mass ratio of solution and angular velocity. The higher speed contributes in thinner thickness under every specific mass ratio of PVDF/DMF solution, and it should be noted that when speed is more than 4000 rpm, the acceleration–thinning effect can be almost negligible (Figure 3.9c). Meanwhile, the increase of rotation speed has a positive effect on β-phase crystallization, which is mainly attributed to the larger shear effect on the PVDF under higher speed (Figure 3.9d). Similarly, the solution with lower PVDF content will obtain a thinner film thickness and higher β-phase content in the same case due to the lower viscosities.

Shaik et al. realized 87% of β-phase with 20 wt% PVDF solution spinning coated on the substrate at 9000 rpm and 100 °C annealing temperature and further proved that the annealing temperature can affect the β-phase content in spin-coated PVDF films, particularly at higher temperature around 95 °C under lower rotation speed (Figure 3.9e) [46]. When the speed was higher than 8000 rpm, it seems like no more β-phase content was achieved with the increase in annealing temperature. Therefore, the inferred conclusion can be obtained that the rotation speed contributed much greater to the β-phase crystallization than the annealing temperature at high speed.

Some attention was paid to the layer effect on the β-phase crystallization in spin-coated PVDF film for assembling the piezoelectric device. For instance, Cardoso et al. observed that a weak effect existed for crystallization when the PVDF layer was improved [48]. As shown in Figure 3.9f, the β-phase content of one layer is always the highest, while the three layers film is lowest at any annealing temperatures. This is because when the first layer was spin coated, the smooth substrate ensured the more dramatic shear and stretching effect on the PVDF, which mainly contributed to the formation of β-phase crystallization. However, when the remaining layers of PVDF were deposited on the previous PVDF substrate, the increased interaction with the polymer solution and higher surface roughness limited the shear effect on the upper layers, inducing inappreciable variation in the electroactive phase content.

3.5 Electrospinning Piezoelectric Film

Electrospinning is a common technique for preparing fibrous membranes through the way of stretching polymer droplets or melts directly by electric field forces. Benefitted from several advantages of simple equipment, controllable spinning process, low cost, and wide materials choices, electrospinning has received extensive attention from many researchers in these years, and it has become the main way to directly and continuously produce nanofiber materials [49–52]. The process of electrospinning is that the squeezed polymer droplets or melts deform into the shape of "Taylor cone" at the spinneret under the high-voltage electric field. Subsequently, the continuous stronger electric field force induces the charged droplet to overcome its own surface tension and eject to form a jet, which is continuously stretched and thinned during the movement. Meanwhile, the solvent evaporates to solidify the jet at the same time, forming micro/nanoscale fiber, which is eventually collected on the receiver to form thin film with the structure of random and nonwoven fabric [53–55].

In 2011, Baji et al. demonstrated that the electrospun PVDF films possessed excellent piezoelectric properties without further poling process and can be directly used as a piezoelectric film, which is because that electrospinning is carried out in a high-voltage electric field and PVDF suffers both stretching and polarization in the spinning process [55]. Moreover, Yee et al. proved that the crystalline phase structure of PVDF films prepared by electrostatic spinning mainly existed in the form of polar β crystallites, which further verified the above results [56]. However, there are also some problems during the process of electrospinning piezoelectric films. For example, the disordered fiber arrangement in prepared PVDF films will result in an offsetting effect, further reducing the overall piezoelectric performance of the fiber films. In this section, four typical electrospinning methods are discussed in detail, and some guidance is summarized for understanding this fabrication strategy.

Firstly, Deng et al. prepared the cowpea-structured PVDF/ZnO nanofibers piezoelectric sensors by electrospinning the dispersed PVDF/ZnO solution directly (Figure 3.10a) [57]. For the smooth and successful proceeding of electrospinning process, the nanoscale ZnO was selected as the additive to enhance the β-phase crystallinity. The design of disordered nanofibers embedded with ZnO contributed to the high piezoelectric performance, such as short response time and high sensitivity, and made the remote control of gestures feasible (Figure 3.10b,c). Impressively, the method of electrode deposition has universal reference significance that sprays flexible MXene (Ti_3C_2) electrodes onto both sides of porous structure, preventing short circuits caused by electron beam deposition. Although the excellent property was obtained, the disordered arrangement of PVDF nanofiber hinders the voltage output of the device. To avoid this negative effect, many researchers attempted to construct a highly aligned structure of PVDF nanofibers [59]. For example, Persano et al. took advantage of the collector with highly linear speed >16 m/s to produce a large-area poly (vinylidene fluoride-trifluoroethylene) (P[VDF–TrFE]) sheet with architectures of aligned arrangements of nanofiber (Figure 3.10d,e) [58]. The polymer chains adopt strongly preferential orientation due to the stretching effect generated by the high rotation speed, enhancing the piezoelectricity along the aligned direction. Figure 3.8f exhibits that this sensor can capture the tiny vibrations of the air caused by 70 dB sound, demonstrating its high sensitivity and application feasibility. Generally, the PVDF electrospinning film could optionally build disordered or aligned architecture to meet different requirements.

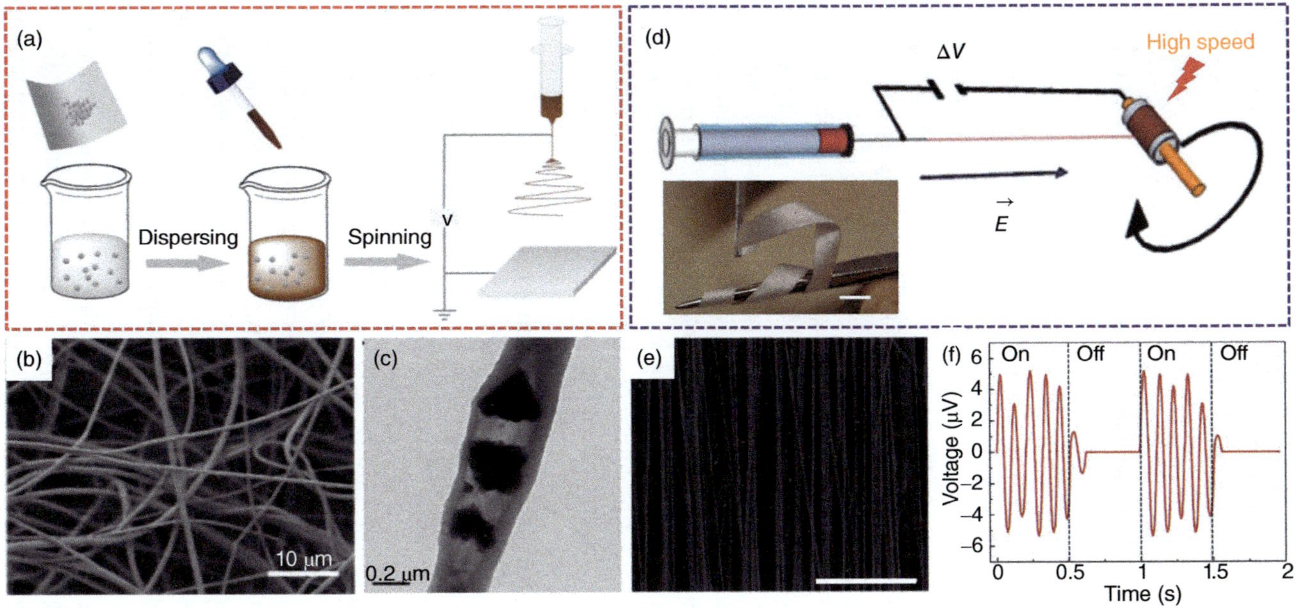

Figure 3.10 (a) The fabrication process of electrospinning random PVDF/ZnO flexible film. Source: Deng et al. [57] / with permission of Elsevier. (b) The SEM image of the nanofibers (NFs). (c) The transmission electron microscopy (TEM) image of a single NF. (d) Schematic illustration of electrospinning highly aligned P(VDF–TrFE) film, the inset is optical photograph. (e) SEM micrograph of highly aligned P(VDF-TrFE) fibers. Scale bar, 10 μm. (f) Output voltage collected from this device exposed to 70 dB sound intensity. Source: Persano et al. [58] / with permission of Springer Nature.

3.6 Summary

Fabrication of PEHs mainly relies on the fabrication processes of various piezoelectric films with different mechanical and electrical properties. For flexible PEHs, on one hand, researchers explored advanced methods to obtain high-quality piezoelectric ceramic thin film combining the high-electromechanical-coupling factor and available flexibility. Piezoelectric thick film attracted more attention because it has similar properties with the piezoelectric bulk materials and compatibility with the mature MEMS fabrication process. On the other hand, researchers developed some novel fabrication processes to enhance the piezoelectric properties of the piezoelectric polymer with intrinsic flexibility. Among these developed fabrication processes, electrospinning has been widely employed to make the flexible PEHs.

References

1 Toprak, A. and Tigli, O. (2014). Piezoelectric energy harvesting: state-of-the-art and challenges. *Applied Physics Reviews* 1: 031104.
2 Erturk, A. and Inman, D.J. (2011). *Piezoelectric Energy Harvesting*, 392. Wiley.
3 Wu, J., Chen, X., Chu, Z. et al. (2016). A barbell-shaped high-temperature piezoelectric vibration energy harvester based on $BiScO_3$–$PbTiO_3$ ceramic. *Applied Physics Letters* 109: 173901.
4 Dhakar, L., Liu, H.C., Tay, F.E.H., and Lee, C. (2013). A new energy harvester design for high power output at low frequencies. *Sensors and Actuators A: Physical* 199: 344–352.
5 Yang, Y., Shen, Q., Jin, J. et al. (2014). Rotational piezoelectric wind energy harvesting using impact-induced resonance. *Applied Physics Letters* 105 (5): 053901.
6 Kuo, Y.-K., Lin, H.-W., Chang, J.-Y. et al. (2012). Polarization effect on the photovoltaic characteristics of $Al_{0.14}Ga_{0.86}N/In_{0.21}Ga_{0.79}N$ superlattice solar cells. *IEEE Electron Device Letters* 33: 1159–1161.
7 Wang, S., Wang, Z.L., and Yang, Y. (2016). A one-structure-based hybridized nanogenerator for scavenging mechanical and thermal energies by triboelectric–piezoelectric–pyroelectric effects. *Advanced Materials* 28: 2881–2887.
8 Alluri, N.R., Saravanakumar, B., and Kim, S.-J. (2015). Flexible, hybrid piezoelectric film $(BaTi_{(1-x)}Zr_xO_3)$/PVDF nanogenerator as a self-powered fluid velocity sensor. *ACS Applied Materials & Interfaces* 7: 9831–9840.
9 Li, H., Tian, C., and Deng, Z.D. (2014). Energy harvesting from low frequency applications using piezoelectric materials. *Applied Physics Reviews* 1: 041301.
10 Aktakka, E. E., Peterson, R. L., and Najafi, K. (2010). A CMOS-compatible piezoelectric vibration energy scavenger based on the integration of bulk PZT films on silicon, Electron Devices Meeting IEEE.
11 Lin, S.C. and Wu, W.J. (2013). Fabrication of PZT MEMS energy harvester based on silicon and stainless-steel substrates utilizing an aerosol deposition method. *Journal of Micromechanics and Microengineering* 23 (12): 125028.

12 Hajati, A. and Kim, S.G. (2011). Ultra-wide bandwidth piezoelectric energy harvesting. *Applied Physics Letters* 99 (8): 083105.

13 Tang, G., Liu, J.Q., Yang, B. et al. (2012). Fabrication and analysis of high-performance piezoelectric MEMS generators. *Journal of Micromechanics and Microengineering* 22.

14 Roundy, S., Wright, P.K., and Rabaey, J. (2003). A study of low level vibrations as a power source for wireless sensor nodes. *Computer Communications* 26: 1131–1144.

15 Morimoto, K., Kanno, I., Wasa, K., and Kotera, H. (2010). High-efficiency piezoelectric energy harvesters of *c*-axis-oriented epitaxial PZT films transferred onto stainless steel cantilevers. *Sensors and Actuators A: Physical* 163: 428–432.

16 Yeo, H.G., Ma, X., Rahn, C., and Trolier-McKinstry, S. (2016). Efficient piezoelectric energy harvesters utilizing (001) textured bimorph PZT films on flexible metal foils. *Advanced Functional Materials* 26: 5940–5946.

17 Yang, Z., Zhou, S., Zu, J., and Inman, D. (2018). High-performance piezoelectric energy harvesters and their applications. *Joule* 2: 642–697.

18 Akedo, J., Park, J.-H., and Kawakami, Y. (2018). Piezoelectric thick film fabricated with aerosol deposition and its application to piezoelectric devices. *Japanese Journal of Applied Physics* 57: 07LA02.

19 Eom, C.-B. and Trolier-McKinstry, S. (2012). Thin-film piezoelectric MEMS. *MRS Bulletin* 37: 1007–1021.

20 Yi, Z., Yang, B., Li, G. et al. (2017). High performance bimorph piezoelectric MEMS harvester via bulk PZT thick films on thin beryllium-bronze substrate. *Applied Physics Letters* 111: 013902.

21 Wang, Z., Miao, J., Tan, C.W., and Xu, T. (2010). Fabrication of piezoelectric MEMS devices-from thin film to bulk PZT wafer. *Journal of Electroceramics* 24: 25–32.

22 Aktakka, E.E., Peterson, R.L., and Najafi, K. (2013). Wafer-level integration of high-quality bulk piezoelectric ceramics on silicon. *IEEE Transactions on Electron Devices* 60: 2022–2030.

23 Yi, Z., Yang, H., Tian, Y. et al. (2018). Self-powered force sensor based on thinned bulk PZT for real-time cutaneous activities monitoring. *IEEE Electron Device Letters* 39: 1226–1229.

24 Zhu, M., Shi, Q., He, T. et al. (2019). Self-powered and self-functional cotton sock using piezoelectric and triboelectric hybrid mechanism for healthcare and sports monitoring. *ACS Nano* 13: 1940–1952.

25 Li, N., Yi, Z., Ma, Y. et al. (2019). Direct powering a real cardiac pacemaker by natural energy of a heartbeat. *ACS Nano* 13: 2822–2830.

26 Yi, Z., Yang, B., Zhang, W. et al. (2020). Batteryless tire pressure real-time monitoring system driven by an ultralow frequency piezoelectric rotational energy harvester. *IEEE Transactions on Industrial Electronics* 68: 3192–3201.

27 Yi, Z., Huang, J., Liu, Z. et al. (2020). Portable, wireless wearable piezoelectric arterial pulse monitoring system based on near-field communication approach. *IEEE Electron Device Letters* 41: 183–186.

28 Yi, Z., Xie, F., Tian, Y. et al. (2020). A battery- and leadless heart-worn pacemaker strategy. *Advanced Functional Materials* 30: 2000477.

29 Krishna, G.M. and Rajanna, K. (2004). Tactile sensor based on piezoelectric resonance. *IEEE Sensors Journal* 4: 691–697.

30 Curry, E.J., Ke, K., Chorsi, M.T. et al. (2018). Biodegradable piezoelectric force sensor. *Proceedings of the National Academy of Sciences of the United States of America* 115: 909–914.

31 Park, D.Y., Joe, D.J., Kim, D.H. et al. (2017). Self-powered real-time arterial pulse monitoring using ultrathin epidermal piezoelectric sensors. *Advanced Materials* 29: 1702308.

32 Ramadan, K.S., Sameoto, D., and Evoy, S. (2014). A review of piezoelectric polymers as functional materials for electromechanical transducers. *Smart Materials and Structures* 23: 26.

33 Jackson, N., Keeney, L., and Mathewson, A. (2013). Flexible-CMOS and biocompatible piezoelectric AlN material for MEMS applications. *Smart Materials and Structures* 22 (11): 115033.

34 Lee, C.-T. and Chiu, Y.-S. (2015). Piezoelectric ZnO-nanorod-structured pressure sensors using GaN-based field-effect-transistor. *Applied Physics Letters* 106 (7): 073502.

35 Zhang, F. and Wang, L. (2002). *Advances in Piezoelectricity*, 531. Science Press.

36 Roscow, J.I., Pearce, H., Khanbareh, H. et al. (2019). Modified energy harvesting figures of merit for stress- and strain-driven piezoelectric systems. *The European Physical Journal Special Topics* 228: 1537–1554.

37 Liu, H., Zhong, J., Lee, C. et al. (2018). A comprehensive review on piezoelectric energy harvesting technology: materials, mechanisms, and applications. *Applied Physics Reviews* 5: 041306.

38 Mahale, B., Bodas, D., and Gangal, S.A. (2017). Study of β-phase development in spin-coated PVDF thick films. *Bulletin of Materials Science* 40: 569–575.

39 Jung, H.J., Chang, J., Park, Y.J. et al. (2009). Shear-induced ordering of ferroelectric crystals in spin-coated thin poly(vinylidene fluoride-*co*-trifluoroethylene) films. *Macromolecules* 42: 4148–4154.

40 Nishiyama, T., Sumihara, T., Sato, E., and Horibe, H. (2017). Effect of solvents on the crystal formation of poly (vinylidene fluoride) film prepared by a spin-coating process. *Polymer Journal* 49: 319–325.

41 Divya, S. and Hemalatha, J. (2017). Study on the enhancement of ferroelectric β phase in P(VDF-HFP) films under heating and poling conditions. *European Polymer Journal* 88: 136–147.

42 Ramasundaram, S., Yoon, S., Kim, K.J., and Lee, J.S. (2008). Direct preparation of nanoscale thin films of poly(vinylidene fluoride) containing β-crystalline phase by heat-controlled spin coating. *Macromolecular Chemistry and Physics* 209: 2516–2526.

43 Cardoso, V.F., Costa, C.M., Minas, G., and Lanceros-Mendez, S. (2012). Improving the optical and electroactive response of poly(vinylidene fluoride-trifluoroethylene) spin-coated films for sensor and actuator applications. *Smart Materials and Structures* 21: 8.

44 Cho, Y., Park, J.B., Kim, B.S. et al. (2015). Enhanced energy harvesting based on surface morphology engineering of P(VDF-TrFE) film. *Nano Energy* 16: 524–532.
45 Ribeiro, C., Costa, C.M., Correia, D.M. et al. (2018). Electroactive poly(vinylidene fluoride)-based structures for advanced applications. *Nature Protocols* 13: 681–704.
46 Shaik, H., Rachith, S.N., Rudresh, K.J. et al. (2017). Towards β-phase formation probability in spin coated PVDF thin films. *Journal of Polymer Research* 24: 6.
47 Cardoso, V.F., Minas, G., Costa, C.M. et al. (2011). Micro and nanofilms of poly(vinylidene fluoride) with controlled thickness, morphology and electroactive crystalline phase for sensor and actuator applications. *Smart Materials and Structures* 20: 4.
48 Cardoso, V.F., Minas, G., and Lanceros-Mendez, S. (2013). Multi layer spin-coating deposition of poly(vinylidene fluoride) films for controlling thickness and piezoelectric response. *Sensors and Actuators A: Physical* 192: 76–80.
49 Park, S.H., Lee, H.B., Yeon, S.M. et al. (2016). Flexible and stretchable piezoelectric sensor with thickness-tunable configuration of electrospun nanofiber mat and elastomeric substrates. *ACS Applied Materials & Interfaces* 8: 24773–24781.
50 Lang, C.H., Fang, J., Shao, H. et al. (2016). High-sensitivity acoustic sensors from nanofibre webs. *Nature Communications* 7: 7.
51 Li, T., Feng, Z.Q., Yan, K. et al. (2018). Pure OPM nanofibers with high piezoelectricity designed for energy harvesting in vitro and in vivo. *Journal of Materials Chemistry B* 6: 5343–5352.
52 Lin, M.F., Xiong, J.Q., Wang, J.X. et al. (2018). Core–shell nanofiber mats for tactile pressure sensor and nanogenerator applications. *Nano Energy* 44: 248–255.
53 Pan, X.M., Wang, Z., Cao, Z. et al. (2016). A self-powered vibration sensor based on electrospun poly(vinylidene fluoride) nanofibres with enhanced piezoelectric response. *Smart Materials and Structures* 25: 8.
54 Siddiqui, S., Lee, H.B., Kim, D.I. et al. (2018). An omnidirectionally stretchable piezoelectric nanogenerator based on hybrid nanofibers and carbon electrodes for multimodal straining and human kinematics energy harvesting. *Advanced Energy Materials* 8: 11.
55 Baji, A., Mai, Y.W., Li, Q., and Liu, Y. (2011). Electrospinning induced ferroelectricity in poly(vinylidene fluoride) fibers. *Nanoscale* 3: 3068–3071.
56 Yee, W.A., Kotaki, M., Liu, Y., and Lu, X.H. (2007). Morphology, polymorphism behavior and molecular orientation of electrospun poly(vinylidene fluoride) fibers. *Polymer* 48: 512–521.
57 Deng, W.L., Yang, T., Jin, L. et al. (2019). Cowpea-structured PVDF/ZnO nanofibers based flexible self-powered piezoelectric bending motion sensor towards remote control of gestures. *Nano Energy* 55: 516–525.

58 Persano, L., Dagdeviren, C., Su, Y. et al. (2013). High performance piezoelectric devices based on aligned arrays of nanofibers of poly(vinylidenefluoride-*co*-trifluoroethylene). *Nature Communications* 4: 10.

59 Sharma, T., Naik, S., Langevine, J. et al. (2015). Aligned PVDF-TrFE nanofibers with high-density PVDF nanofibers and PVDF core–shell structures for endovascular pressure sensing. *IEEE Transactions on Biomedical Engineering* 62: 188–195.

4

Cantilever Piezoelectric Energy Harvesters

4.1 Introduction

Energy harvesting from ambient energy for activating low-powered electronics has emerged during the last decades as an enabling technology for self-powered devices [1–5]. The piezoelectric effect is commonly used for energy harvesters or sensors due to its advantages of simple configuration and high-energy-conversion efficiency [6–8]. Currently, most investigations for power generation of vibration-based piezoelectric energy harvesters (PEHs) focus on high-output power, low resonant frequency, and good reliability [9–12]. Among these studies, PEHs are commonly based on a cantilever structure with a large proof mass at the free end, which is contributed for two reasons [13, 14]. First, for a given force input, the cantilever configuration can result in the highest average strain, which is beneficial to obtain high-output power. Second, the cantilever configuration has a low resonant frequency for a given size, which is easy to meet the low-frequency ambient vibrations.

Most cantilever-based PEHs consist of the piezoelectric and supporting layers with the same or custom length. However, the subtle changes in cantilever configuration can generate great influence on the output power, especially on microscale [15]. To learn the effect of the piezoelectric layer length on the output performance, Gao et al. [16] have recently explored the relationship between the length ratio of the nonpiezoelectric layer to the piezoelectric layer and the output voltage under a concentrated static force at the cantilever tip. The results indicated that the optimal voltage was obtained when the piezoelectric layer and the nonpiezoelectric layer have the same length under constant tip displacement. However, in view of most of the practical application scenarios, PEHs are expected to capture the ambient vibration energy, not the static force or displacement. Therefore, a further study for such cantilever PEHs under the excitation of harmonic vibration was carried out by Gao et al. [17]. The results indicated that the maximum power under the optimal resistance occurred at the same length of the piezoelectric layer and the nonpiezoelectric layer. However, the additional mechanical impedance from the defect of the theoretical model influenced their calculation results, leading to that the output power and beam displacement being underestimated. Moreover, the nonpiezoelectric layer thickness has been studied as the only influence factor of the optimal nonpiezoelectric/piezoelectric layer length

Flexible Piezoelectric Energy Harvesters and Sensors, First Edition. Bin Yang, Zhiran Yi, and Chengkuo Lee.

ratio. Thus, it cannot be applied to the PEHs with a concentrated proof mass at the tip of cantilever. Patel et al. [18] focused on the optimization of the microscale cantilever PEHs under the constant frequency through investigating the influence of geometrical parameters. The theoretical analysis is based on the combination of a distributed parameter electromechanical model and finite element analysis. Yu et al. [19] presented a vibration-based MEMS piezoelectric energy harvester, where the influence of piezoelectric patch length and width, and thickness on the output voltage was also discussed. Sunithamani et al. [20] presented enhancing the performance of energy harvester by optimizing the PZT length. Although these works indicated that performance was improved by optimizing the piezoelectric layer geometry, the optimization schemes are still not systematic. Furthermore, the effect of other key design parameters on the optimal piezoelectric layer design and its underlying mechanism is still not clear. In addition, the natural frequency of 244.4 Hz is higher than the frequency of most natural vibrational sources for practical applications. Therefore, it is significant to optimize the piezoelectric layer length of the low-frequency PEHs with a proof mass for high performance.

The main purpose of this chapter is to determine the optimal PZT layout length, investigate the effects of geometric and material parameters on this length, and reveal the foundational optimization mechanism for the piezoelectric cantilever harvester with a tip-proof mass, which contributes to high-conversion efficiency at low-frequency harmonic vibrations. An optimal piezoelectric layer length for the PEH under the base motion excitation was obtained to achieve a high-power density via a systematic theoretical analysis and finite element analysis. The effects of the entire length, width, and elastic modulus of supporting layer; thickness of piezoelectric layer; and height of mass proof on the optimal length of piezoelectric layer were also investigated. Experimental results are in agreement with the above analyses. A nonuniform cantilever piezoelectric harvester with the optimal piezoelectric layer length is fabricated by the thinned PZT thick film, which performs high-power density (30 mW/cm^3) and good stability (3.3 million cycles) at the resonant frequency of 46 Hz. The comparison with previous reports demonstrates that the optimal functional layer length is an effective way of improving the strain-induced piezoelectric harvesters, which also can be considered a promising method for such strain-induced harvesters or sensors, such as electrochemically driven mechanical energy harvester [21].

4.2 Optimized Cantilever PEH

The results of theoretical analysis of different lengths of piezoelectric layer coverings show that the length of piezoelectric layer arrangement has a large impact on the output performance of the cantilever beam vibration energy harvester. In this section, the output performance of the cantilever-beam-type vibration energy harvester with different piezoelectric layer arrangement lengths is studied by the preparation process of the piezoelectric thin-film composite structure with metal substrate, and the results of the theoretical analysis are compared and analyzed.

4.2.1 Structure Design and Fabrication

As shown in Figure 4.1, six different single-crystal cantilever beam vibration energy harvester structures with effective piezoelectric layer arrangement lengths of 8, 6.5, 5, 3, and 1 mm are designed, respectively, with beryllium bronze as the substrate material and PZT as the piezoelectric material, with a total cantilever beam length of 15 mm, a width of 2.5 mm, and a mass bulk size of $5 \times 2.5 \times 0.2\ mm^3$.

Figure 4.2 shows the vibration energy harvester test platform, mainly with vibration control and dynamic signal analyzer (m + p VibPilot), shaker, power amplifier,

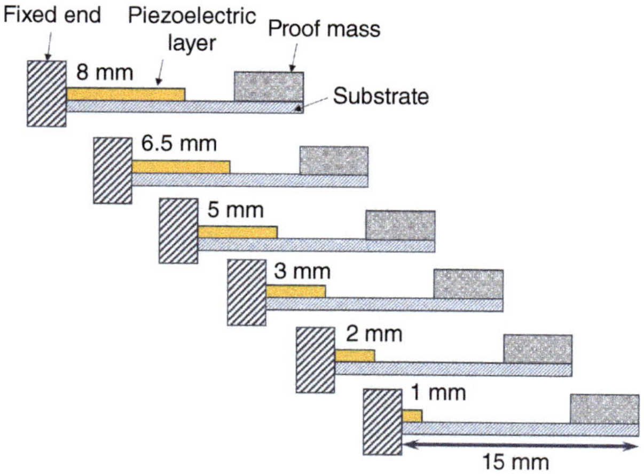

Figure 4.1 Schematic illustration of cantilever piezoelectric vibration energy harvester with different layout lengths of the piezoelectric layer.

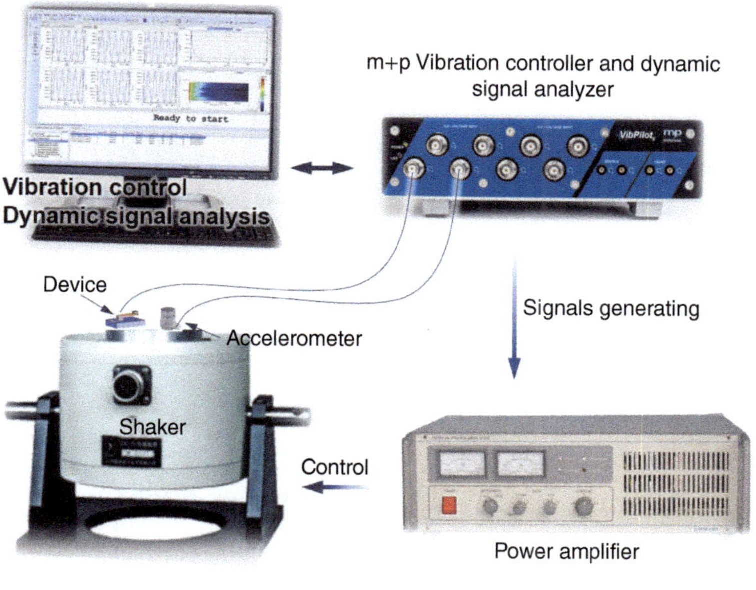

Figure 4.2 Test platform for vibration energy harvester.

acceleration sensor, and control and data acquisition computer components, the computer through the control software to set the excitation acceleration and frequency, dynamic signal analyzer output excitation signal through the power amplifier amplification to excite the shaker to generate vibration. The acceleration sensor monitors the vibration state of the shaker to maintain the set acceleration and frequency, and the test device is fixed to the vibration pallet of the shaker.

4.2.2 Output Performance

Figure 4.3 shows the closed-circuit output voltage performance of the cantilever vibration energy harvester with the piezoelectric layer arrangement lengths of 1, 3, 5, 6.5, and 8 mm for 0.1g and 1.0g excitation acceleration amplitudes, respectively. The cantilever beam vibration energy harvester with a piezoelectric layer arrangement length of 3 mm at a load resistance of 1000 kΩ exhibits the maximum closed-circuit output voltage.

By calculating the output power as shown in Figure 4.4, it can be seen that the piezoelectric layer arrangement length of 3 mm has the maximum output power achieved. It is known that the resonant frequency of the cantilever beam vibration energy harvester increases gradually as the length of the piezoelectric layer arrangement increases from 1 to 8 mm, and therefore, the input vibration energy also increases gradually. At the same time, it can be seen that as the arrangement length increases, the best matching load resistance gradually decreases, which is more conducive to the effective output of electrical energy. However, for the piezoelectric layer arrangement lengths of 5, 6.5, and 8 mm, the piezoelectric layer arrangement length of 3 mm has the highest energy-conversion efficiency. For the vibration energy harvester with the piezoelectric layer arrangement length of 1 mm,

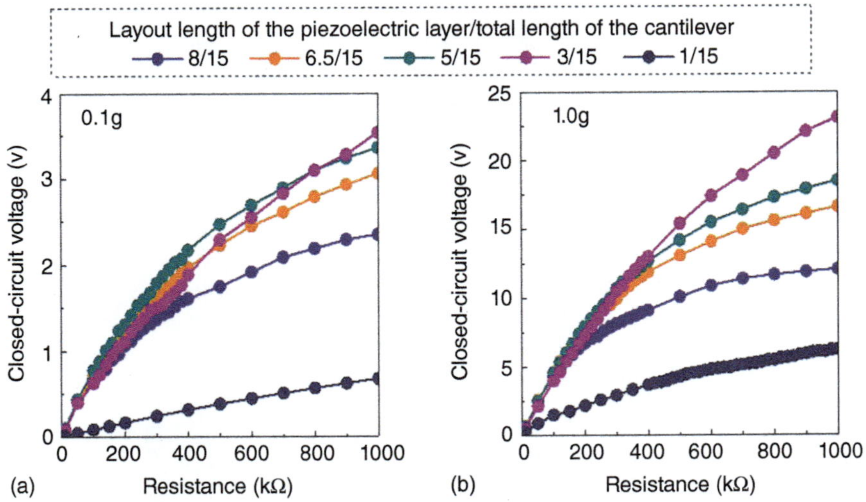

Figure 4.3 Closed-circuit voltage of cantilever piezoelectric energy harvester (PEH) with the partial layout length of the piezoelectric layer under the applied acceleration amplitude of 0.1g (a) and 1.0g (b).

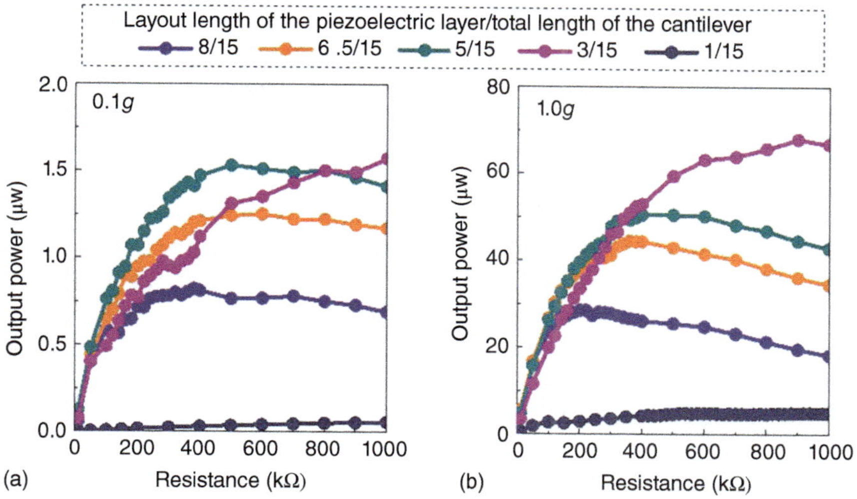

Figure 4.4 Output power of cantilever piezoelectric energy harvester (PEH) with the partial layout length of the piezoelectric layer under the applied acceleration amplitude of 0.1g (a) and 1.0g (b).

the output power is influenced by the high matching load resistance and the output power is lower.

To verify the theoretical analysis results, Figure 4.5 shows the comparison between the FEA results and the corresponding experimental data. Figure 4.5a gives the relationship between the output power and the load resistance for the cantilever beam vibration energy harvester (total effective length 15 mm) with the piezoelectric layer arrangement lengths of 3, 5, and 8 mm. The experimental and simulation results show that the output power is highly dependent on the

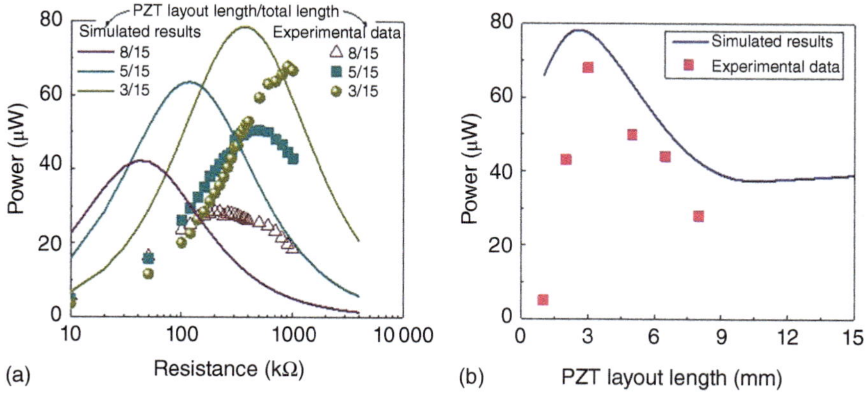

Figure 4.5 Output powers of the cantilever piezoelectric energy harvester (PEH) with different layout lengths of the piezoelectric layer, (a) output powers of the harvesters as a function of the external resistance. (b) The maximum output power is a function of the PZT layout length. The applied acceleration amplitude is 1.0g. Curves represent simulated results and the dot represents experimental data.

load resistance, and the best matching resistance and maximum output power are decreasing with the increase of the arrangement length of piezoelectric layer from 3 to 8 mm, the best matching resistance and the maximum. The simulated analysis and the experimental results are in good agreement with each other. In addition, a difference in the best-fit resistance value between the simulated and experimental results is also observed, which is mainly attributed to the preparation process error and measurement equipment impedance. Figure 4.5b reflects the variation of the maximum output power of the cantilever vibration energy harvester with different piezoelectric layer section arrangement lengths. However, the theoretical and experimental results reflect the same influence law of the piezoelectric layer arrangement length, which verifies the validity of the proposed optimization analysis method and reveals the important influence law of the piezoelectric layer arrangement length on the design of cantilever-beam-type vibration energy harvester, which provides key technical support for the design of high-efficiency piezoelectric cantilever beam vibration energy harvester.

The normalized power density NPD is commonly used to measure the advantages and disadvantages of piezoelectric vibration energy harvesters of different structures, sizes, materials, and vibration modes [12, 22]. It is calculated as

$$\mathrm{NPD} = \frac{\mathrm{Power}_{\mathrm{average}}}{\mathrm{Volume}_{\mathrm{eff}} \times \mathrm{Acceleration}^2 \times \mathrm{Frequency}} \tag{4.1}$$

where $\mathrm{Power}_{\mathrm{average}}$ is the average power density, $\mathrm{Volume}_{\mathrm{eff}}$ is the effective volume of the device, Acceleration is the excitation acceleration, and Frequency is the excitation frequency. Table 4.1 collates the performance comparison of unimorph cantilever beam vibration energy harvester with partial arrangement of piezoelectric layer; through the table, it can be seen that as the proportion of piezoelectric layer decreases, its intrinsic frequency and volume is gradually reduced, the reduced volume is the part of the piezoelectric layer reduced, the recent matching resistance shows an increasing trend, the singularity appears when the proportion of arrangement length is 2/15, which is not in line with the theoretical results, and

Table 4.1 Comparison for unimorph cantilever vibration energy harvesters via different layout lengths of the PZT layer (acceleration is 1*g*).

Type	PZT ratio of the length	Volume (mm^3)	Frequency (Hz)	Matched resistance (kΩ)	Power (mW)	Power density (mW/cm^3)	Normalized power density (mW/cm^3 g^2 Hz)
Unimorph	8/15	5.38	123.1	200	0.028	5.3	0.0431
	6.5/15	5.19	95.3	360	0.044	8.5	0.0891
	5/15	5	80	500	0.05	10.1	0.126
	3/15	4.75	67	900	0.068	14.3	0.213
	1/15	4.5	38.8	1120	0.005	1.1	0.029

it is thought to be the experimental preparation. The output power, power density, and normalized power density reach the maximum at the PZT length ratio of 3/15, reaching 0.213 mW/cm^3 g^2 Hz, which is consistent with the theoretical analysis conclusion, indicating that this piezoelectric layer arrangement ratio is the optimal design of this unimorph cantilever beam vibration energy harvester.

4.3 Bimorph PEH

To further increase the power density of the cantilever beam vibration energy harvester, the preparation method of the double-crystal cantilever beam vibration energy harvester is studied based on the preparation process of double-crystal piezoelectric thin-film multilayer composite structure, and the key technology of piezoelectric thin film to enhance the output performance of the piezoelectric vibration energy harvester is explored.

4.3.1 Structure Design and Fabrication

As shown in Figure 4.6, the total length of the cantilever beam of the bimorph cantilever beam vibration energy harvester is 18 mm, the width is 3 mm, 50-μm-thick beryllium bronze is used as the substrate layer, the fixed end is about 2 mm long for bonding between the cantilever beam and the base, the effective length of the cantilever beam is controlled at 16 mm, the end is fixed with a tungsten mass bulk of 5 mm in length, 3 mm in width and 1.5 mm in thickness, and the upper and lower piezoelectric sheets are designed with a thickness of 1.5 mm. The design thickness of the lower piezoelectric sheet is 50 μm.

The finite element simulation (FEA) analysis of the designed double-crystal cantilever beam vibration energy harvester was carried out using COMSOL Multiphysics, and the mesh division in the simulation analysis is shown in Figure 4.7a. The calculation results found that its first-order modal response frequency is at 61.038 Hz (as shown in Figure 4.7b), the open-circuit output maximum peak-to-peak voltage can reach 66.9 V (as shown in Figure 4.7c), and Figure 4.7d shows that the maximum effective power of 630 μW can be output at the matching resistance of 180 kΩ.

As shown in Figure 4.8a, the piezoelectric layer fully covered bimorph cantilever beam vibration energy harvester is prepared using a bimorph piezoelectric thin-film multilayer composite structure preparation process, which mainly includes surface

Figure 4.6 Design of bimorph piezoelectric vibration energy harvester with an equal length between the piezoelectric layer and the cantilever.

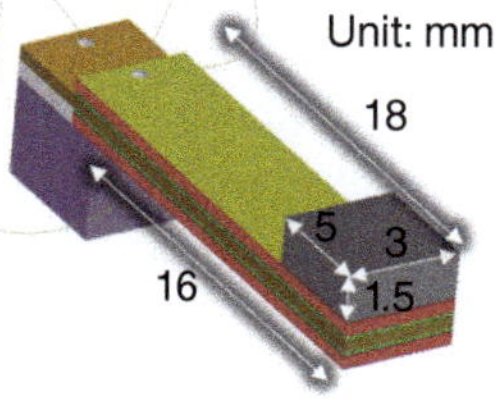

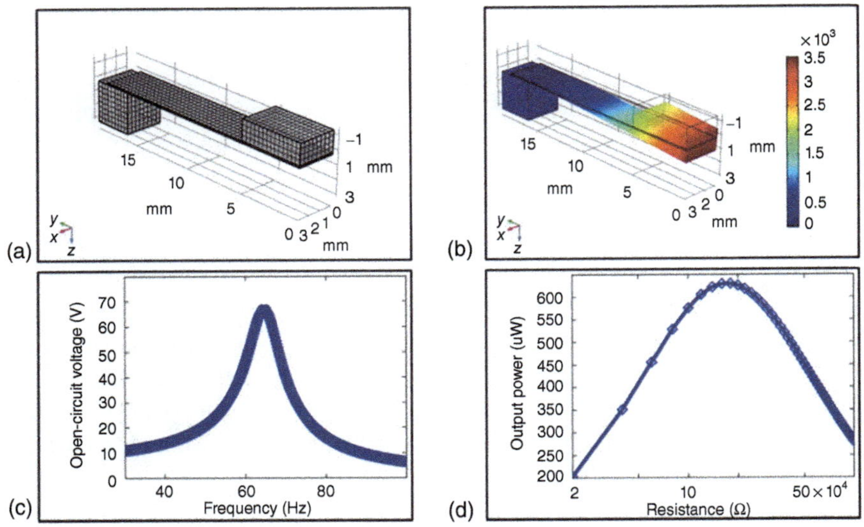

Figure 4.7 Simulation of bimorph piezoelectric vibration energy harvester with the equal length between the piezoelectric layer and the cantilever by FEA, (a) mesh, (b) Ffrst-order resonant mode, (c) output voltage, and (d) output power of the cantilever piezoelectric energy harvester under the acceleration amplitude of 1.0*g*.

polishing, hot-press bonding, mechanical thinning, sputtering electrode, laser cutting, and cantilever beam fixation and lead process. In the hot-press bonding process, two equal area 300-μm-thick piezoelectric ceramic sheets of PZT (C-6, Fuji ceramics) are bonded to both sides of 50-μm-thick beryllium bronze simultaneously by silver conductive adhesive, and the upper and lower piezoelectric sheets have opposite polarization directions. Considering that the modulus of beryllium bronze is close to that of pure copper and has better fatigue resistance than copper, beryllium bronze was used as the substrate material in the preparation of the bimorph cantilever beam. Figure 4.8b shows the bimorph cantilever beam energy harvester prepared by this process, and the cantilever beam structure is encapsulated by Plexiglas.

The preparation process of the piezoelectric bimorph multilayer composite structure is very similar to that of the piezoelectric single-crystal multilayer composite structure, with differences mainly in the bonding and thinning processes, which are described below in detail.

(1) Bonding of piezoelectric layer and metal foil
Piezoelectric and metal foil bonding: take the polished metal foil and two piezoelectric sheets of the same size, clean them with acetone, alcohol and deionized water, respectively, for 5 minutes by ultrasonic cleaning, and then dry them in an oven at 80 °C. Apply a layer of conductive silver glue on the two polished sides of the metal foil and attach the side of the piezoelectric sheet with electrodes to it, gently press it to make the conductive glue uniform, and wipe off the excess conductive glue around it with a cotton ball; and then, after fine-tuning the attachment position under the microscope, the wafer is laminated with the cut dust-free cloth using polyimide tape to play a simple fixing role, and then

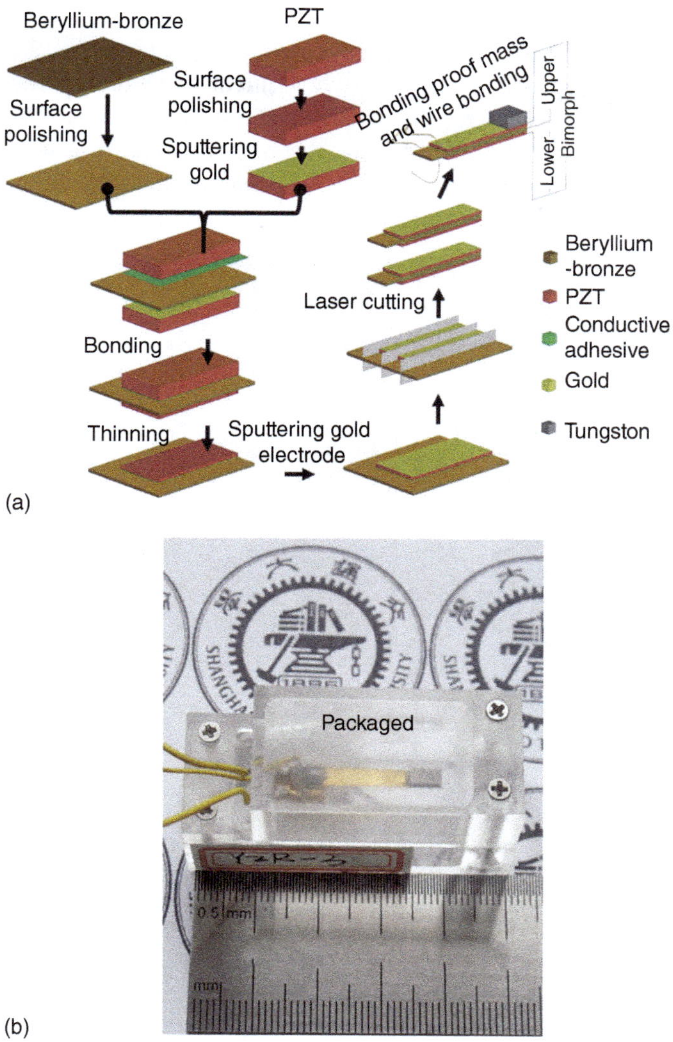

Figure 4.8 (a) Scheme illustration and micromachining processes of the proposed bimorph piezoelectric energy harvester, (b) photograph of the packaged device. Source: Yi et al. [23].

covered with a piece of dust-free cloth of the same size on the top layer, placed in the jig, fixed the jig, applied a certain pressure to the wafer, and put the whole jig into a constant temperature (175 °C) vacuum oven for 3.5 hours for hot-pressing bonding.

(2) Piezoelectric layer thinning
Considering that the thickness of the bonded double wafer is large, direct thinning will cause instability of the grinding head; it is necessary to fix silicon sacrificial wafers of equal thickness in four directions symmetrically around the wafer, and the remaining steps are the same as those for single-wafer thinning, except that thinning of two wafers is required.

Figure 4.9a shows an SEM photograph of the piezoelectric bimorph layer cross section, which shows that the actual thickness of the upper bulk PZT piezoelectric ceramic is reduced to approximately 53 μm after the mechanical grinding and polishing process, and the actual thickness of the lower PZT layer is reduced to approximately 76 μm using the same process, for a total effective volume of the device of approximately 30.6 mm^3. Figure 4.9b shows the impedance analysis of the fully covered bimorph cantilever beam energy

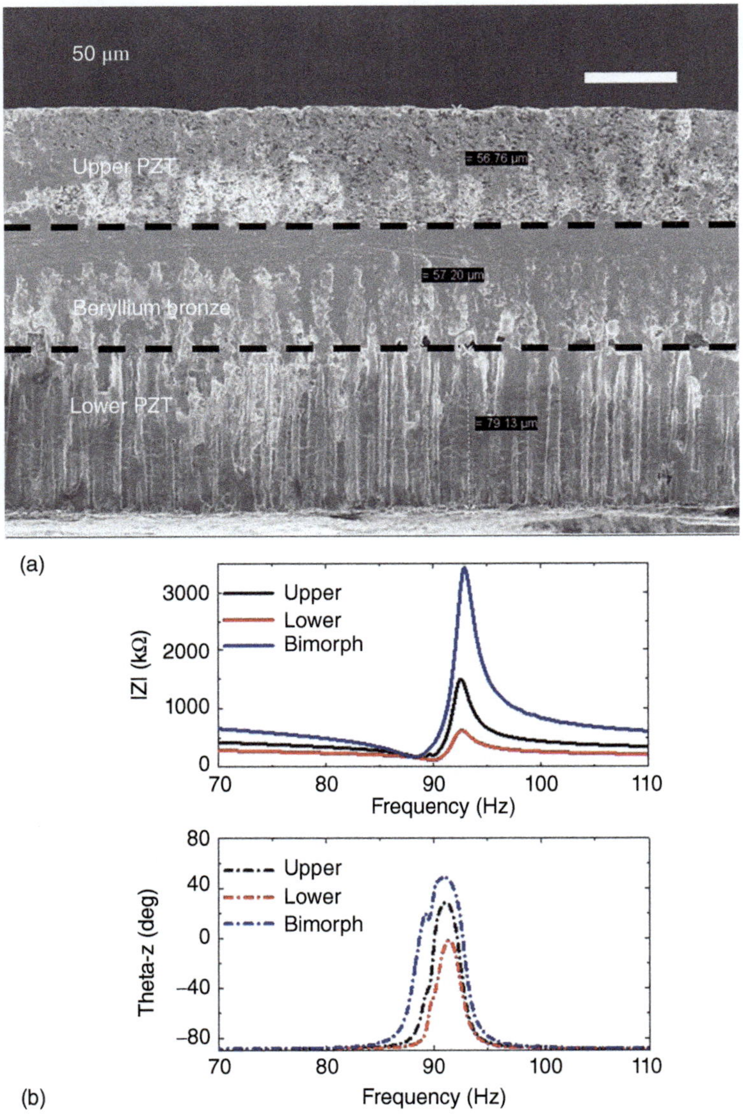

Figure 4.9 Cross-sectional morphology of the composite and impedance analysis, (a) cross-sectional scanning electron microscope (SEM) image of the bimorph PZT layer. (b) Impedance spectrum of the device measured under free condition. Source: (a) Yi et al. [23] / AIP Publishing, (b) Yi et al. [23].

harvester. The measured impedance curves show a resonant frequency of 89.6 Hz and an antiresonant frequency of 92.4 Hz, with a best-fit impedance of 325 kΩ at resonance. Similar measurements were made between the bottom and middle electrodes and between the middle and top electrodes. Similar measurements were performed between the bottom and middle electrodes and between the middle and top electrodes, where the resonant and antiresonant frequencies were kept constant and the best matching impedances were 123 and 235 kΩ, respectively, which varied mainly due to the difference between the thicknesses of the upper and lower PZT films. The peak of the phase angle curve is about 91.2 Hz, and their quality factor Q values are calculated to be 23, 38, and 65 based on the phase angle measurements of the bimorph overall and separate upper piezoelectric layer and bottom piezoelectric layer, respectively; so, the bimorph type has a lower quality factor Q compared to the unimorph piezoelectric cantilever beam energy harvester. A high Q factor is reported to be known to imply lower energy loss, which is a good choice for energy converters, but a high Q-factor also means that the output behavior of the device is more sensitive to the ambient vibration frequency, and a small deviation of the vibration frequency from the resonant frequency of the device can cause a great reduction in output power, and a lower Q-factor helps to maintain high-power output over a wide-frequency band.

4.3.2 Output Performance

The output performance of the full-coverage dual-crystal cantilever beam energy harvester is characterized by the built excitation test rig, which mainly includes the exciter, power amplifier, acceleration sensor, and vibration-control system. Figure 4.10a shows the variation of the open-circuit output voltage of the energy harvester with the excitation frequency in the range of 40–130 Hz for different excitation acceleration amplitudes from 0.1 to 3.5g. The curves show that when the excitation acceleration amplitudes are 0.1, 0.5, 1.0, 2.0, 3.0, and 3.5g; the peak-to-peak values of the open-circuit output voltage are 9.6, 32.2, 49.8, 75, 94, and 105 , where $g = 9.81\ \text{m/s}^2$ is the gravitational constant. Also, it can be found that the resonant frequency of the maximum output voltage has a significant bias toward lower frequencies as the excitation acceleration amplitude increases, which is mainly due to the nonlinear softening characteristics of Young's modulus of the PZT piezoelectric ceramic. When the energy harvester is in resonance, the output power is mainly influenced by the size of the load resistance, and the maximum output power can be obtained with the best matching load resistance. The output characteristics of the device in the closed-circuit case are analyzed by connecting the load resistance in series with this device and between 1 and 1000 kΩ.

Figure 4.10b shows the relationship curve between the peak-to-peak output voltage of the energy harvester and the resistance of the external load resistance of the loop under the excitation acceleration of 0.1–3.5g. It can be seen that under a certain excitation acceleration amplitude, the peak-to-peak output voltage increases rapidly in the lower load resistance range and level off gradually, and the closed-circuit

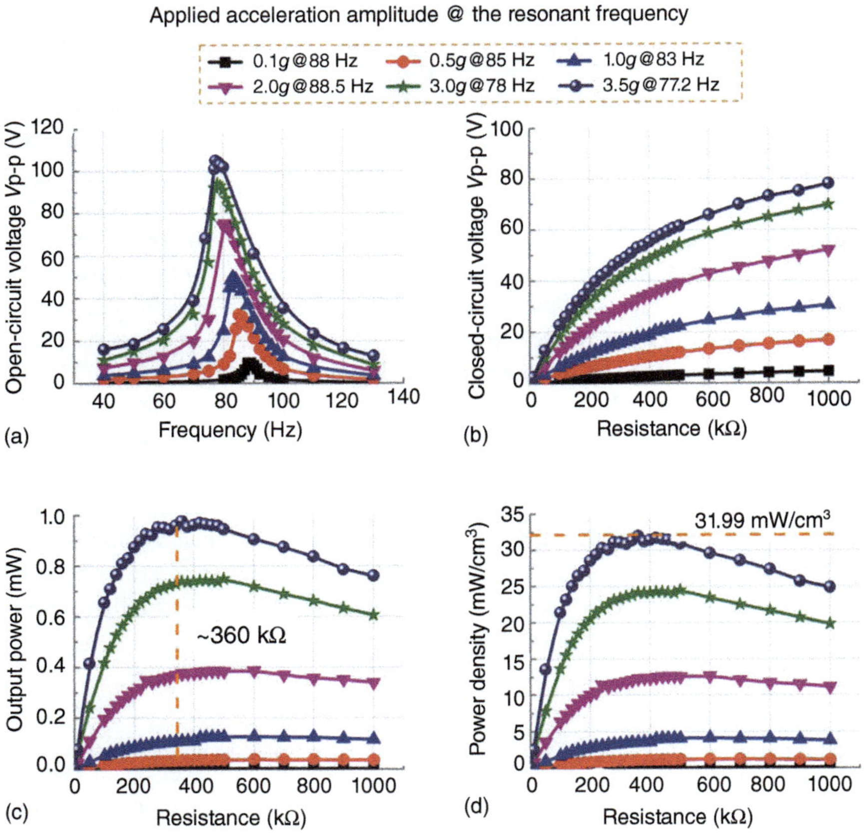

Figure 4.10 Output performance of the bimorph piezoelectric energy harvester. (a) Frequency dependence of the open output voltages under different applied accelerations amplitude varied from 0.1 to 3.5*g*, (b) the closed-circuit output voltage, (c) the output power, and (d) the power density versus the load resistances under different applied accelerations amplitude. Source: Yi et al. [23].

output voltage will gradually converge to the open-circuit output voltage as the load resistance continues to increase. As shown in Figure 4.10c, the maximum effective power is calculated by the formula $P_{eff} = V_{pp}^2/(8 \cdot \text{Resistance})$, and the best matching resistance is about 360 kΩ, at which time, when the excitation acceleration amplitude is 3.5g, the peak voltage of the closed-circuit output, is 53.1 V, and the effective power can reach about 0.979 mW, and the output power increases with the excitation The output power increases with the increase of the excitation acceleration amplitude. Figure 4.10d depicts the effective power density of the energy harvester at different excitation acceleration amplitudes, and the maximum effective power density of 31.99 mW/cm^3 is achieved at 3.5g excitation acceleration amplitude. Figure 4.11 shows the variation of the loop current with load resistance calculated from the closed-circuit output voltage at different resistances, and it can be seen that the loop current is about 150 μA at 3.5g excitation acceleration amplitude when the maximum power output is reached (i.e. resistance at 360 kΩ).

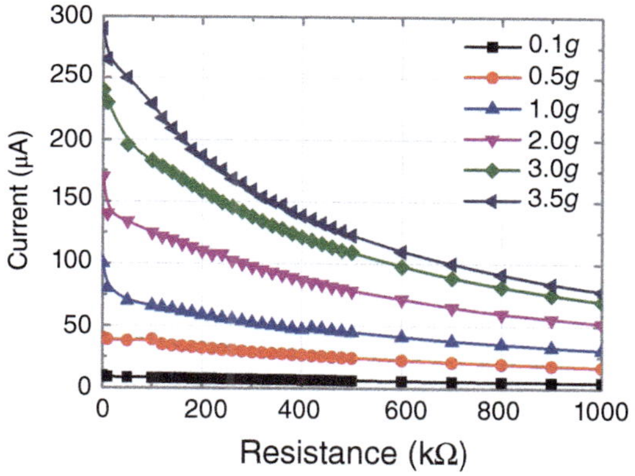

Figure 4.11 The closed-circuit current as a function of the resistance. Source: Yi et al. [23].

In addition, to understand the contribution of the upper and lower layers with different thicknesses of piezoelectric wafers in terms of output performance, the closed-circuit output voltages of the upper monolithic wafer, the lower monolithic wafer, and the biaxial wafer as a whole were measured at different excitation acceleration amplitudes in the experiments, as shown in Figure 4.12a. When the excitation acceleration amplitude is 3.0g, the maximum effective output power generated by the upper and lower monolithic wafers and the biaxial wafer is 0.563, 0.396, and 0.741 mW at matched load resistances around 200, 120, and 360 kΩ,

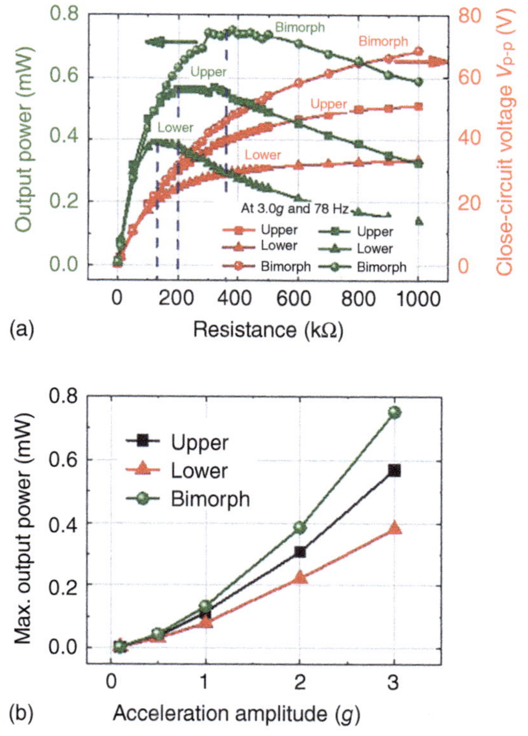

Figure 4.12 Output performance of the different PZT layers. (a) Output voltage (red) and power (green) as a function of external load resistances for upper, lower, and bimorph PZT layers of the piezoelectric harvester, respectively. Labels color changed with the inner color. (b) Maximum output power versus accelerations amplitude. Source: Yi et al. [23].

respectively, and here the matched load resistance (360 kΩ) of the biaxial crystal is in good agreement with the impedance analysis test results (325 kΩ). Theoretically, when two voltage sources are connected in series, the output voltage should generally increase proportionally, but the effective output power calculated from the measured output voltage of the bimorph is lower than the sum of the output power of the upper and lower monocrystal layers, which is mainly due to the fact that the best matching resistance of the bimorph composite piezoelectric layer is slightly larger than the best matching resistance of the two monocrystal layers, and the mismatch of the best load resistance makes the output power slightly lower than the

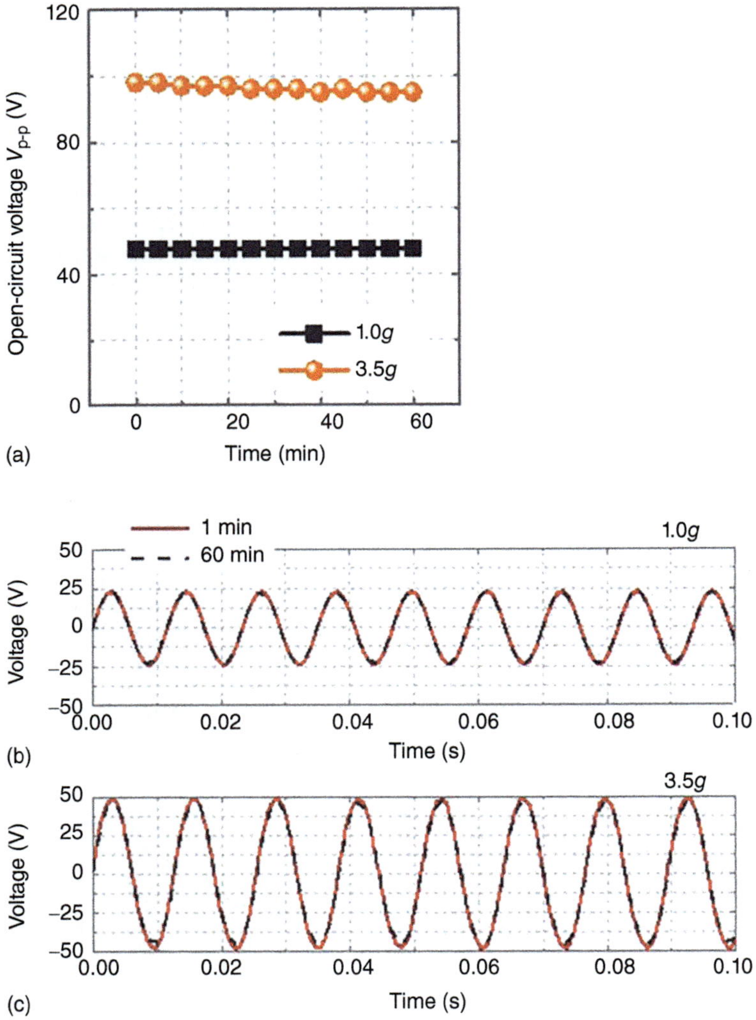

Figure 4.13 Stability testing of the B-PEH. (a) The open-circuit voltage recorded during the stability test for 60 min at 1.0 and 3.5*g* accelerations amplitude at its resonant frequency. (b) and (c) Comparison of the open-circuit voltage curve between 1 and 60 min under 1.0 and 3.5*g*, respectively. Source: Yi et al. [23].

sum of the output effective power. The relationship between the maximum output effective power and the excitation acceleration amplitude is shown in Figure 4.12b. The piezoelectric bilayer has higher output performance than two separate PZT monocrystalline layers, and the increase in output power with increasing acceleration amplitude shows a quadratic change, which is consistent with the general model theory of velocity-damped oscillators used for electromechanical coupling conversion.

To investigate the stability of the full-coverage double-crystal cantilever beam energy harvester, it was experimentally subjected to excitation at resonant frequencies and high-acceleration amplitudes to evaluate the reliability of its structure and stability of its output. Figure 4.13a depicts the open-circuit voltage output at 1.0 and 3.5g excitation acceleration amplitude for 60 minutes. Figure 4.13b shows the time domain output waveform with 1.0g applied, indicating that the dual-crystal cantilever beam energy harvester shows good stability during this time. Figure 4.13c shows that the output voltage under high acceleration of 3.5g decreases slightly after onehour, which is mainly attributed to the fact that the cantilever beam under high acceleration enters the fatigue state more quickly.

Considering the actual use effect of the generated electrical energy, as shown in Figure 4.14a, the energy harvester output is converted from AC to DC by rectifier to directly supply the series of connected green LEDs without any storage device, and twenty-one LEDs can be lit at the same time under 3.0g excitation

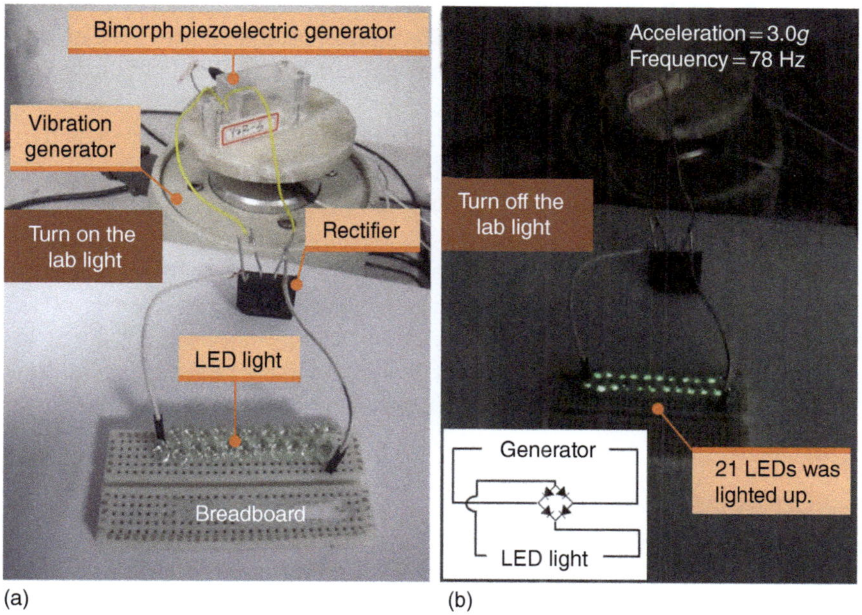

Figure 4.14 Application of the bimorph piezoelectric harvester. (a) Photograph of connect circuit for lighting LEDs via the bimorph piezoelectric harvester. (b) The serial twenty-one LEDs were lighted up by the bimorph piezoelectric harvester under a 78 Hz frequency and 3.0*g* applied acceleration amplitude. Source: Yi et al. [23]/AIP Publishing.

Table 4.2 Performance comparison of recently reported PEHs.

Piezo-electric materials	Substrate	Volume (mm^3)	Accele-ration (g)	Fre-quency (Hz)	Power (mW)	Power density (mW/cm^3)	Normalized power density (mW/cm^3 g^2 Hz)	References
Epitaxial PZT	Steel	4.88	0.5	126	0.0053	1.086	0.034	Morimoto et al. [24]
Thinned bulk PZT	Silicon	27	1.5	154	0.205	7.59	0.022	Aktakka et al. [25]
Thinned bulk PZT	Silicon	47.82	1.0	96	0.0824	1.72	0.018	Janphuang et al. [26]
Deposited PZT	Ni	8	0.5	320	0.00015	0.019	0.00024	Gupta et al. [27]
Mn-doped KNN	Silicon	2	1.0	132	0.00362	1.81	0.014	Won et al. [28]
Thinned PZT	Bronze	30.6	1.0	83	0.125	4.092	0.0493	Yi et al. [23]
			3.5	77.2	0.979	31.99	0.0337	

acceleration amplitude and 78 H excitation frequency (Figure 4.14b). According to the experimental results of the open-circuit test, the open-circuit peak value is 94 V. The effective voltage of each green LED cannot be less than 2.0 V. The output power can be better conformed to the actual application, and the slight difference is mainly affected by the brightness of these LEDs.

In addition, the performance comparison of piezoelectric energy harvesters that have been reported in recent years is shown in Table 4.2. The piezoelectric bimorph PZT thick-film energy harvester with beryllium bronze foil substrate has higher output power and higher normalized power density. In summary, the flexible substrate-based piezoelectric composite layer cantilever beam vibration energy harvester has higher normalized power density characteristics and lower intrinsic frequency than the silicon-based substrate vibration energy harvester, and is able to operate under high-acceleration excitation conditions and has better adaptability to the random shock loads that often occur in environmental vibration sources.

4.4 Optimized Bimorph PEH

To further increase the normalized power density of the cantilever beam vibration energy harvester, a bimorph cantilever beam vibration energy harvester with different piezoelectric layer arrangement lengths was designed and prepared by the preparation technique of bimorph piezoelectric thin-film multilayer composite structure to study its output characteristics and verify the relevant theoretical analysis conclusions.

4.4.1 Structure Design and Fabrication

Figure 4.15 gives a schematic diagram of the affective dimensions of the device structure, considering that the cantilever beam partially covered by the piezoelectric layer has a more obvious performance in improving the cantilever beam energy harvester. In this section, with reference to the idea of unimorph structure optimization, the double-crystal cantilever beam vibration energy harvester with partial piezoelectric layer coverage is designed as shown in the figure. The effective length of the piezoelectric layer to the total effective length of the cantilever beam is 4/18 mm, the thickness of the piezoelectric layer is 50 μm, and the substrate is beryllium bronze, and two tungsten mass bulks ($6 \times 9 \times 0.2\ mm^3$) are fixed symmetrically at the end. The total effective volume of the device is $32\ mm^3$.

A physical view of the partially piezoelectric layer-covered bimorph cantilever beam energy harvester prepared according to this dimension is shown in Figure 4.16. First, the bulk PZT was polished and the bottom electrode (20/180 nm Cr/Au) was sputtered after polishing. Next, the 50-μm-thick beryllium bronze was polished by a mechanical polishing process and bonded to the polished surface of the bulk PZT using conductive epoxy resin as an interlayer at 175 °C. The piezoelectric layer of the bulk PZT wafer bonded with the beryllium bronze substrate is then thinned to 50 μm by mechanical thinning method and polished, still sputtering Cr/Au as the top electrode of the piezoelectric layer. The top and bottom electrodes are separated by a hard mask prior to sputtering the metal. Then, the piezoelectric wafer is patterned in the structure of a 3-mm-wide cantilever beam using a UV-laser method, and a tungsten bulk is bonded at the end of the cantilever beam to further reduce the resonant frequency of the device. Finally, the cantilever beam is fixed to the PCB and the top and bottom film electrodes of the piezoelectric layer are led out using wires for testing.

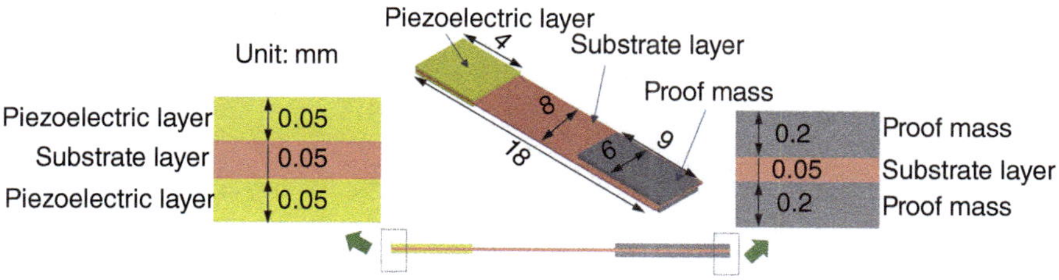

Figure 4.15 Schematic diagram of the structural design of the piezoelectric energy harvester.

Figure 4.16 Photograph of the cantilever piezoelectric energy harvester with the optimal layout length of the piezoelectric layer. Source: Hu et al. [29]/Elsevier.

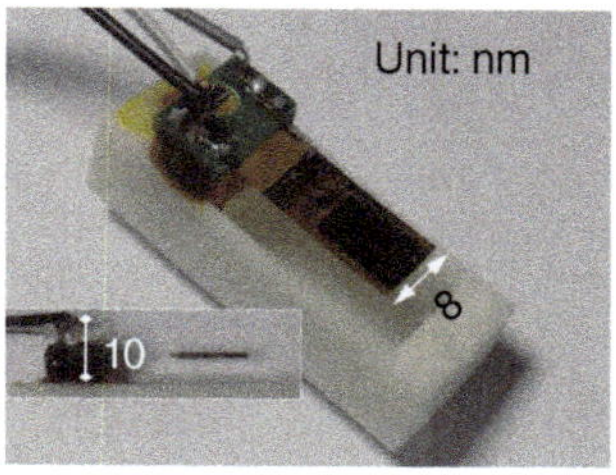

4.4.2 Output Performance

Figure 4.17 show the open-circuit voltage output frequency domain response and the closed-circuit voltage variation curve with external resistance for the double-crystal cantilever beam energy harvester covered by this partial piezoelectric layer at different accelerations from 0.1 to 2.0g. The performance of the closed-circuit output power and power density subject to the variation of external resistance is also solved. Figure 4.17a shows the peak value of the maximum open-circuit voltage output of the device increasing from 14.78 to 38.97 V as the acceleration increases from 0.5 to 2.0g. The resonant frequency of the maximum output voltage also shifts significantly to lower frequencies as the applied acceleration increases. The output power has a high dependence on the load resistance, and the optimal matching load resistance is obtained at the maximum output power. Figure 4.17b shows the curve of the peak-to-peak output voltage at acceleration from 0.5 to 2.0g with the change of

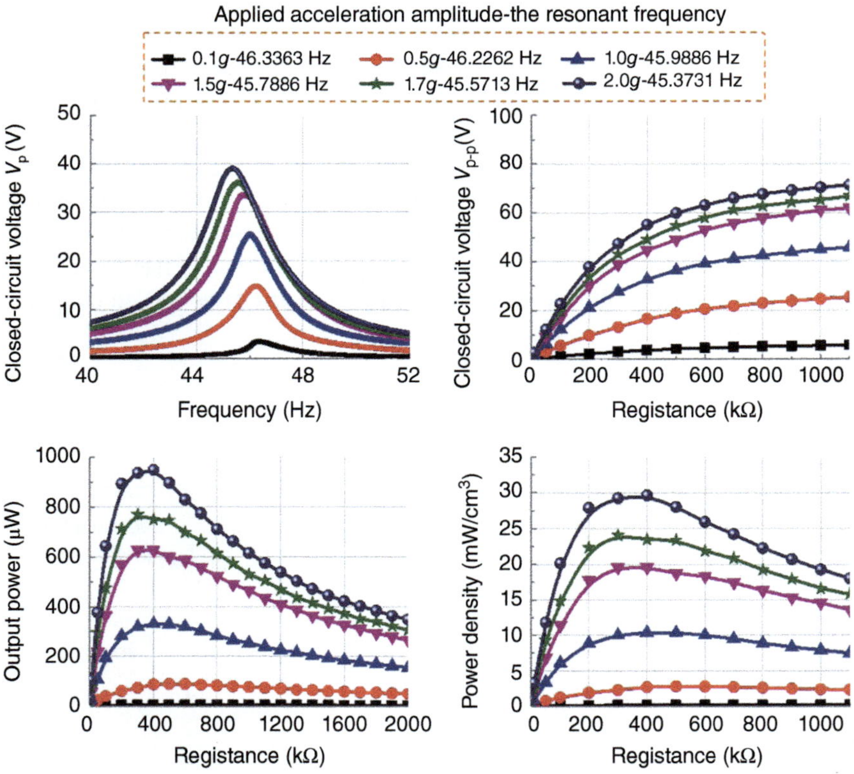

Figure 4.17 Output performance of the piezoelectric energy harvester. (a) Frequency dependence of the open output voltages under different applied accelerations amplitude varied from 0.1 to 2.0*g*. (b) The closed-circuit output voltage, (c) the output power, and (d) the power density versus the load resistances under different applied accelerations amplitude. Source: Hu et al. [29].

external load resistance. The output voltage increases rapidly at lower resistances and gradually stabilizes at higher load resistance ranges, the closer the maximum closed-circuit output voltage is to the maximum open-circuit output voltage. The load resistance varies from 1 to 2000 kΩ, with an optimum load resistance of approximately 400 kΩ, as shown in Figure 4.17c. At this optimum point, the peak voltage is 27.75 V at an acceleration of 2.0g, and the effective power reaches about 0.963 mW. Also, the output power increases with the amplitude of the applied acceleration. Figure 4.17d shows the curve of the effective power density with external resistance for different applied accelerations, reaching a maximum effective power density of 30.1 mW/cm^3 at 2.0g.

To evaluate the output performance stability of this double-crystal cantilever beam, the fatigue experimental condition was chosen to keep it in resonance for a long time under the acceleration of 2.0g. The output voltage waveform was recorded by the vibration dynamic monitoring system to observe its changes. Figure 4.18 records the output voltage value of the device for 20 hours, and the results show that the output voltage of the device maintains good stability during this time period.

4.4.2.1 Comparative Analysis of Normalized Power Density

Figure 4.19 shows the performance of the piezoelectric energy harvesters reported in recent years, as well as the performance comparison of the piezoelectric bimorph vibration energy harvesters with unoptimized and optimized piezoelectric layer arrangement lengths. The optimized device always shows obvious performance superiority under the same acceleration.

Table 4.3 summarizes the continuous improvement methods and their performance of the cantilever beam vibration energy harvesters for low-stiffness metal foil substrates in the study, which continuously achieve lower frequency and higher power density vibration energy harvesters. The performance of each energy harvester under 1.0g excitation acceleration is given in the table, considering the influence of avoiding acceleration on the analysis results. The optimized piezoelectric bimorph exhibits higher output power, lower intrinsic frequency, and higher power density compared with the optimized unimorph cantilever beam vibration energy harvester, and the normalized power density is slightly larger,

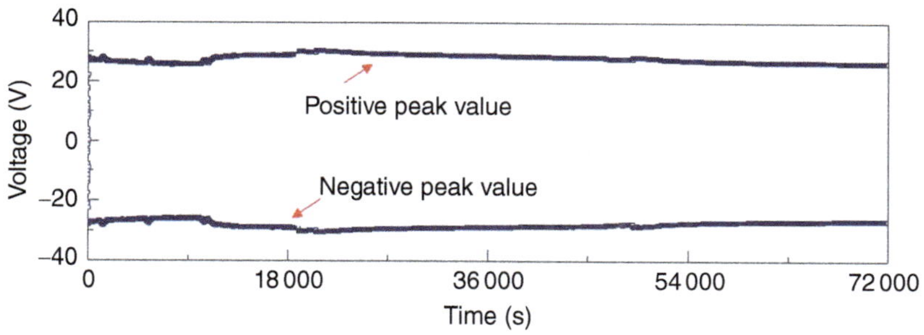

Figure 4.18 Output stability of the cantilever piezoelectric energy harvester with the optimal layout length of the piezoelectric layer. Source: Hu et al. [29].

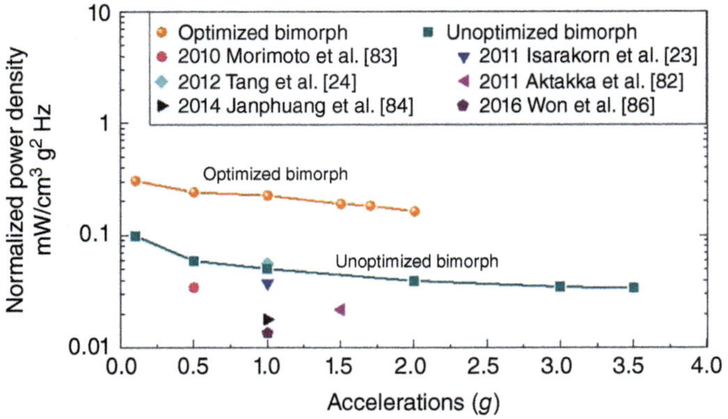

Figure 4.19 Comparison of normalized power density between recently reported PEHs and the presented bimorph piezoelectric energy harvesters with/without the optimal PEH in this chapter. Source: Refs. [24–26, 28–31].

Table 4.3 summary of cantilever vibrational energy harvesters.

Configuration								
Types	**Length ratio of the PZT layer**	**Volume (mm^3)**	**Frequency (Hz)**	**Acceleration (*g*)**	**Matched resistance (kΩ)**	**Power (mW)**	**Power density (mW/cm^3)**	**Normalized power density (mW/cm^3 g^2 Hz)**
Single crystal	3 : 15	4.75	67	1.0	900	0.068	14.3	0.213
Bimorph	16 : 16	30.6	83	1.0	360	0.125	4.092	0.0493
			77.2	3.5		0.979	31.99	0.0337
	4 : 18	32	45.9	1.0	400	0.332	10.38	0.226
			45	2.0		0.963	30.08	0.167

indicating the optimized bimorph cantilever beam vibration energy harvesters. The optimized double-crystal cantilever beam vibration energy harvester has performance advantages and prospect of better meeting practical application requirements.

This chapter focuses on the process development and device preparation of the high-voltage electrical characteristic thin film prepared by MEMS mechanical thinning technology, investigates the energy-conversion characteristics of the flexible cantilever-beam-type piezoelectric vibration energy harvester, explores the power density enhancement method through experimental methods guided by the theoretical analysis results, analyzes the feasibility of the theoretical method, and comes up with the design and preparation of the low-frequency piezoelectric vibration energy harvester with high-power density, and preparation scheme for high-power density low-frequency piezoelectric vibration energy harvester.

4.5 Summary

The method of preparing flexible beryllium bronze foil substrate piezoelectric composite layer based on low-temperature vacuum bonding process and mechanical thinning process is proposed, and the unimorph piezoelectric vibration energy harvester with different arrangement lengths of piezoelectric film in the length direction of cantilever beam is designed under the guidance of the theoretical analysis results in the previous chapter, and the preparation process is explored, and the output performance of the energy harvester with different piezoelectric film arrangement lengths is investigated to verify the optimal. The effectiveness of the optimized piezoelectric film arrangement length in improving the device output performance is verified.

A bimorph composite layer structure based on flexible beryllium bronze substrate is proposed to further enhance the output performance of the energy harvester, and the preparation process of the bimorph cantilever beam energy harvester on flexible beryllium bronze foil substrate is investigated to investigate its output performance, which is three times higher in power density and about one-third higher in normalized power density than the reported silicon-based energy harvesters of the same type.

The design and preparation of high-power density piezoelectric vibrating energy harvesters are investigated by combining the design scheme of bimorph structure and optimized arrangement of piezoelectric layers, and the output performance of bimorph piezoelectric vibrating energy harvesters with optimized piezoelectric layer length is investigated, and the power density of bimorph vibrating energy harvesters with equal length of piezoelectric layer and substrate layer is nearly doubled and the normalized power density is about four times higher than that of bimorph vibrating energy harvesters with equal length of piezoelectric layer and substrate layer, and the high-output power characteristics of nearly milliwatt level are maintained. Compared with the unimorph vibrating energy harvester with optimized piezoelectric layer length, the normalized power density is slightly increased, but the output power is significantly increased, which has more important practical application value.

References

1 Jeong, C.K. et al. (2015). A hyper-stretchable elastic-composite energy harvester. *Advanced Materials* 27: 2866–2875.
2 Abdelkefi, A. (2016). Aeroelastic energy harvesting: a review. *International Journal of Engineering Science* 100: 112–135.
3 Selvan, K.V. and Ali, M.S.M. (2016). Micro-scale energy harvesting devices: review of methodological performances in the last decade. *Renewable & Sustainable Energy Reviews* 54: 1035–1047.
4 Ma, T. et al. (2018). Development of inline hydroelectric generation system from municipal water pipelines. *Energy* 144: 535–548.

5 Jasim, A. et al. (2018). Laboratory testing and numerical simulation of piezoelectric energy harvester for roadway applications. *Applied Energy* 224: 438–447.

6 Lee, H.S. et al. (2014). Flexible inorganic piezoelectric acoustic nanosensors for biomimetic artificial hair cells. *Advanced Functional Materials* 24: 6914–6921.

7 Wu, N. et al. (2015). Cellular polypropylene piezoelectret for human body energy harvesting and health monitoring. *Advanced Functional Materials* 25: 4788–4794.

8 Shaikh, F.K. and Zeadally, S. (2016). Energy harvesting in wireless sensor networks: a comprehensive review. *Renewable & Sustainable Energy Reviews* 55: 1041–1054.

9 Kim, D.H. et al. (2017). In vivo self-powered wireless transmission using biocompatible flexible energy harvesters. *Advanced Functional Materials* 27: 1700341.

10 Wang, X. et al. (2017). A frequency and bandwidth tunable piezoelectric vibration energy harvester using multiple nonlinear techniques. *Applied Energy* 190: 368–375.

11 Wu, J., Shi, H., Zhao, T. et al. (2016). High-temperature $BiScO_3$-$PbTiO_3$ piezoelectric vibration energy harvester. *Advanced Functional Materials* 26: 7186–7194.

12 Yeo, H.G., Ma, X., Rahn, C., and Trolier-McKinstry, S. (2016). Efficient piezoelectric energy harvesters utilizing (001) textured bimorph PZT films on flexible metal foils. *Advanced Functional Materials* 26: 5940–5946.

13 Roundy, S. and Wright, P.K. (2004). A piezoelectric vibration based generator for wireless electronics. *Smart Materials and Structures* 13: 1131–1142.

14 Toprak, A. and Tigli, O. (2014). Piezoelectric energy harvesting: state-of-the-art and challenges. *Applied Physics Reviews* 1: 031104.

15 Takei, R., Makimoto, N., Okada, H. et al. (2016). Design of piezoelectric MEMS cantilever for low-frequency vibration energy harvester. *Japanese Journal of Applied Physics* 55: 06GP14.

16 Gao, X., Shih, W.-H., and Shih, W.Y. (2009). Induced voltage of piezoelectric unimorph cantilevers of different nonpiezoelectric/piezoelectric length ratios. *Smart Materials and Structures* 18: 125018.

17 Gao, X., Shih, W.H., and Shih, W.Y. (2010). Vibration energy harvesting using piezoelectric unimorph cantilevers with unequal piezoelectric and nonpiezoelectric lengths. *Applied Physics Letters* 97: 233503.

18 Patel, R., McWilliam, S., and Popov, A.A. (2011). A geometric parameter study of piezoelectric coverage on a rectangular cantilever energy harvester. *Smart Materials and Structures* 20: 085004.

19 Yu, H., Zhou, J., Deng, L., and Wen, Z. (2014). A vibration-based MEMS piezoelectric energy harvester and power conditioning circuit. *Sensors* 14: 3323–3341.

20 Sunithamani, S., Lakshmi, P., and Eba Flora, E. (2013). PZT length optimization of MEMS piezoelectric energy harvester with a non-traditional cross section: simulation study. *Microsystem Technologies* 20: 2165–2171.

21 Kim, S. et al. (2016). Electrochemically driven mechanical energy harvesting. *Nature Communications* 7: 10146.

22 Zhang, Y. et al. (2018). Micro electrostatic energy harvester with both broad bandwidth and high normalized power density. *Applied Energy* 212: 362–371.

23 Yi, Z. et al. (2017). High performance bimorph piezoelectric MEMS harvester via bulk PZT thick films on thin beryllium-bronze substrate. *Applied Physics Letters* 111: 013902. https://doi.org/10.1063/1.4991368.

24 Morimoto, K., Kanno, I., Wasa, K., and Kotera, H. (2010). High-efficiency piezoelectric energy harvesters of c-axis-oriented epitaxial PZT films transferred onto stainless steel cantilevers. *Sensors and Actuators A: Physical* 163: 428–432.

25 Aktakka, E.E., Peterson, R.L., and Najafi, K. (2011) Thinned-PZT on SOI process and design optimization for piezoelectric inertial energy harvesting," *2011 16th International Solid-State Sensors, Actuators and Microsystems Conference*, pp. 1649–1652, https://doi.org/10.1109/TRANSDUCERS.2011.5969857, http://ieeexplore.ieee.org/ielx5/5958191/5969118/05969857.pdf?tp=&arnumber=5969857&isnumber=5969118.

26 Janphuang, P., Lockhart, R., Uffer, N. et al. (2014). Vibrational piezoelectric energy harvesters based on thinned bulk PZT sheets fabricated at the wafer level. *Sensors and Actuators A: Physical* 210: 1–9.

27 Gupta, R., Tomar, M., Rammohan, S. et al. (2016). Multiferroic cantilever for power generation using dual functionality. *Applied Physics Letters* 109: 193901.

28 Won, S.S., Lee, J., Venugopal, V. et al. (2016). Lead-free Mn-doped $(K_{0.5},Na_{0.5})NbO_3$ piezoelectric thin films for MEMS-based vibrational energy harvester applications. *Applied Physics Letters* 108: 232908. https://doi.org/10.1063/1.4953623.

29 Hu, Y., Yi, Z., Dong, X. et al. (2019).High power density energy harvester with non-uniform cantilever structure due to high average strain distribution. *Energy* 169: 294–304.

30 Isarakorn, D. et al. (2011). The realization and performance of vibration energy harvesting MEMS devices based on an epitaxial piezoelectric thin film. *Smart Materials and Structures* 20: 025015.

31 Tang, G. et al. (2012). Fabrication and analysis of high-performance piezoelectric MEMS generators. *Journal of Micromechanics and Microengineering* 22: 065017.

5

Free-Vibration Nonlinear Piezoelectric Energy Harvesters

5.1 Introduction

The theoretical analysis and experimental study of the vibration energy harvester based on the flexible piezoelectric composite layer preparation technology were introduced previously, and the superior performance of the piezoelectric bulk-thinning process based on the flexible substrate, in enhancing the output performance of the energy harvester and reducing its intrinsic frequency, can be found. However, the available bandwidth of the high-performance output for vibration energy harvesters based on the cantilever beam structure is narrow, and the power output characteristics are highly dependent on the excitation frequency. The future development of intelligent piezoelectric energy harvesters (PEHs) should have the characteristics of high output response in a broad bandwidth, self-tunable capability in the output, and adaptability to different excitation intensities.

Researchers have obtained effective broadband electrical energy output by mechanical deformation-limiting method based on a stopper [1] or linear superposition method [2], but the mechanical deformation-limiting method mainly obtains broadband output response by losing part of the high electromechanical coupling output, while the linear superposition method relies on the outputs superposition of multiple linear vibration energy harvesters with near-resonant frequencies to exhibit broadband characteristics in the final output performance, which makes the energy-harvesting system more complex.

The nonlinear piezoelectric energy-harvesting technique is increasingly becoming another representative method to achieve broadband piezoelectric energy conversion. Nonlinear piezoelectric wideband vibration energy harvesters consist of two main types [3] – one with broadband characteristics achieved by external magnetic coupling, and the other with buckled nonlinear characteristics due to the device structure itself. The former requires the creation of an external magnetic field by rigid magnets, while the latter is relatively simple in structure and easy to miniaturize and integrate for preparation.

Therefore, this chapter focuses on discussing the nonlinear vibration characteristics of the flexible piezoelectric buckled structure and investigating the mechanism and optimal control method of broadband piezoelectric energy conversion based on the buckled nonlinear structural characteristics. Meanwhile, this chapter also delves

Flexible Piezoelectric Energy Harvesters and Sensors, First Edition. Bin Yang, Zhiran Yi, and Chengkuo Lee.

into the nonlinear vibration characteristics of flexible buckled vibration PEH and develops a broadband PEH with a stacked double-buckled structure. The control methods of nonlinear softening and hardening vibration effects are studied, and the broadband vibration energy-harvesting strategy with a joint mechanism of nonlinear softening and hardening effects is investigated to provide a reference for realizing the enhancement of nonlinear stable bandwidth.

5.2 Free-Vibration Nonlinear Mechanism

5.2.1 Theoretical Analysis

Here, we harnessed the flexible buckled-bridge PEH [4–8] to give an example of nonlinear vibration mechanism used for piezoelectric energy harvesting. The vibration energy harvester based on the piezoelectric buckled structure uses a flexible polyethylene terephthalate (PET) polymer material as the support layer, and the degree of structural buckling can be characterized by the buckled angle θ. Figure 5.1a shows a schematic diagram of a piezoelectric buckled-bridge structure, which consists of a PET support sheet and a piezoelectric unimorph structure polarized along the Z-direction (here "unimorph" refers to a piezoelectric composite structure formed by a single piezoelectric layer and a substrate layer). The unimorph structure is prepared by bonding a piezoelectric thinned film (Figure 5.1b) on a metal substrate of beryllium bronze. A mass bulk is set in the center of the buckling bridge to form a concentrated inertial force in the center of the bridge when the vibration excitation occurs to induce buckling deformation. The upper and lower electrodes of the piezoelectric layer are connected in series with an external load resistor R through conduction lead to form a closed-loop circuit. The output characteristics of this piezoelectric buckled-bridge-type vibration energy harvester are analyzed by calculating the voltage change across the load resistor. In addition, assuming that the fixed base of the energy harvester vibrates only vertically, the vibration displacement of the fixed base with respect to the ground (displacement

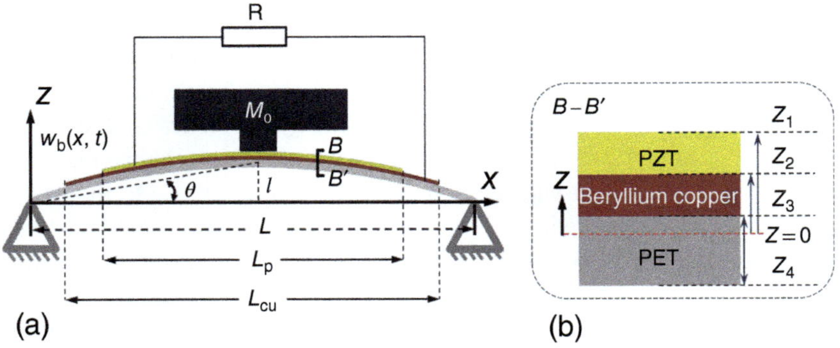

Figure 5.1 Schematic illustration of the piezoelectric buckled-bridge model. (a) The simplified structural model for the theory analysis. (b) Layer numbering used for this laminated structure. The plane $z = 0$ is the mechanical neutral plane of the piezoelectric composite.

is zero) can be expressed by $w_b(t)$, and the flexural deformation of the piezoelectric buckled bridge can be expressed by $w(x, t)$, while the potential difference in the z-direction is considered to be caused not only by the horizontal strain of the piezoelectric buckling bridge structure but also as an overall expression containing the strain effects in other directions. To obtain the displacement equations of this piezoelectric buckled bridge, the first-order composite plate theory [9] derivation can be used, and the Lagrangian function is calculated according to the energy conservation theorem.

By designing different lengths of each layer of the structure to form an end constraint with variable stiffness, on the one hand, due to only the PET layer at the fixed end, with low-bending stiffness, it is easy to achieve resonance under low-frequency conditions. The fatigue resistance of the polymer in contrast to ceramic, metal, and other materials, can guarantee better stability of the device. On the other hand, the end gradient stiffness can avoid the strain concentration at the fixed end so that the piezoelectric layer cannot generate an effective strain.

The piezoelectric instanton equation in the form of stress-potential shift based on the strain and electric field tensor as independent variables is known as [10]:

$$\begin{cases} T_x = c_{xx}^E S_x - e_{zx}^T E_z \\ D_z = e_{zx} S_x + \varepsilon_{zz}^S E_z \end{cases} \tag{5.1}$$

Some stress or electric field components can be eliminated depending on the geometric characteristics or the electrodeposition and expanded in matrix form as follows:

$$\begin{bmatrix} T_1 \\ T_2 \\ T_3 \\ T_4 \\ T_5 \\ T_6 \\ D_1 \\ D_2 \\ D_3 \end{bmatrix} = \begin{bmatrix} c_{11}^E & c_{12}^E & c_{13}^E & & & & 0 & 0 & -e_{31} \\ c_{12}^E & c_{11}^E & c_{13}^E & & 0 & & 0 & 0 & -e_{31} \\ c_{13}^E & c_{13}^E & c_{33}^E & & & & 0 & 0 & -e_{33} \\ & & & c_{55}^E & 0 & 0 & 0 & -e_{15} & 0 \\ & 0 & & 0 & c_{55}^E & 0 & -e_{15} & 0 & 0 \\ & & & 0 & 0 & c_{66}^E & 0 & 0 & 0 \\ 0 & 0 & 0 & 0 & e_{15} & 0 & \varepsilon_{11}^T & 0 & 0 \\ 0 & 0 & 0 & e_{15} & 0 & 0 & 0 & \varepsilon_{11}^T & 0 \\ e_{31} & e_{31} & e_{33} & 0 & 0 & 0 & 0 & 0 & \varepsilon_{33}^T \end{bmatrix} \begin{bmatrix} S_1 \\ S_2 \\ S_3 \\ S_4 \\ S_5 \\ S_6 \\ E_1 \\ E_2 \\ E_3 \end{bmatrix} \tag{5.2}$$

For a piezoelectric thin plate subjected to two-dimensional strain fluctuations, the positive and transverse shear stress components in the thickness direction of the piezoelectric ceramic are negligible according to Kirchhoff-love theory: $T_3 = T_4 = T_5 = 0$. Also consider that the electric field exists only on the upper and lower surfaces of the piezoelectric layer covered by electrodes, i.e. along the z-axis direction ($E_1 = E_2 = 0$), then the instanton equation can be simplified as:

$$\begin{cases} T_1 = c_{11}^E S_1 + c_{12}^E S_2 - e_{31} E_3 \\ T_2 = c_{12}^E S_1 + c_{11}^E S_2 - e_{31} E_3 \\ T_6 = c_{66}^E S_6 \\ D_3 = e_{31} S_1 + e_{31} S_2 + \varepsilon_{33}^T E_3 \end{cases} \tag{5.3}$$

For a thin slender plate, the strain in the y-axis direction S_2 and the strain in the xy-direction S_6 can be neglected with respect to the strain in the x-axis direction, so the present constitutive equation can be further simplified as

$$\begin{cases} T_1 = c_{11}^E S_1 - e_{31} E_3 \\ T_2 = c_{12}^E S_1 - e_{31} E_3 \\ D_3 = e_{31} S_1 + \varepsilon_{33}^T E_3 \end{cases} \tag{5.4}$$

The analysis of the piezoelectric layer strain, defining the intermediate layer displacements as (u, v, w), along the x-, y-, and z-axis directions, respectively, can be calculated by the Kirchhoff hypothesis, after the superposition of the underlying excitation displacements $w_b(t) = \frac{\text{acc}}{(2\pi f)^2}\sin(2\pi ft)$ can be derived as

$$u_m = \begin{bmatrix} u(x,z,t) \\ v(x,z,t) \\ w(x,z,t) \end{bmatrix} = \begin{bmatrix} u^0(x,t) - z\dfrac{\partial w^0(x,t)}{\partial x} \\ 0 \\ w^0(x,t) + w_b(t) \end{bmatrix} \tag{5.5}$$

where $u^0(x, t)$ and $w^0(x, t)$ are for the transverse motion displacement and longitudinal motion displacement of the neutral axis relative to the ground at point x and moment t, respectively, where the transverse motion displacement is small and can be neglected, so the main strain can be expressed as

$$\begin{aligned} S_{xx}(x,z,t) &= \frac{\partial u_m}{\partial x} = \frac{\partial u^0(x,t)}{\partial x} + \frac{1}{2}\left(\frac{\partial w^0(x,t)}{\partial x}\right)^2 - z\frac{\partial^2 w^0(x,t)}{\partial x^2} \\ &\approx \frac{1}{2}\left(\frac{\partial w^0(x,t)}{\partial x}\right)^2 - z\frac{\partial^2 w^0(x,t)}{\partial x^2} \end{aligned} \tag{5.6}$$

5.2.2 Energy Equation

For this system, the total potential energy [10, 11] is

$$U = \frac{1}{2}\left(\int_{V_s} S^t T \, dV_s + \int_{V_p} S^t T \, dV_p\right) \tag{5.7}$$

where S is the engineering strain tensor, T is the engineering stress tensor, and the subscripts s and p represent the substrate and piezoelectric layer, respectively, and the integration is performed for each part of the material volume. For the substrate layer is isotropic, it is known by Hooke's law: $T_{xx}(x, z, t) = E_s S_{xx}(x, z, t)$, where E_s is the elastic modulus of the substrate layer material. For the piezoelectric ceramic layer, the equation for the stress component is $T_{xx}(x,z,t) = T_1 = c_{11}^E S_{xx}(x,z,t) + e_{31}\frac{v(t)}{h_p}$, where the electric field can be calculated from the voltage $v(t)$ across the load resistor, i.e. $E_3(t) = -v(t)/h_p$, where h_p is the thickness of the piezoelectric layer.

Thus, the total potential energy of the system is

$$U = \frac{1}{2}\int_{V_s} E_s (S_{xx})^2 dV_s + \frac{1}{2}\int_{V_p} \left[\left(c_{11}^E + c_{12}^E\right)(S_{xx})^2 + 2e_{31}\frac{v(t)}{h_p} S_{xx}\right] dV_p \tag{5.8}$$

For this system (neglecting the moment of inertia of the mass bulk), the total kinetic energy is

$$T = \frac{1}{2}\left(\int_{V_s} \rho_s \left(\frac{\partial u_m}{\partial t}\right)^2 dV_s + \int_{V_p} \rho_p \left(\frac{\partial u_m}{\partial t}\right)^2 dV_p\right)$$
$$= \frac{1}{2}\int_{V_s} \rho_s \left[\left(z\frac{\partial^2 w^0(x,t)}{\partial t \partial x}\right)^2 + \left(\frac{\partial w^0(x,t)}{\partial t} + \frac{\partial w_b(t)}{\partial t}\right)^2\right] dV_s$$
$$+ \frac{1}{2}\int_{V_p} \rho_p \left[\left(z\frac{\partial^2 w^0(x,t)}{\partial t \partial x}\right)^2 + \left(\frac{\partial w^0(x,t)}{\partial t} + \frac{\partial w_b(t)}{\partial t}\right)^2\right] dV_p$$
$$+ \frac{1}{2}M_0\left[\frac{\partial w^0(L/2,t)}{\partial t} + \frac{\partial w_b(t)}{\partial t}\right]^2 \tag{5.9}$$

The electrical energy available in the piezoelectric layer is

$$W_{ie} = \frac{1}{2}\int_{V_p} E^t D dV_p = -\frac{1}{2}\int_{V_p} \frac{v(t)}{h_p}\left\{e_{31}\left[\frac{1}{2}\left(\frac{\partial w^0(x,t)}{\partial x}\right)^2 - z\frac{\partial^2 w^0(x,t)}{\partial x^2}\right]\right\} dV_p$$
$$+ \frac{1}{2}C_p v^2(t) \tag{5.10}$$

where $C_p = w\varepsilon_{33}^T L_p/h_p$ is the equivalent capacitance of this piezoelectric buckling bridge and w is the width of the buckling bridge. Define the zeroth, first, and second-order moments of the substrate and piezoelectric layer cross sections at any point x as

$$(A_s, H_s, I_s) = \iint_s (1, z, z^2) dydz = \left[wt_s, \frac{1}{2}w\left(t_{k+1}^2 - t_k^2\right), \frac{1}{3}w\left(t_{k+1}^3 - t_k^3\right)\right]$$
$$(A_p, H_p, I_p) = \iint_p (1, z, z^2) dydz$$

The term related to the piezoelectric constant is defined as

$$B_p = \iint_p \frac{e_{31}}{h_p} dydz, \quad J_p = \iint_p \frac{e_{31}}{h_p} z dydz$$

The total potential energy can be expressed as:

$$U = \frac{1}{2}\int_0^L E_s\left\{A_s\left(\frac{1}{2}\left(\frac{\partial w^0(x,t)}{\partial x}\right)^2\right)^2 + I_s\left(\frac{\partial^2 w^0(x,t)}{\partial x^2}\right)^2\right.$$
$$\left. -2H_s\left[\frac{1}{2}\left(\frac{\partial w^0(x,t)}{\partial x}\right)^2\right]\frac{\partial^2 w^0(x,t)}{\partial x^2}\right\} dx$$
$$+ \frac{1}{2}\int_0^L \left(c_{11}^E + c_{12}^E\right)\left\{A_p\left(\frac{1}{2}\left(\frac{\partial w^0(x,t)}{\partial x}\right)^2\right)^2\right.$$

$$+I_{\mathrm{p}}\left(\frac{\partial^2 w^0(x,t)}{\partial x^2}\right)^2 - 2H_{\mathrm{p}}\left[\frac{1}{2}\left(\frac{\partial w^0(x,t)}{\partial x}\right)^2\right]\frac{\partial^2 w^0(x,t)}{\partial x^2}\Bigg\}\mathrm{d}x$$

$$+\frac{1}{2}\int_0^L\left\{B_{\mathrm{p}}v(t)\left[\frac{1}{2}\left(\frac{\partial w^0(x,t)}{\partial x}\right)^2\right] - J_{\mathrm{p}}v(t)\frac{\partial^2 w^0(x,t)}{\partial x^2}\right\}\mathrm{d}x \tag{5.11}$$

The total kinetic energy can be expressed as:

$$\begin{aligned}T = \frac{1}{2}\int_0^L \rho_{\mathrm{s}}\Bigg\{&I_{\mathrm{s}}\left(\frac{\partial^2 w^0(x,t)}{\partial t\partial x}\right)^2 + A_{\mathrm{s}}\left(\frac{\partial w^0(x,t)}{\partial t}\right)^2\\&+2A_{\mathrm{s}}\frac{\partial w^0(x,t)}{\partial t}\frac{\partial w_{\mathrm{b}}(t)}{\partial t} + A_{\mathrm{s}}\left(\frac{\partial w_{\mathrm{b}}(t)}{\partial t}\right)^2\Bigg\}\mathrm{d}x\\&+\frac{1}{2}\int_0^L \rho_{\mathrm{p}}\Bigg\{I_{\mathrm{p}}\left(\frac{\partial^2 w^0(x,t)}{\partial t\partial x}\right)^2 + A_{\mathrm{p}}\left(\frac{\partial w^0(x,t)}{\partial t}\right)^2\\&+2A_{\mathrm{p}}\frac{\partial w^0(x,t)}{\partial t}\frac{\partial w_{\mathrm{b}}(t)}{\partial t} + A_{\mathrm{p}}\left(\frac{\partial w_{\mathrm{b}}(t)}{\partial t}\right)^2\Bigg\}\mathrm{d}x\\&+\frac{1}{2}M_0\left(\frac{\partial w^0\left(\frac{L}{2},t\right)}{\partial t} + \frac{\partial w_{\mathrm{b}}(t)}{\partial t}\right)^2\end{aligned} \tag{5.12}$$

The electrical energy available in the piezoelectric layer can be expressed as:

$$W_{ie} = -\frac{1}{2}\int_0^L v(t)\left\{B_{\mathrm{p}}\left[\frac{1}{2}\left(\frac{\partial w^0\left(\frac{L}{2},t\right)}{\partial x}\right)^2\right] - J_{\mathrm{p}}\frac{\partial^2 w^0(x,t)}{\partial x^2}\right\}\mathrm{d}x + \frac{1}{2}C_{\mathrm{p}}v^2(t) \tag{5.13}$$

Without considering the loss of mechanical dissipation, the expanded Hamilton's principle for considering the electrical energy of the piezoelectric layer is $\int_{t_1}^{t_2}(\delta T - \delta U + \delta W_{ie} + \delta W_{\mathrm{virtual}})\mathrm{d}t = 0$. $\delta W_{\mathrm{virtual}}$ is the virtual work of nonconservative mechanical forces and charges, and since the work done by the underlying excitation is already reflected in the total kinetic energy term, the mechanical damping effect is introduced in the subsequent process, and thus the only nonconservative virtual work is caused by the charge output, i.e. $\delta W_{\mathrm{virtual}} = Q(t)\delta v(t)$.

The energy equation is discretized according to the hypothetical modal method with modal number N. The vibration deflection is expanded with a finite term as follows:

$$w^0(x,t) = \sum_{r=1}^{N} a_r(t)\psi_r(x) \tag{5.14}$$

where $\psi_r(x)$ and $a_r(t)$ are called kinematic tolerance trial functions that satisfy the respective fundamental boundary conditions. It is known that the first-order mode of this bridge-type vibration energy harvester plays a major role in energy conversion, so the total kinematic displacement can be approximated as the first-order modal kinematic displacement: $w^0(x, t) = a(t)\psi(x)$, where the initial displacement function of the buckled bridge can be expressed as: $\psi(x) = h_0 \frac{1-\cos\left(\frac{2\pi x}{L}\right)}{2}$.

Then, the total potential energy can be expressed as:

$$U = a^4 U_1 + a^3 U_2 + a^2 U_3 + a^2 v(t) U_4 + a v(t) U_5 \tag{5.15}$$

where

$$U_1 = \frac{1}{8}\int_0^L \left(E_s A_s + c_{11}^E A_p + c_{12}^E A_p\right) \psi'^4 \mathrm{d}x$$

$$U_2 = -\int_0^L \left(E_s H_s + c_{11}^E H_p + c_{12}^E H_p\right) \psi'^2 \psi'' \mathrm{d}x$$

$$U_3 = \frac{1}{2}\int_0^L \left[\left(E_s I_s + c_{11}^E I_p + c_{12}^E I_p\right) (\psi'')^2\right] \mathrm{d}x$$

$$U_4 = \frac{1}{4}\int_0^L [B_p \psi'^2] \mathrm{d}x$$

$$U_5 = -\frac{1}{2}\int_0^L J_p \psi'' \mathrm{d}x$$

The first-order partial differentiation of the total potential energy with respect to displacement and voltage is:

$$\frac{\partial U}{\partial a_i} = 4a^3 U_1 + 3a^2 U_2 + 2a U_3 + 2av U_4 + v(t) U_5$$

$$\frac{\partial U}{\partial v_i} = a^2 U_4 + a U_5$$

Total kinetic energy is:

$$T = (\dot{a})^2 T_1 + \dot{a} F_1 + F_2 \tag{5.16}$$

where

$$T_1 = \frac{1}{2}\int_0^L [(\rho_s I_s + \rho_p I_p)(\psi')^2 + (\rho_s A_s + \rho_p A_p)(\psi)^2] \mathrm{d}x + \frac{1}{2} M_0 \left(\psi\left(\frac{L}{2}\right)\right)^2$$

$$F_1 = \int_0^L (\rho_s A_s + \rho_p A_p)\psi \frac{\partial w_b(t)}{\partial t} \mathrm{d}x + M_0 \psi\left(\frac{L}{2}\right) \frac{\partial w_b(t)}{\partial t} = F \frac{\partial w_b(t)}{\partial t}$$

$$F = \int_0^L (\rho_s A_s + \rho_p A_p)\psi \, \mathrm{d}x + M_0 \psi\left(\frac{L}{2}\right)$$

$$F_2 = \frac{1}{2}\int_0^L (\rho_s A_s + \rho_p A_p)\left(\frac{\partial w_b(t)}{\partial t}\right)^2 \mathrm{d}x + \frac{1}{2} M_0 \left(\frac{\partial w_b(t)}{\partial t}\right)^2$$

Partial differential form of kinetic energy:

$$\frac{\partial T}{\partial \dot{a}_i} = 2T_1 \dot{a} + F_1, \frac{\partial T}{\partial a_i} = 0, \frac{\partial T}{\partial v_i} = 0, \frac{\mathrm{d}}{\mathrm{d}t}\left(\frac{\partial T}{\partial \dot{a}_i}\right) = 2\ddot{a} T_1 + \dot{F}_1, \frac{\mathrm{d}}{\mathrm{d}t}\left(\frac{\partial T}{\partial \dot{v}_i}\right) = 0$$

The electrical energy possessed by the piezoelectric layer is:

$$W_{ie} = a^2 v(t) W_1 + a v(t) W_2 + \frac{1}{2} C_p v^2(t) \tag{5.17}$$

where

$$W_1 = -\frac{1}{4}\int_0^L B_p\psi'^2 \mathrm{d}x = -U_4$$

$$W_2 = \frac{1}{2}\int_0^L J_p\psi'' \mathrm{d}x = -U_5$$

$$\frac{\partial W_{ie}}{\partial a_i} = 2avW_1 + vW_2$$

$$\frac{\partial W_{ie}}{\partial v_i} = a^2W_1 + aW_2 + C_p v$$

The Lagrange electromechanical coupling equation based on the extended Hamilton's principle is as follows:

$$\begin{cases} \dfrac{\mathrm{d}}{\mathrm{d}t}\left(\dfrac{\partial T}{\partial \dot{a}_i}\right) - \dfrac{\partial T}{\partial a_i} + \dfrac{\partial U}{\partial a_i} - \dfrac{\partial W_{ie}}{\partial a_i} = 0 \\ \dfrac{\mathrm{d}}{\mathrm{d}t}\left(\dfrac{\partial T}{\partial \dot{v}_i}\right) - \dfrac{\partial T}{\partial v_i} + \dfrac{\partial U}{\partial v_i} - \dfrac{\partial W_{ie}}{\partial v_i} = Q \end{cases} \tag{5.18}$$

Thus, a system of differential equations for the displacement and voltage of the electromechanical coupling can be derived as follows:

$$\begin{cases} 2\ddot{a}T_1 + 4a^3U_1 + 3a^2U_2 + 2aU_3 + 4avU_4 + 2vU_5 = -\dot{F}_1 \\ 4a\dot{a}U_4 + 2\dot{a}U_5 - C_p\dot{v} = v/R \end{cases} \tag{5.19}$$

According to the parameters shown in Table 5.1, the Runge–Kutta method is used to numerically solve the above higher-order differential equations and analyze the output characteristics of the displacement at the center point and the voltage of the vibration process under different excitation accelerations and different action frequencies, respectively.

As shown in Figure 5.2, the perturbation displacement curves at the center of the buckled bridge are shown from 50 to 130 Hz at 0.1, 0.2, 0.3, 0.4, 0.5, and 1.0g acceleration amplitude, respectively. When the excitation acceleration amplitude is 0.1g, the maximum displacement perturbation is found at 90 Hz excitation frequency, and the resulting displacement perturbation gradually decreases in an approximately symmetrical order with increasing and decreasing frequency. When the excitation acceleration amplitude increases, the calculated disturbance displacement amplitude also gradually increases, and at 0.4g the disturbance displacement appears to obvious change, and the maximum displacement occurs at the frequency of 100 Hz; and when the frequency decreases from 100 Hz, the disturbance displacement gradually decreases with a smaller amplitude, but when the frequency increases from 100 to 110 Hz, the disturbance displacement appears a sudden and significant decrease, and when the frequency is increased from 100 to 110 Hz, the disturbance displacement decreases abruptly and then continues to decrease gradually. When the excitation acceleration amplitude increases to 1.0g, the maximum perturbation displacement appears at 110 Hz, with obvious asymmetry at both ends of the frequency point, where the peak displacement is

Table 5.1 Parameters for the theoretical analysis of the piezoelectric buckled-bridge energy harvester.

Parameters	Denotations	Value	Unit
Young's modulus of PET	E_{PET}	4	GPa
Density of PET	ρ_{PET}	1380	kg/m^3
Thickness of PET	h_{PET}	0.1	mm
Length of PET	L	25	mm
Young's modulus of beryllium copper	E_{copper}	128	GPa
Poisson ratio of beryllium copper		0.3	
Density of beryllium copper	ρ_{copper}	8250	kg/m^3
Length of beryllium copper	L_{copper}	20	mm
Thickness of beryllium copper	h_{copper}	0.05	mm
Elastic modulus of PZT	C_{11} C_{12}	127.2 80.2	GPa GPa
Density of PZT	ρ_{PZT}	7500	kg/m^3
Thickness of PZT	h_{PZT}	0.05	mm
Piezoelectric constant	e_{31}	−6.2	C/m^2
Density of tungsten	$\rho_{Tungsten}$	19 200	kg/m^3
Length of piezoelectric layer	L_p	16	mm
Dielectric constant of piezoelectric layer	ε_{33}	1433	
Vacuum permittivity	ε_0	8.85×10^{-12}	F/m
Width of buckled bridge	w	5	mm
Height of initial buckled bridge	h_0	0.5	mm
Load resistance	R	10	MΩ
Volume of mass	V_{mass}	0.2×10^{-6}	m^3

located, and the basic change trend remains the same as that at 0.5g excitation acceleration amplitude.

Figure 5.3 shows the waveform curves of the theoretically calculated disturbance displacement at different frequencies for a magnified excitation acceleration amplitude of 1.0g. The output disturbance displacement amplitude has obvious periodic fluctuations, and the bending deviates from the displacement volatility characteristic of the input simple harmonic vibration. As the frequency increases from 50 to 110 Hz, the disturbance displacement amplitude gradually increases and the periodic fluctuation of the waveform becomes more and more obvious. When the frequency continues to increase to 120 Hz, the disturbance displacement suddenly decreases and the periodic fluctuation of the waveform is relatively weakened.

Figure 5.4 reflects the variation between the amplitude of the extracted disturbance displacement curve and the applied frequency under different excitation accelerations. It can also be seen from this figure that the frequency response of

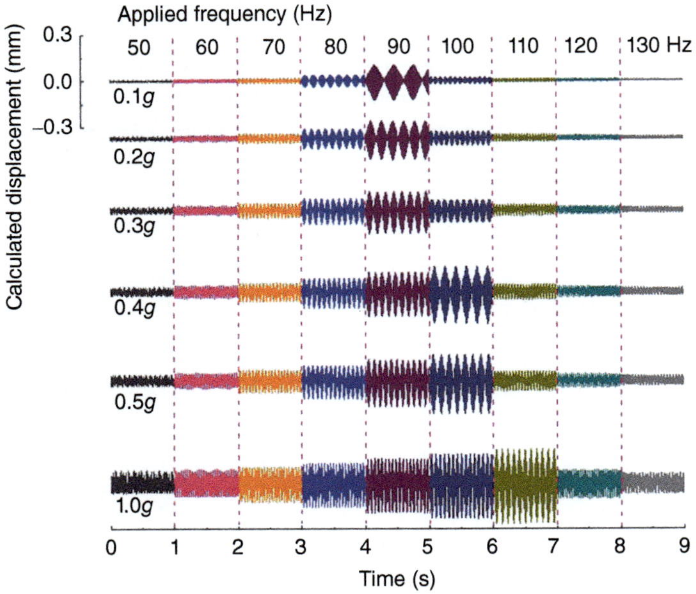

Figure 5.2 Calculated displacement waveforms of the central point in the buckled bridge under the 0.1, 0.2, 0.3, 0.4, 0.5, and 1.0*g* acceleration amplitude with frequencies varied from 50 to 130 Hz.

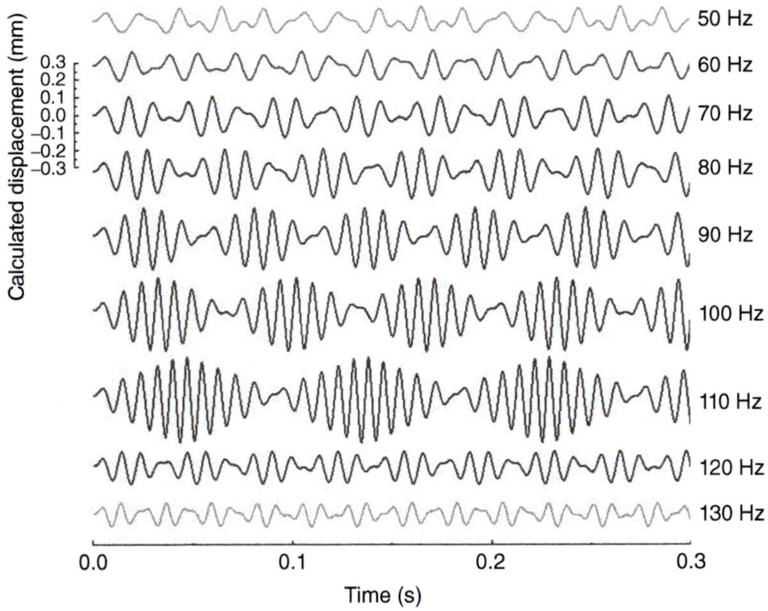

Figure 5.3 Enlarged calculated displacement waveforms of the central point of the buckled bridge under the acceleration amplitude of 1.0*g* with varied frequencies.

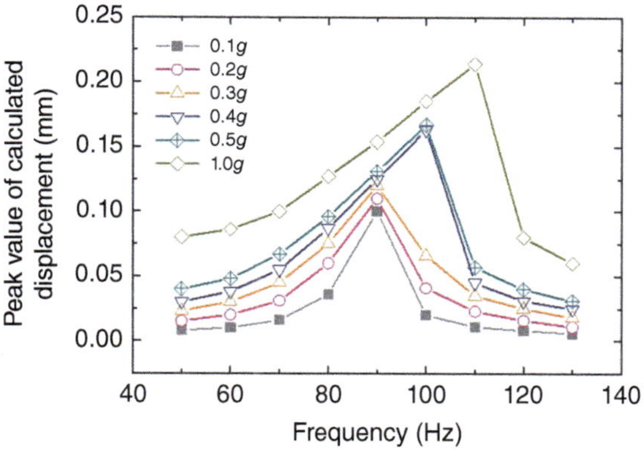

Figure 5.4 Calculated peak displacement as a function of the applied frequency under 0.1, 0.2, 0.3, 0.5, and 1.0*g* acceleration amplitude.

the disturbance displacement amplitude has a peak at 90 Hz when the acceleration amplitude is 0.1–0.3g. Currently, the frequency-domain curve is relatively symmetrical, which means that the nonlinear vibration response is not obvious, and the peak is mainly due to the applied frequency close to the inherent frequency of the structure, so it means that the inherent frequency of the buckled-bridge structure is about 90 Hz. When the acceleration amplitude is 0.4g, the frequency-domain response starts to be obviously biased toward high frequency, and the peak value appears at 100 Hz. As the excitation acceleration continues to increase, the peak frequency-domain response appears at 110 Hz at 1.0g. This indicates that the vibration excitation of 0.4g acceleration amplitude makes the buckled-bridge structure produce obvious nonlinear vibration response, and according to the result that the frequency response curve is biased toward high frequency under high excitation conditions, we can judge the structure. The vibration belongs to the nonlinear vibration type with monostatic hardening effect, which is beneficial for the energy harvester to enhance the width of the frequency band in the high-frequency band.

Figure 5.5 shows the output voltage curves of this buckled-bridge-type piezoelectric vibration energy harvester for different excitation accelerations and different action frequencies corresponding to the disturbance displacement curves. It can be seen from the curves that the output voltage amplitude increases as the excitation acceleration increases, while the applied frequency of the output voltage generating the maximum peak has a tendency to favor high frequency. When the excitation acceleration amplitude is 0.1–0.3g, the applied frequency of the maximum output voltage amplitude is at 90 Hz, when the excitation acceleration amplitude increases to 0.4 and 0.5g, the applied frequency of the maximum output voltage amplitude appears at 100 Hz, and the applied frequency of the maximum output voltage amplitude appears at 110 Hz at 1.0g. Meanwhile, it can be seen that when the excitation acceleration amplitude is greater than 0.4g, the output voltage curve decreases significantly after the maximum output voltage amplitude, and a jump in amplitude occurs, and the phenomenon becomes more and more obvious with the increase of excitation acceleration.

Figure 5.6 shows the output voltage waveforms of the buckled-bridge-type PEH at different frequencies calculated theoretically for an amplified excitation

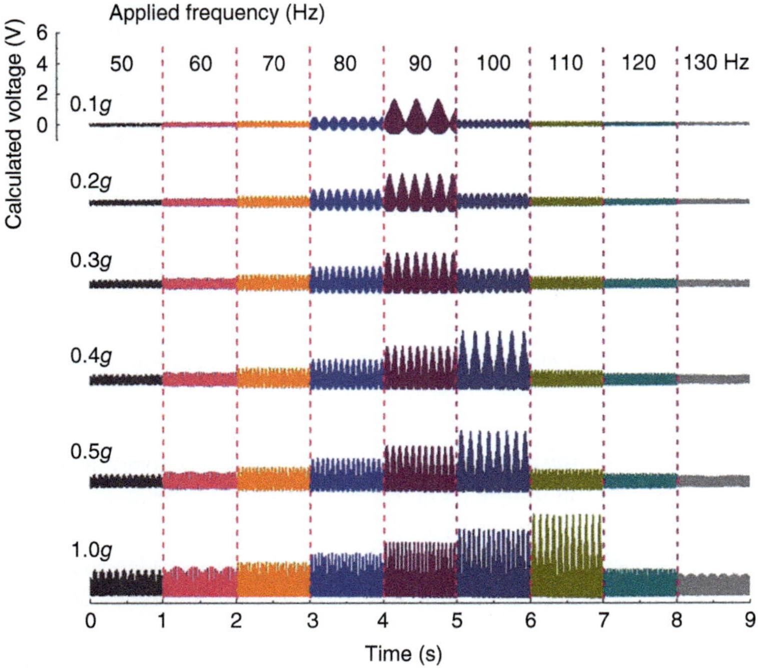

Figure 5.5 Calculated voltage waveforms of the piezoelectric buckled bridge energy harvester under the 0.1, 0.2, 0.3, 0.4, 0.5, and 1.0*g* acceleration amplitude with frequency varied from 50 to 130 Hz.

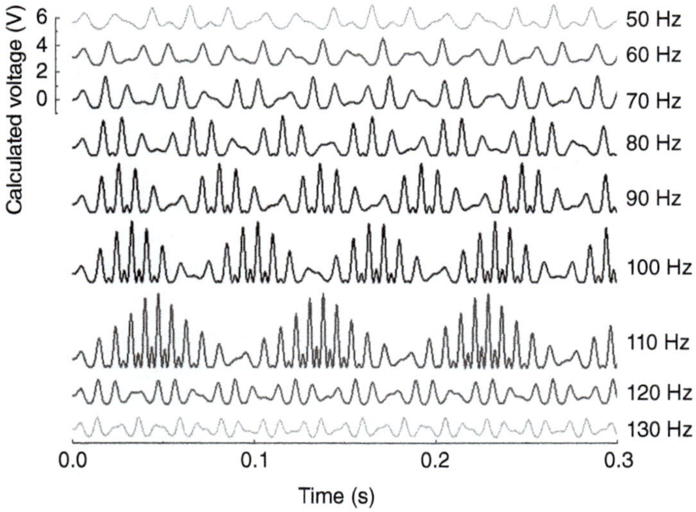

Figure 5.6 Enlarged calculated voltage waveforms of the buckled-bridge piezoelectric energy harvester under the acceleration amplitude of 1.0*g* with varied frequencies.

acceleration amplitude of 1.0g. Similar to the displacement waveform, the output voltage amplitude also shows periodic fluctuation characteristics, which still do not follow the input simple harmonic vibration waveform. As the applied frequency increases from 50 to 110 Hz, the periodic fluctuations become more and more pronounced, and the output voltage amplitude also increases gradually, with the maximum voltage output peak at 110 Hz. When the applied frequency is greater than 110 Hz, the output voltage amplitude suddenly decreases a lot, i.e. the amplitude jumps, and the large periodic fluctuation of the voltage curve is weakened. At the same time, unlike the displacement fluctuation characteristics, the output voltage waveform is biased to the positive voltage side, and the negative voltage increases insignificantly with the increase of excitation acceleration, mainly due to the peak positive voltage change, which is mainly due to the asymmetric deformation of the buckled structure with respect to the piezoelectric neutral layer.

Figure 5.7 shows the output voltage amplitude at different action frequencies for different excitation acceleration amplitudes, reflecting the variation curve of the output voltage of the buckled-bridge-type vibration energy harvester with the applied frequency. It can be seen from this figure that the peak output voltage is mainly due to the resonance of the structure at small excitation acceleration, indicating that the intrinsic frequency of the structure is about 90 Hz. When the excitation acceleration increases to 0.4g, the output voltage starts to favor the high-frequency band, indicating that the vibration form of the energy harvester belongs to the monostable hardening effect.

5.2.3 Finite Element Analysis

Finite element analysis (FEA) is generally used for mechanical and electrical analysis of PEHs with linear vibration. For nonlinear energy harvesters, it is usually difficult to analyze their force-electric characteristics by means of finite element simulation because of instability and chaos in their frequency-domain output. Considering that the finite element method is more suitable for calculating the steady-state characteristics, this section adopts the steady-state device output time-domain characteristics at different frequencies to study the frequency-domain

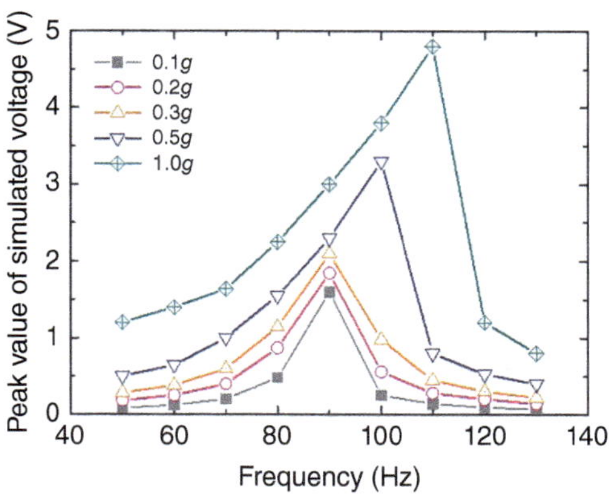

Figure 5.7 Calculated peak voltage as a function of the applied frequency under 0.1, 0.2, 0.3, 0.4, 0.5, and 1.0*g* acceleration amplitude.

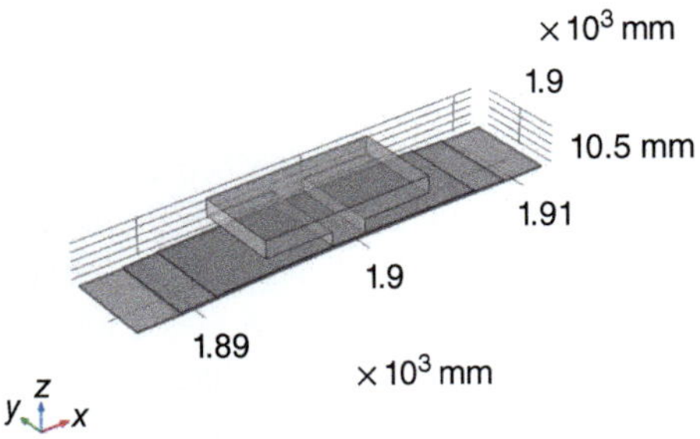

Figure 5.8 Three-dimensional model for the FEA.

characteristics of the output voltage, which of course only contains the output characteristics of the stable region of the nonlinear energy-harvesting output process and is also the most practically significant output characteristic in the development of this nonlinear vibration PEH.

Figure 5.8 shows a three-dimensional (3D) finite element simulation model built according to the structural and dimensional parameters in Figure 5.1 and Table 5.1. The simulation analysis was performed with the commercial multi-physics field-coupling analysis software COMSOL Multiphysics 5.4. By fixing both ends of the suspension bridge and applying inertial loads in the z-direction, as shown in the figure, the boundary conditions, such as mechanical boundary, charge conservation, grounding and termination boundary, and external resistance, are set for this PEH using the multi-physics field-coupling boundary of solid mechanics, electrostatic field, electric circuit, and piezoelectric effect.

Figure 5.9 shows the mesh dissection used in the simulation analysis. Figure 5.9a shows the full view of the mesh dissection, and Figure 5.9b shows the side enlarged view of the mesh dissection. It can be seen that the triangles used on the sides are used for the lateral lattice division, and then the lattice is swept in the width direction to achieve the overall structural lattice dissection. The cell size is predefined as the normal section with the following parameters – maximum cell size of 2.5 mm, minimum cell size of 0.45 mm, maximum cell growth rate of 1.5, curvature factor of 0.6, and resolution of 0.5 for narrow areas. The time-domain calculation time range is also set from 0 to 0.12 second with an interval of 0.0001 second. The tolerance

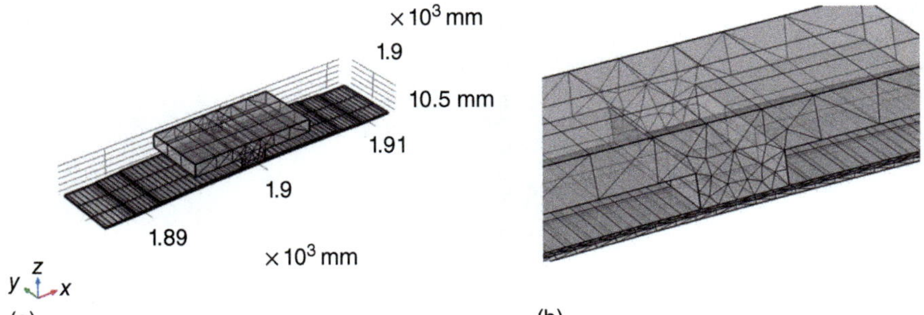

Figure 5.9 Mesh for the FEA. (a) Schematic diagram for FEA mesh, (b) the later view of the FEA mesh.

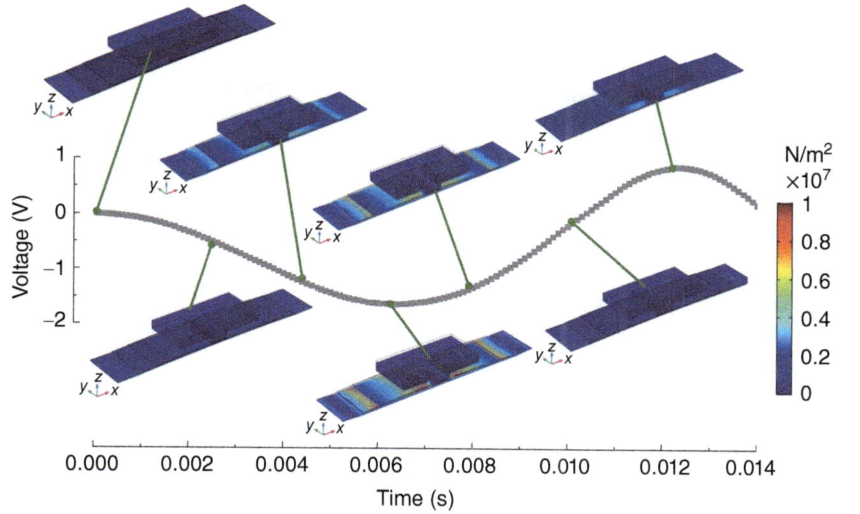

Figure 5.10 Stress distribution and output voltage as a function of the time under the acceleration amplitude of 1.0*g* and the applied frequency of 84 Hz.

is set for physical field control, and geometric nonlinearity and full coupling are set for the solution, and the system preset MUMPS solver is used to start the calculation.

Figure 5.10 shows the stress distribution and output voltage variation with time for the buckled-bridge structure analyzed in real-time simulation under the conditions of 1.0g excitation acceleration amplitude and 84 Hz applied frequency. It can be seen that the central mass bulk setup generates a central concentrated force under the inertial load, which causes large stress in the piezoelectric layer during the vibration process. The buckled-bridge end is fixed with variable stiffness, and the flexible polymer support layer is used as the fixed end to reduce the bending stiffness of the fixed end. The variable stiffness design prevents excessive stress concentration at the fixed end and improves the reliability of the vibrating structure.

Figure 5.11 shows the output voltage waveforms of the simulation analysis at different action frequencies for 1.0g excitation acceleration amplitude. It can be seen that the output voltage does not follow the fluctuating form of the simple harmonic vibration load applied at the boundary, and the output voltage amplitude has a fluctuating characteristic with time, and the fluctuating characteristic is more obvious at about 80 Hz, which also reaches the maximum output voltage amplitude. This fluctuation is also a characteristic of the output voltage of the nonlinear vibration energy harvester and is the biggest problem faced by the finite element simulation when analyzing its output voltage frequency-domain curve, and this fluctuation makes the frequency-domain curve difficult to calculate accurately.

Figure 5.12 shows the output voltage as a function of the applied frequency of the simulation analysis under different excitation accelerations (0.1, 0.2, 0.3, 0.4, 0.5, and 1.0g), which displays that the simulated output voltage is period change in the peak value at the time from 0 to 0.12 second under a fixed excitation frequency. The

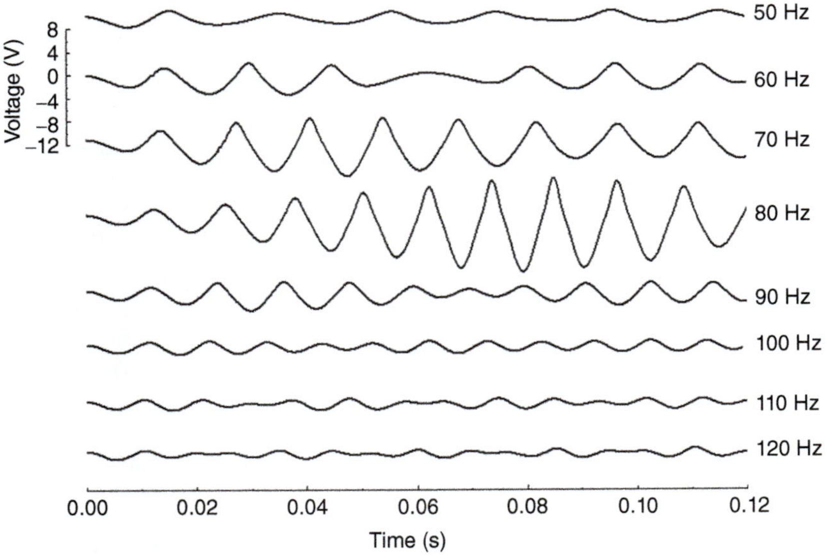

Figure 5.11 Simulated voltage as a function of the time at the different applied frequencies.

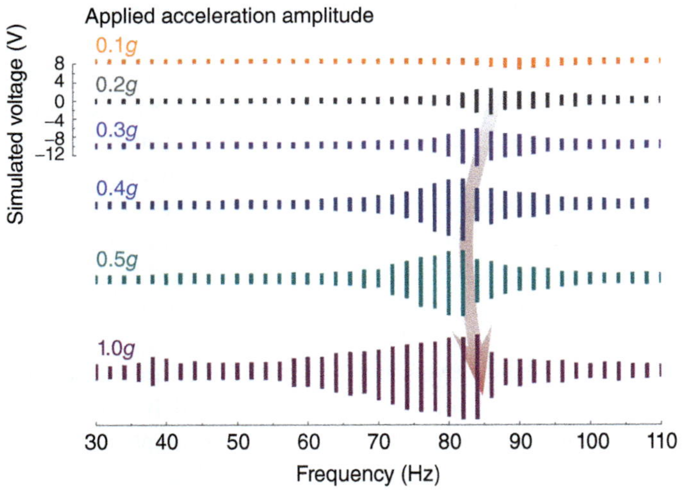

Figure 5.12 Simulated voltage as a function of the applied frequency under 0.1, 0.2, 0.3, 0.4, 0.5, and 1.0*g* acceleration amplitude.

frequency-domain characteristics are analyzed by the distribution of the output voltage values in the time domain at different frequencies. It makes the results more effective and more accurate. As the acceleration increases, the output voltage amplitude increases, and the output frequency point that produces the largest amplitude is also shifted from 88 Hz at 0.1g to 82 Hz at 0.3g and then to 84 Hz at 1.0g. The frequency-domain characteristics of the output voltage change significantly from 0.3 to 0.4g. After 0.4g, there is an obvious amplitude jump. The results are in good agreement with the theoretical analysis. For the maximum output voltage amplitude, the applied frequency in the low-acceleration band is biased toward the low frequency mainly due to the softening effect of the piezoelectric material layer stiffness, and in the high-frequency band, the applied frequency starts to bias toward the high frequency mainly due to the hardening effect of the buckled-bridge-type nonlinear structure.

Additionally, there are numerous ultralow-frequency mechanical kinetic energies known to exist in the environment, and how to efficiently carry out the conversion of these energies is also one of the challenges faced by the current piezoelectric energy-harvesting technology. Through previous analysis and experimental studies, it can be found that the performance of PEHs has a large dependence on the excitation frequency. The conversion mechanism of free-vibration energy harvesters is to use large perturbations in the case of structural resonance to induce large strains in the structure to obtain efficient energy conversion, while it is more difficult to achieve structural resonance of micromachined devices under ultralow-frequency excitation conditions. The forced vibration form can maintain the device with the action of the vibration source to generate sufficient structural strain to achieve efficient energy conversion. Mechanism of the forced vibration source applied in the nonlinear buckled PEH will be introduced in Chapter 6.

5.3 Design Method for Custom Nonlinear Characteristics

Buckling bridge as a typical nonlinear structure is widely used to realize wideband piezoelectric energy harvesting [3, 4, 11–15]. However, its frequency hysteresis region contains two output states of high energy and low energy, and they correspond to unstable output and stable output, respectively [16–18]. In the face of actually fixed frequency range vibration sources, a high-energy stable output is real. Therefore, how to improve the broadband stable output characteristics of nonlinear vibration energy harvesters becomes an important challenging problem for its research. Therefore, this section discusses a strategy based on a combination of hardened and softened nonlinear vibration energy harvesters to obtain a stable, wider bandwidth output, which enhances the practical application of nonlinear energy harvesters. Two buckled drum structure PEHs with different nonlinear effects are fabricated, and by controlling the buckled angle, the energy harvesters with low- and high-resonant frequencies are made to have hardened and softened nonlinear effect characteristics, respectively, and when the two frequency response curves overlap together with a certain acceleration amplitude, the low energy jumps in the output are avoided and a continuous ultrawide bandwidth energy output can be obtained.

The three-dimensional schematic of the drum-shaped PEH is shown in Figure 5.13a. The low stiffness tympanic membrane structure mainly consists of PET, stainless-steel foil, and lead zirconate titanate (PZT) film. The four support edges of a piezoelectric buckled drum were fixed on a 3D-printed rigid base, and a mushroom-shaped mass bulk was assembled at the center of the drum to provide

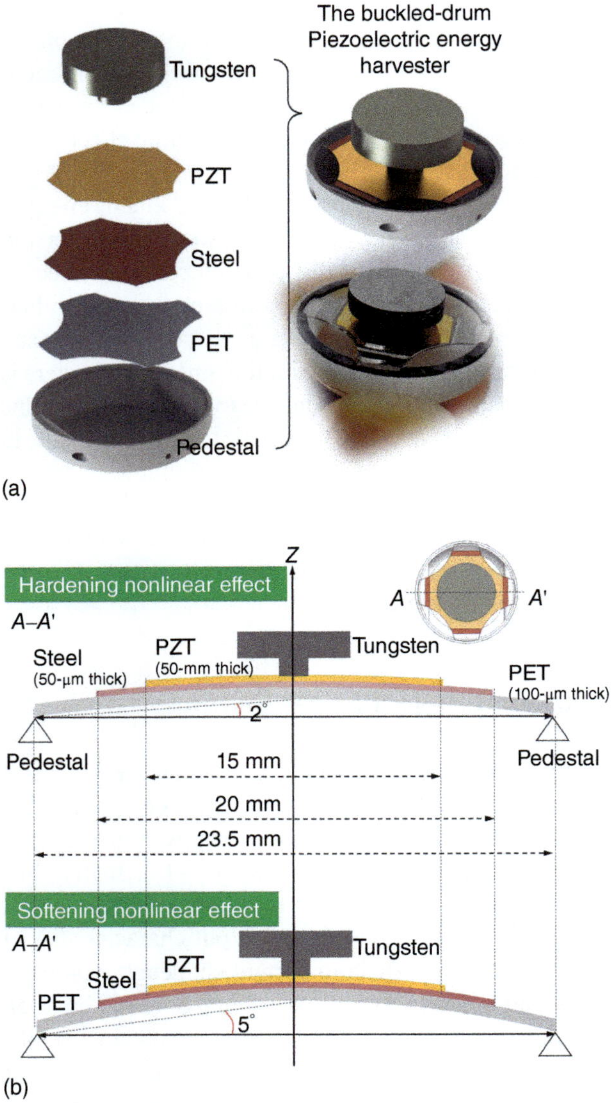

Figure 5.13 Structural design of the buckled drum piezoelectric energy harvester. Source: From Yi et al. [7] / with permission of IEEE. (a) Schematic illustration and photograph of the harvester, (b) schematic diagram for the hardening nonlinear mechanism (buckled angle of 2°) and the softening nonlinear mechanism (buckled angle of 5°) buckled-drum harvester.

a concentrated force. Different buckled angles were prepared by controlling the preset heights of different center points during the fixation process, with buckled angles of 2° and 5°, respectively, as shown in Figure 5.13b.

The open-circuit output voltages of these two PEHs in the frequency from 60 to 160 Hz exhibit two opposite nonlinear responses at different applied acceleration amplitudes of 0.1, 0.5, and 1.0g [7]. The bandwidth increases rapidly as the excitation

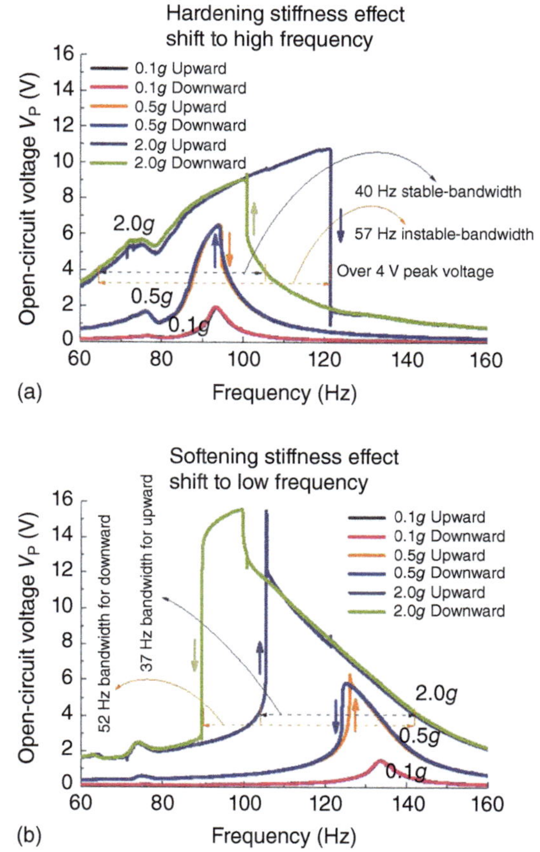

Figure 5.14 Open-circuit output voltage performance of the hardening and softening nonlinear mechanism buckled-drum PEHs at the acceleration amplitudes of 0.1, 0.5, and 2.0*g*. Source: Yi et al. [7] / with permission of IEEE. (a) Hardening stiffness effect at the buckled angle of 2°. (b) Softening stiffness effect at the buckled angle of 5°. Source: Yi et al. [7] / with permission of IEEE.

acceleration amplitude increases, and the hysteresis characteristics of the forward and negative sweep curves become more pronounced. At high excitation, the buckled angle of 2° exhibits a nonlinear hardening effect, as shown in Figure 5.14a, and the output voltage broadens toward high frequencies, while the buckled angle of 5° exhibits a nonlinear softening effect and the output voltage broadens toward low frequencies, as shown in Figure 5.14b. In addition, the frequency shift at the flexion angle of 5° is more obvious than that at the flexion angle of 2°. This is mainly attributed to the inversion of nonlinear effects at high-excitation accelerations for large flexion angles.

5.4 Flexible Buckled-Bridge PEH

In Section 5.2, the potential and output characteristics of the flexible buckled PEH for wideband energy harvesting are explained by theoretical analysis and simulation analysis, and the energy conversion mechanisms under different vibration loads are revealed, as well as the solutions to the challenges of ultralow frequency. This section introduces a free-vibration-type flexible buckled PEH [4], which is fabricated

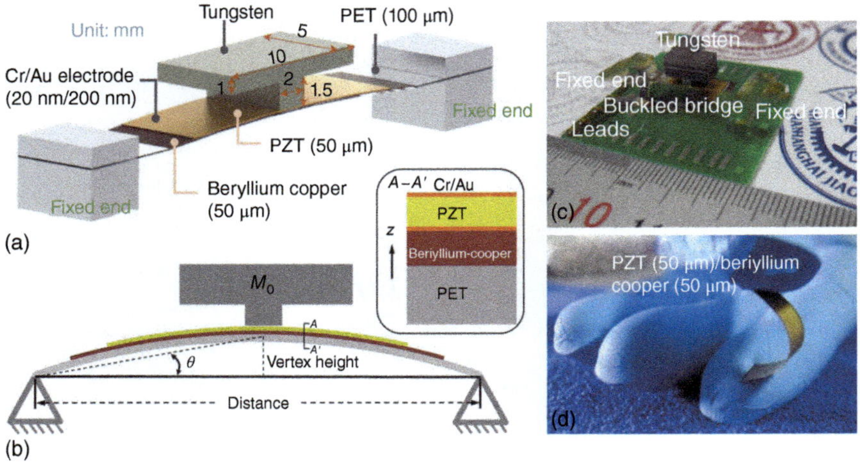

Figure 5.15 Structural design of the flexible buckled piezoelectric energy harvester. (a) Three-dimensional model illustration of the proposed flexible harvester. (b) The planar graph of the device's cross section. (c) Photograph of the fabricated device. (d) Evaluation of flexibility. Source: From Yi et al [4] / with permission of AIP Publishing.

by the piezoelectric thick film with mechanical-thinning technique and discusses its output characteristics and performance in broadband.

Figure 5.15a shows a three-dimensional model of a flexible buckled-bridge-type piezoelectric vibration energy harvester (PEH) [4] based on a thinned PZT thick film. The buckled bridge mainly consists of a piezoelectric ceramic layer, a metal layer, and a polymer layer. Piezoelectric composite structure is end-fixed at a size of 25 mm pitch of both ends. The symmetry of the structure is maintained, as shown in Figure 5.15b, using the PET layer as a flexible support layer for the piezoelectric composite structure to reduce the vibration process. Low-bending stiffness is used to achieve a low-resonant frequency. In addition, to induce large deformation of the buckled structure, increase the piezoelectric energy conversion efficiency and further reduce the resonant frequency of the system, a T-shaped tungsten mass bulk (top structure: $10 \times 1 \times 5\ mm^3$, bottom structure: $2 \times 1.5 \times 5\ mm^3$) is built in the center of the constructed buckled-bridge-type structure, as shown in Figure 5.15c, for the prepared piezoelectric vibration energy harvester of this buckled bridge type. Figure 5.15d demonstrates the good flexibility of the bonded piezoelectric composite.

Figure 5.16a shows that the hysteresis characteristic of the output voltage curve can be clearly observed under high-acceleration excitation, and the peak output voltage is biased toward the high-frequency side, which is a typical hardening nonlinearity phenomenon [19, 20]. The detailed analysis of the frequency-domain experiments showed that when the excitation acceleration amplitude was about 0.7g, there were significant hysteresis phenomena in the frequency-domain curves obtained from both the forward and negative sweep tests. The same conclusion can be obtained by analyzing the frequency at which the maximum peak voltage appears (Figure 5.16b) and the variation pattern of the output maximum peak voltage value

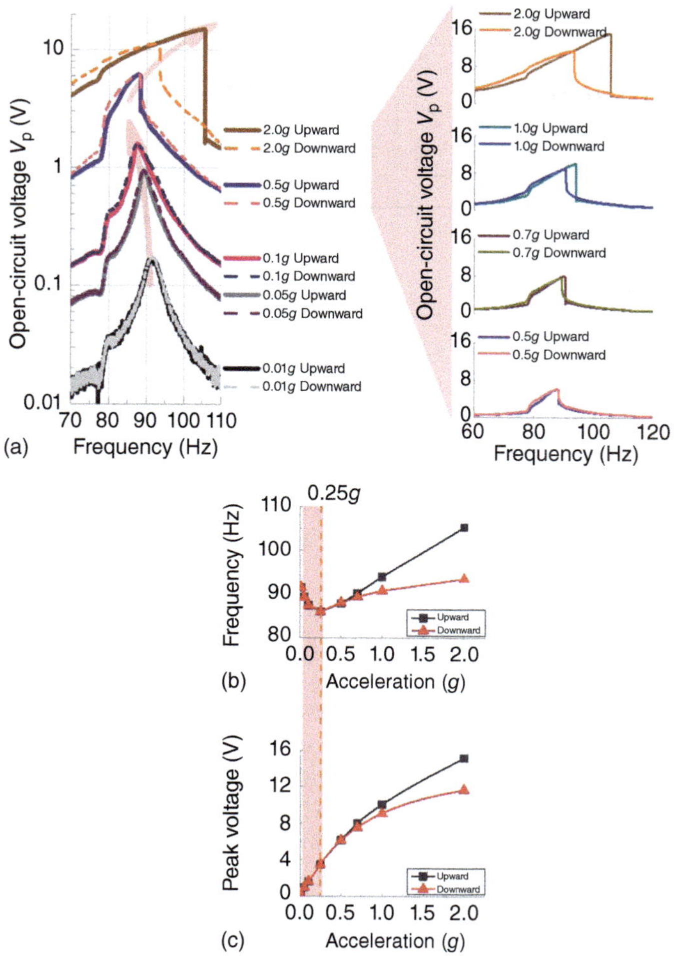

Figure 5.16 Output performance of the buckled bridge piezoelectric energy harvester. (a) Frequency dependence of the open output voltages under different applied accelerations amplitude varied from 0.01 to 2.0*g*. Enlarged graph: detailed results of frequency sweeping at the 0.5, 0.7, 1.0, and 2.0*g* acceleration amplitudes. (b) The resonant frequency for peak voltage and (c) output peak voltage (upward and down sweeping frequency) as a function of the acceleration amplitude. Source: Yi et al. [4] / with permission of AIP Publishing.

(Figure 5.16c) as the excitation acceleration amplitude increases. In the region of low applied acceleration, the response frequency of the maximum peak voltage appears to be gradually shifted to low frequency with increasing applied acceleration amplitude, which is mainly attributed to the softening nonlinear effect of the piezoelectric material due to the increase of the elastic flexibility of the PZT under large stress [21]. When the applied acceleration amplitude is greater than about 0.25g, the response frequency is gradually shifted to high frequency, which is mainly caused by the hardening of the buckled bridge, caused by the nonlinear properties of the buckled bridge playing a major role. For excitation acceleration amplitudes greater

Table 5.2 Performance comparison of recently reported bridge PEHs.

Materials	Volume of piezoelectric layer (mm^3)	Acceleration (*g*)	Frequency (Hz)	Power (μW)	Normalized power density (mW/(cm^3 g^2 Hz))	References
Poly (vinylidene fluoride-trifluoroethylene) (PVDF)	1.664	3	30	2.9	0.0064	Jung and Yun [25]
Macro fibre composites (MFC)	457.2	0.5	68	100	0.0129	Gafforelli et al. [26]
PVDF	131.2	14 m/s	—	61.8	—	Zhang et al. [27]
PZT	60	—	3	5	—	Jiang et al. [14]
PZT	42.24	—	14	48	—	Xie et al. [28]
PZT	4	2	105.3	600	0.3561	Yi et al. [4]

than 0.7*g*, an increasingly wide hysteresis region can be observed between the frequency-domain curves obtained from the forward and negative sweep tests. In the low-excitation case, such as the general piezoelectric vibration energy harvesters [22–24], the peak output voltage is linear with respect to the applied acceleration amplitude, with nonlinearity occurring for acceleration amplitudes above 0.25g.

Above all, the flexible polymer-supported piezoelectric buckled bridge-based nonlinear vibration energy harvester has outstanding output power of 0.6 mW and output power density of 150 mW/cm^3, as displayed in Table 5.2, at an excitation acceleration amplitude of 2.0g and an excitation frequency of 105.3 Hz, with a normalized power density of 0.36 mW/(cm^3 g^2 Hz), and this piezoelectric thin-film buckled bridge has good reliability.

5.5 Flexible Buckled-Bridges-Stacked PEH

Nonlinear vibration mechanisms are an effective way to broaden the frequency band of energy harvesters, and buckled-bridge-type structures as a typical nonlinear structure have greater potential in the study of broadband PEHs [13, 15, 27, 29–31]. To further enhance the output performance of buckled-bridge-type energy harvesters, a vertical integration method is proposed in this section to discuss the fabrication of two identical buckled-bridge-type structures. The vertical integration fabrication is achieved by rigidly connecting two identical buckled-bridge-type structures through the middle mass bulk to enhance their output characteristics while expanding the three-dimensional space utilization of the device.

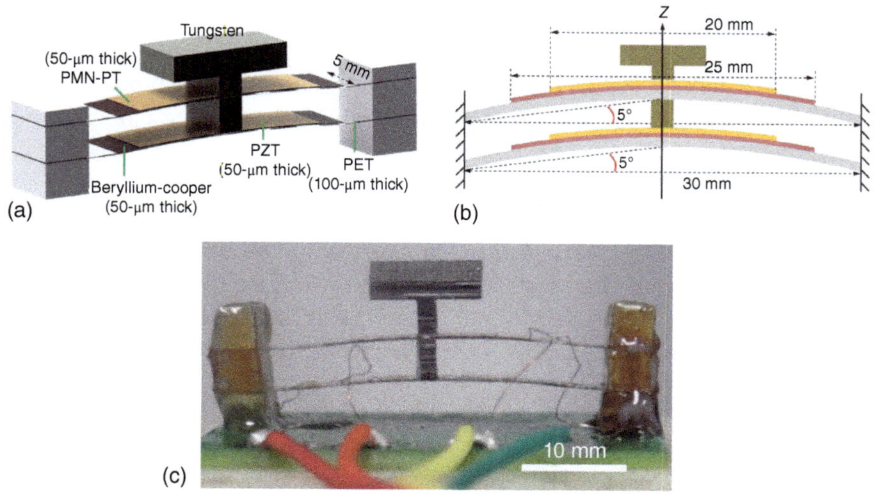

Figure 5.17 Schematic of "工" shaped piezoelectric energy harvester. (a) Three-dimensional model for this PEH. (b) Cross-sectional view of the buckled-bridge PEH. (c) Photograph of the fabricated harvester. Source: From Yi et al. [32] / with permission of IEEE.

As shown in Figure 5.17a, the buckled bridge [32] mainly consists of PET, beryllium bronze, and PZT/lead magnesium niobate-lead titanate (PMN-PT) layers, and the upper and lower buckled bridges with the same bending angle (5°). The dual-buckled bridges are connected by tungsten mass bulks to form an "工"-shaped structure. A "T" shaped tungsten mass bulk is bonded at the center of the uppermost buckled bridge to provide a concentrated inertial force at the center by occupying a smaller effective area. The cross section of the "工"-shaped PEH is shown in Figure 5.17b, with the length of the piezoelectric layer being 20 mm and the width being 5 mm, and the lengths of the beryllium bronze and PET layers being 25 and 30 mm, respectively. The fabricated device is shown in Figure 5.17c.

The frequency-domain responses of two layers connected in parallel and series exhibit a significant softening effect and hysteresis characteristics of the forward and negative sweep at high accelerations [32]. According to the output performance and vibration state characteristics of the two buckling bridge elements, they maintain a high degree of consistency, which is mainly because they form a resonant overall structure through rigid connection. Based on this vertical stacking integration strategy, the energy collector is expected to achieve higher energy output in a limited chip area. The low stiffness fixed end is conducive to the low-frequency resonance of the buckling bridge, causing large deformation in the piezoelectric layer and improving the output performance. For the further development of this type of energy collector, the influence of the fixed end on the resonance frequency of the whole device needs to be considered, and the specific actual vibration source and space size requirements need to be considered to further reduce the stiffness of the fixed end to adjust the resonance frequency of the structure, to achieve more efficient energy conversion.

5.6 Summary

In this chapter, we have mainly introduced the free-vibration nonlinear mechanism and discussed a broadband vibration energy-harvesting strategy using a joint mechanism of nonlinear softening and hardening effects. The softening and hardening effect control method of the flexible buckled wideband PEH is studied, and the combined high-frequency offset of the nonlinear hardening effect and low-frequency offset characteristics of the nonlinear softening effect are proposed to realize stable wideband vibration energy conversion, and the effectiveness of the method is investigated by experimental means, and its important theoretical guidance value in enhancing the nonlinear vibration energy harvester in practical applications is elaborated.

The output performance of a flexible buckled PEH was discussed, and its broadband characteristics are investigated to develop a broadband PEH with a stacked double-buckled structure. The design method for a dual-buckled-structure-stacked wideband piezoelectric energy harvesting is developed to illustrate its application potential in high-density integration.

References

1 Liu, H.C., Lee, C., Kobayashi, T. et al. (2012). Piezoelectric MEMS-based wideband energy harvesting systems using a frequency-up-conversion cantilever stopper. *Sensors and Actuators A: Physical* 186: 242–248.

2 Hu, Y., Mou, F.X., Yang, B. et al. (2018). A broadband E-shaped piezoelectric energy harvester based on vortex-shedding induced vibration from low velocity liquid flow. *AIP Advances* 8 (12): 125214.

3 Harne, R.L. and Wang, K.W. (2013). A review of the recent research on vibration energy harvesting via bistable systems. *Smart Materials and Structures* 22: 023001.

4 Yi, Z., Hu, Y., Ji, B. et al. (2018). Broad bandwidth piezoelectric energy harvester by a flexible buckled bridge. *Applied Physics Letters* 113: 183901.

5 Yang, B., Yi, Z., Tang, G., and Liu, J. (2019). A gullwing-structured piezoelectric rotational energy harvester for low frequency energy scavenging. *Applied Physics Letters* 115: 063901.

6 Hu, Y., Yi, Z.R., Dong, X.X. et al. (2019). High power density energy harvester with non-uniform cantilever structure due to high average strain distribution. *Energy* 169: 294–304.

7 Zhiran, Y., Dong, X., Zhang, W. et al. (2019). Controllable nonlinear effect for stable and broad bandwidth piezoelectric energy harvester. In: *2019 20th International Conference on Solid-State Sensors, Actuators and Microsystems & Eurosensors XXXIII (TRANSDUCERS & EUROSENSORS XXXIII)*, 2511–2514. IEEE.

8 Yi, Z., Xie, F., Tian, Y.W. et al. (2020). A battery- and leadless heart-worn pacemaker strategy. *Advanced Functional Materials* 30 (25): 2000477.

9 Reddy, J.N. (2003). *Mechanics of Laminated Composite Plates and Shells: Theory and Analysis*, 858. CRC Press.

10 Erturk, A. and Inman, D.J. (2011). *Piezoelectric Energy Harvesting*. John Wiley & Sons, Ltd.

11 Cottone, F., Gammaitoni, L., Vocca, H. et al. (2012). Piezoelectric buckled beams for random vibration energy harvesting. *Smart Materials and Structures* 21: 035021.

12 Cottone, F., Vocca, H., and Gammaitoni, L. (2009). Nonlinear energy harvesting. *Physical Review Letters* 102: 080601.

13 Zhu, Y. and Zu, J.W. (2013). Enhanced buckled-beam piezoelectric energy harvesting using midpoint magnetic force. *Applied Physics Letters* 103: 041905.

14 Jiang, X.-Y., Zou, H.-X., and Zhang, W.-M. (2017). Design and analysis of a multi-step piezoelectric energy harvester using buckled beam driven by magnetic excitation. *Energy Conversion and Management* 145: 129–137.

15 Li, H.T., Qin, W.Y., Zu, J., and Yang, Z. (2018). Modeling and experimental validation of a buckled compressive-mode piezoelectric energy harvester. *Nonlinear Dynamics* 92: 1761–1780.

16 Arrieta, A.F., Delpero, T., Bergamini, A.E., and Ermanni, P. (2013). Broadband vibration energy harvesting based on cantilevered piezoelectric bi-stable composites. *Applied Physics Letters* 102: 173904.

17 Mallick, D., Amann, A., and Roy, S. (2016). Surfing the high energy output branch of nonlinear energy harvesters. *Physical Review Letters* 117: 197701.

18 Udani, J.P. and Arrieta, A.F. (2017). Sustaining high-energy orbits of bi-stable energy harvesters by attractor selection. *Applied Physics Letters* 111: 213901.

19 Jordan, D.W. and Smith, P. (2007). *Nonlinear Ordinary Differential Equations*, vol. 73, 164–166. Oxford University Press Inc.

20 Erturk, A. and Inman, D.J. (2011). Broadband piezoelectric power generation on high-energy orbits of the bistable Duffing oscillator with electromechanical coupling. *Journal of Sound and Vibration* 330: 2339–2353.

21 Shen, D., Park, J.H., Ajitsaria, J. et al. (2008). The design, fabrication and evaluation of a MEMS PZT cantilever with an integrated Si proof mass for vibration energy harvesting. *Journal of Micromechanics and Microengineering* 18 (5): 055017.

22 Elfrink, R., Kamel, T.M., Goedbloed, M. et al. (2009). Vibration energy harvesting with aluminum nitride-based piezoelectric devices. *Journal of Micromechanics and Microengineering* 19 (9): 094005.

23 Kamel, T.M., Elfrink, R., Renaud, M. et al. (2010). Modeling and characterization of MEMS-based piezoelectric harvesting devices. *Journal of Micromechanics and Microengineering* 20 (10): 105023.

24 Yeo, H.G., Ma, X., Rahn, C., and Trolier-McKinstry, S. (2016). Efficient piezoelectric energy harvesters utilizing (001) textured bimorph PZT films on flexible metal foils. *Advanced Functional Materials* 26: 5940–5946.

25 Jung, S.-M. and Yun, K.-S. (2010). Energy-harvesting device with mechanical frequency-up conversion mechanism for increased power efficiency and wideband operation. *Applied Physics Letters* 96: 111906.

26 Gafforelli, G., Corigliano, A., Xu, R., and Kim, S.-G. (2014). Experimental verification of a bridge-shaped, nonlinear vibration energy harvester. *Applied Physics Letters* 105: 203901.

27 Zhang, J., Zhang, J., Shu, C., and Fang, Z. (2017). Enhanced piezoelectric wind energy harvesting based on a buckled beam. *Applied Physics Letters* 110: 183903.

28 Xie, Z., Kwuimy, C.A.K., Wang, Z., and Huang, W. (2018). A piezoelectric energy harvester for broadband rotational excitation using buckled beam. *AIP Advances* 8: 015125.

29 Van Blarigan, L., Danzl, P., and Moehlis, J. (2012). A broadband vibrational energy harvester. *Applied Physics Letters* 100: 253904.

30 Van Blarigan, L. and Moehlis, J. (2016). Dynamic analysis of a buckled asymmetric piezoelectric beam for energy harvesting. *Chaos* 26: 033107.

31 Li, H., Qin, W., and Deng, W. (2018). Coherence resonance of a magnet-induced buckled piezoelectric energy harvester under stochastic parametric excitation. *Journal of Intelligent Material Systems and Structures* 29: 1620–1631.

32 Yi, Z., Tian, Y., Dong, X. et al. (2019). Vertically integrated double buckled-bridge for softening nonlinear piezoelectric energy harvester. In: *2019 20th International Conference on Solid-State Sensors, Actuators and Microsystems & Eurosensors XXXIII (TRANSDUCERS & EUROSENSORS XXXIII)*, 1435–1438. https://doi.org/10.1109/TRANSDUCERS.2019.8808441.

6

Forced-Vibration Nonlinear Piezoelectric Energy Harvesters

6.1 Introduction

The previous chapter illustrated the experimental study of the broadband characteristics of the free-vibration type flexible buckled piezoelectric energy harvester and completed the evaluation of its performance through a laboratory-based test, which also verified the potential of the flexible buckled piezoelectric structure prepared by mechanical-thinning technology for efficient energy harvesting. This chapter focuses on the application study of flexible buckled piezoelectric energy harvesting technology based on forced excitation, such as the rotational motion environment of automotive wheels, and investigates the self-powered systems.

6.2 Forced-Vibration Nonlinear Mechanism

A schematic diagram for the theoretical analysis model of the piezoelectric buckled bridge under external loading force [1] is shown in Figure 6.1. The loading force is applied at the center of the buckled bridge, and the upper and lower electrode layers of the piezoelectric layer are led through external wires to collect the electrical charge and form a closed circuit with an external resistor R, as appeared in Figure 6.1a. The piezoelectric buckled bridge consists of a polyethylene terephthalate (PET) polymer support layer and a piezoelectric unimorph sheet layer polarized along the Z-axis as shown in Figure 6.1b.

Based on the theoretical analysis of the previously mentioned free-vibration energy harvester based on flexible buckled bridge, the piezoelectric instanton equation is shown as follows:

$$\begin{cases} T_x = c_{xx}^E S_x - e_{zx} E_z \\ D_z = e_{zx} S_x + \varepsilon_{zz}^S E_z \end{cases} \tag{6.1}$$

where T_x, S_x is the mechanical stress and strain tensor, c_x^E is the elastic modulus of the piezoelectric layer under a constant electric field, ε_{zz}^S is the dielectric constant of the piezoelectric material under strain-free conditions, e_{zx} is the piezoelectric constant, E_z is the electric field component in the z-axis direction, and D_Z is the potential shift along the z-axis direction.

Flexible Piezoelectric Energy Harvesters and Sensors, First Edition. Bin Yang, Zhiran Yi, and Chengkuo Lee.

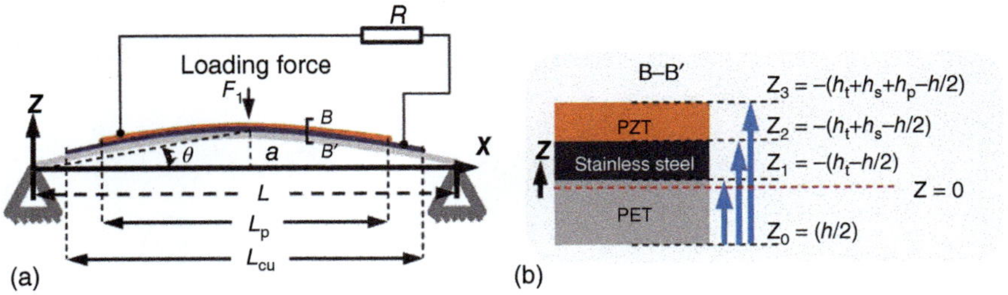

Figure 6.1 Illustration schematic of the piezoelectric buckled bridge for theory analysis. (a) Coordinate system for the simplified structural model. (b) Layer numbering used for this laminated structure. Source: Yi et al. [1] / with permission of IEEE.

The total potential energy of the system is:

$$U = a^4U_1 + a^3U_2 + a^2U_3 + a^2v(t)U_4 + av(t)U_5 \tag{6.2}$$

where,

$$U_1 = \frac{1}{8}\int_0^L \left(E_{\text{PET}}A_{\text{PET}} + E_{\text{Steel}}A_{\text{Steel}} + c_{11}^E A_{\text{p}}\right) \psi'^4 \text{d}x$$

$$U_2 = -\int_0^L \left(E_{\text{PET}}H_{\text{PET}} + E_{\text{Steel}}H_{\text{Steel}} + c_{11}^E H_{\text{p}}\right) \psi'^2 \psi'' \text{d}x$$

$$U_3 = \frac{1}{2}\int_0^L \left[\left(E_{\text{PET}}I_{\text{PET}} + E_{\text{Steel}}I_{\text{Steel}} + c_{11}^E I_{\text{p}}\right) (\psi'')^2\right] \text{d}x$$

$$U_4 = \frac{1}{4}\int_0^L [B_{\text{p}}\psi'^2]\,\text{d}x, \qquad U_5 = -\frac{1}{2}\int_0^L J_{\text{p}}\psi''\text{d}x$$

Partial differential results for the total potential energy:

$$\frac{\partial U}{\partial a_i} = 4a^3U_1 + 3a^2U_2 + 2aU_3 + 2avU_4 + vU_5 \tag{6.3}$$

$$\frac{\partial U}{\partial v_i} = a^2U_4 + aU_5 \tag{6.4}$$

where L is the length of the arrangement along the x-axis, ρ is the material density, b is the width of the buckled bridge, and in addition

$$(A_{\text{PET}}, H_{\text{PET}}, I_{\text{PET}}) = \iint_S (1, z, z^2)\text{d}y\text{d}z$$
$$= \left[bt_{\text{PET}}, \frac{1}{2}b\left(Z_1^2 - Z_0^2\right), \frac{1}{3}b\left(Z_1^3 - Z_0^3\right)\right]$$

$$(\text{A}_{\text{Steel}}, H_{\text{Steel}}, I_{\text{Steel}}) = \iint_S (1, z, z^2)\text{d}y\text{d}z$$
$$= \left[bt_{\text{Steel}}, \frac{1}{2}b\left(Z_2^2 - Z_1^2\right), \frac{1}{3}b\left(Z_2^3 - Z_1^3\right)\right]$$

$$(A_{\text{p}}, H_{\text{p}}, I_{\text{p}}) = \iint_S (1, z, z^2)\text{d}ydz = \left[bt_{\text{p}}, \frac{1}{2}b\left(Z_3^2 - Z_2^2\right), \frac{1}{3}b\left(Z_3^3 - Z_2^3\right)\right]$$

The total kinetic energy of the system is

$$K = (\dot{a})^2 T_1 \tag{6.5}$$

where

$$T_1 = \frac{1}{2}\int_0^L \left[\left(\rho_{\mathrm{PET}}A_{\mathrm{PET}} + \rho_{\mathrm{Steel}}A_{\mathrm{Steel}} + \rho_{\mathrm{p}}A_{\mathrm{p}}\right)(\psi)^2\right] \mathrm{d}x,\ \frac{\partial K}{\partial \dot{a}_i} = 2T_1\dot{a},\ \frac{\partial K}{\partial a_i} = T_2,\ \frac{\partial K}{\partial \dot{v}_i} = 0,$$
$$\frac{\partial T}{\partial v_i} = 0,\ \frac{d}{dt}\left(\frac{\partial K}{\partial \dot{a}_i}\right) = 2\ddot{a}T_1,\ \frac{d}{dt}\left(\frac{\partial K}{\partial \dot{v}_i}\right) = 0.$$

Electrical energy inside the piezoelectric layer:

$$W_{ie} = a^2 v(t) W_1 + a v(t) W_2 + \frac{1}{2}C_{\mathrm{p}}v^2(t) \tag{6.6}$$

where

$$W_1 = -\frac{1}{2}\int_0^L B_{\mathrm{p}}\psi'^2 \mathrm{d}x = -U_4,$$
$$W_2 = \frac{1}{2}\int_0^L J_{\mathrm{p}}\psi'' \mathrm{d}x = -U_5,$$
$$\frac{\partial W_{ie}}{\partial a_i} = 2avW_1 + vW_2,$$
$$\frac{\partial W_{ie}}{\partial v_i} = a^2W_1 + aW_2 + C_{\mathrm{p}}v.$$

where $B_{\mathrm{p}} = \iint_p \frac{e_{31}}{h_{\mathrm{p}}}\mathrm{d}y\mathrm{d}z$, $J_{\mathrm{p}} = \iint_p \frac{e_{31}}{h_{\mathrm{p}}}z\mathrm{d}y\mathrm{d}z$.

According to the reported theoretical model about the buckled-bridge type vibration energy harvester, the transverse strain component [2, 3] along the x-direction is: $S_{xx}(x,z,t) = \frac{1}{2}\left(\frac{\partial w^0(x,t)}{\partial x}\right)^2 - z\frac{\partial^2 w^0(x,t)}{\partial x^2}$. For the substrate, stress tensor can be expressed as $T_{xx}(x, z, t) = E_s S_{xx}(x, z, t)$ and the piezoelectric layer stress tensor can be expressed as:

$$T_{xx}(x,z,t) = c_{11}^E S_{xx}(x,z,t) + e_{31}\frac{v(t)}{h_{\mathrm{p}}} \tag{6.7}$$

where $E_3(t) = -v(t)/h_{\mathrm{p}}$, $v(t)$ is the instantaneous voltage between the two electrodes of the piezoelectric layer, and h_{p} is the thickness of the piezoelectric layer. Therefore, the total potential energy of the system can be expressed as:

$$\begin{aligned} U = {} & \frac{1}{2}\int_{V_{\mathrm{PET}}} E_{\mathrm{PET}}\left[\frac{1}{2}\left(\frac{\partial w^0(x,t)}{\partial x}\right)^2 - z\frac{\partial^2 w^0(x,t)}{\partial x^2}\right]^2 \mathrm{d}V_{\mathrm{PET}} \\ & + \frac{1}{2}\int_{V_{\mathrm{Steel}}} E_s\left[\frac{1}{2}\left(\frac{\partial w^0(x,t)}{\partial x}\right)^2 - z\frac{\partial^2 w^0(x,t)}{\partial x^2}\right]^2 \mathrm{d}V_{\mathrm{Steel}} \\ & + \frac{1}{2}\int_{V_{\mathrm{p}}} \left\{ c_{11}^E\left[\frac{1}{2}\left(\frac{\partial w^0(x,t)}{\partial x}\right)^2 - z\frac{\partial^2 w^0(x,t)}{\partial x^2}\right]^2 \right. \\ & \left. + e_{31}\frac{v(t)}{h_{\mathrm{p}}}\left[\frac{1}{2}\left(\frac{\partial w^0(x,t)}{\partial x}\right)^2 - z\frac{\partial^2 w^0(x,t)}{\partial x^2}\right]\right\} \mathrm{d}V_{\mathrm{p}} \end{aligned} \tag{6.8}$$

The total kinetic energy can be expressed as:

$$K = \frac{1}{2}\int_{V_{\mathrm{PET}}} \rho_{\mathrm{PET}}\left(\frac{\partial w^0(x,t)}{\partial t}\right)^2 \mathrm{d}V_{\mathrm{PET}} + \frac{1}{2}\int_{V_{\mathrm{Steel}}} \rho_{\mathrm{Steel}}\left(\frac{\partial w^0(x,t)}{\partial t}\right)^2 \mathrm{d}V_{\mathrm{Steel}} + \frac{1}{2}\int_{V_{\mathrm{p}}} \rho_{\mathrm{p}}\left(\frac{\partial w^0(x,t)}{\partial t}\right)^2 \mathrm{d}V_{\mathrm{p}} \tag{6.9}$$

The internal electrical energy of the piezoelectric layer is:

$$W_{ie} = \frac{1}{2}\int_{V_{\mathrm{p}}} \boldsymbol{E}^t\boldsymbol{D}\mathrm{d}V_{\mathrm{p}} = -\frac{1}{2}\int_{V_{\mathrm{p}}} \frac{v(t)}{h_{\mathrm{p}}}\left\{e_{31}\left[\frac{1}{2}\left(\frac{\partial w^0(x,t)}{\partial x}\right)^2 - z\frac{\partial^2 w^0(x,t)}{\partial x^2}\right]\right\}\mathrm{d}V_{\mathrm{p}} + \frac{1}{2}C_{\mathrm{p}}v^2(t) \tag{6.10}$$

From the extended Hamilton's principle:

$$\int_{t_1}^{t_2} (\delta K - \delta U + \delta W_{ie} + \delta W_{\mathrm{virtual}})\mathrm{d}t = 0 \tag{6.11}$$

where the virtual work is mainly the work done by the charge, $\delta W_{\mathrm{virtual}} = Q(t)\delta v(t)$.

In turn, the Lagrange equation for electromechanical coupling can be derived as follows:

$$\frac{\mathrm{d}}{\mathrm{d}t}\left(\frac{\partial K}{\partial \dot{v}_i}\right) - \frac{\partial K}{\partial v_i} + \frac{\partial U}{\partial v_i} - \frac{\partial W_{ie}}{\partial v_i} = Q \tag{6.12}$$

Thus, $a^2U_4 + aU_5 - a^2W_1 - aW_2 - C_{\mathrm{p}}v = Q$.

Taking the derivative of time for this equation yields: $4a\dot{a}U_4 + 2\dot{a}U_5 - C_{\mathrm{p}}\dot{v} = \frac{v}{R}$.

It is known that when the time is zero, the voltage output is also zero, i.e. $v_{t=0} = 0$, so the instantaneous voltage output from the device can be expressed as:

$$v = \mathrm{e}^{-\frac{t}{C_{\mathrm{p}}R}}\left(\int_0^t \frac{4a\dot{a}U_4 + 2\dot{a}U_5}{C_{\mathrm{p}}}\mathrm{e}^{\frac{t}{C_{\mathrm{p}}R}}\mathrm{d}t\right) \tag{6.13}$$

The perturbation displacement of the neutral layer can be expressed as:

$$w(x,t) = \psi(x)\cdot a(t) = \frac{1-\cos\left(\frac{2\pi x}{L}\right)}{2} \times \begin{cases} \sigma\sin(k\pi f t), & \frac{n}{f} \le t \le \frac{1}{kf} + \frac{n}{f} \\ 0, & \left(1-\frac{1}{k}\right)\frac{1}{f} + \frac{n}{f} \le t \le \frac{1}{f} + \frac{n}{f} \end{cases} \tag{6.14}$$

where σ is the amplitude of extrusion deformation borne by the buckled bridge, it is known that the design radius of the hemispherical tooth and the deformation amplitude generated by the buckled bridge are equal, f is the frequency of the applied force, n is a natural number, k is the influence factor of the gear, and the number of gear design, so the output instantaneous power can be expressed as

$$P = \frac{1}{16}\frac{\pi^6 R(fbkh_{\mathrm{p}}e_{31})^2\sigma^4}{L^2\left[(2\pi LRfk\varepsilon_0\varepsilon_{33}b)^2 + h_{\mathrm{p}}{}^2\right]^2}\left(\mathrm{e}^{-\frac{h_{\mathrm{p}}t}{b\varepsilon_0\varepsilon_{33}LR}}\right)^2$$

$$\times \left\{ e^{\frac{h_p t}{b\varepsilon_0\varepsilon_{33}LR}} [2\pi LRfk\varepsilon_0\varepsilon_{33}b\cos(2k\pi ft) - h_p \sin(2k\pi ft)] - 2\pi LRfk\varepsilon_0\varepsilon_{33}b \right\}^2 \tag{6.15}$$

The output instantaneous power is a periodic function of time, defined $H = (2\pi LRk\varepsilon_0\varepsilon_{33}b)^2$, the compression displacement σ is 0.5 mm, the influence factor $k = 1.25$ for the gear (12 teeth), and the amplitude of the power can be expressed as (Figure 6.2a)

$$P_{\text{peak}} \propto \frac{f^4}{\left(Hf^2 + {h_p}^2\right)^2} \tag{6.16}$$

In general, the tire rotation frequency in normal driving of most motor vehicles is less than 20 Hz. As shown in Figure 6.2b, the frequency point where the maximum output power increases fastest is defined as f_0. If the frequency is less than f_0, the output power is low and dependent on frequency, and conversely, the dependence of the output power on frequency will be relatively weaker in the high-frequency range. Figures 6.2c and d give the calculated instantaneous output voltage and the corresponding output power waveforms at 100 Hz. With this dependence in mind, the gear design provides the ability to convert from low to high frequency for a buckled-bridge-type piezoelectric rotary motion energy harvester with a frequency applied to the device that is twelve times the actual rotational frequency, allowing the device to attenuate the frequency dependence of the output power in ultra-low frequency applications.

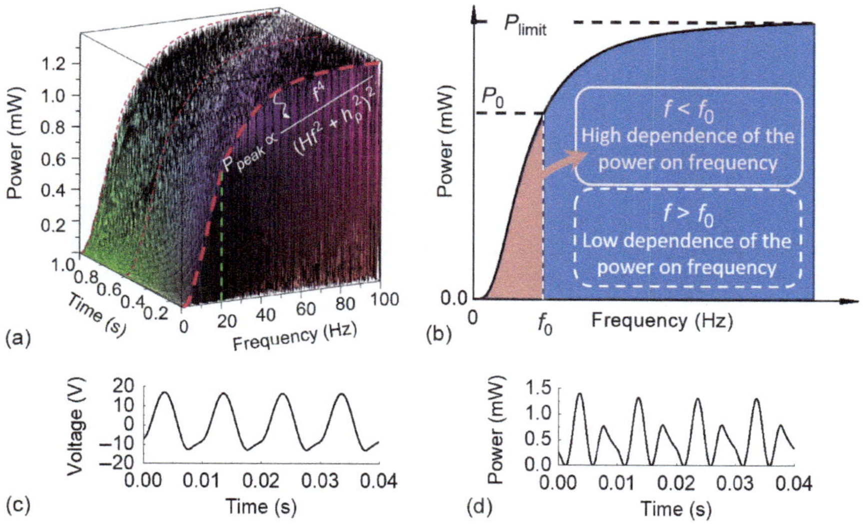

Figure 6.2 Theory analysis. (a) The power as a function of the applied frequency. (b) Dependence of the power on the applied frequency. (c) The calculated output voltage as a function of the time at 100 Hz. (d) The corresponding calculated output power as a function of the time. Source: Yi et al. [1] / with permission of IEEE.

6.3 Self-Powered Tire-Pressure-Monitoring Systems

More than 1.2 million people die on the road and as many as 50 million others are injured each year worldwide [4], more than half of which are caused by traffic accidents due to tire problems. Currently, tire pressure monitoring has become the most effective way to prevent tire problems, integrated sensor module, microprogram control unit, communication module, and battery tire pressure monitoring systems (TPMS) emerged for real-time assessment of tire safety is very necessary to ensure the safety of driving. In recent years, lots of countries have successively included TPMS in the strict regulatory framework to improve road safety, and it is foreseeable that TPMS will be more widely used in vehicles in the future. However, unfortunately, all existing TPMS are powered by batteries, and due to the limited lifetime of chemical batteries, the extensive use of battery-powered TPMS will bring new safety risks and ongoing environmental pollution problems, so it is important to develop a TPMS with sustainable power supply capability to avoid these problems. Among them, energy harvesting technology is expected to achieve this goal. The power consumption of the most used TPMS chips in the market is usually 0.45 mW on average to achieve a data transfer rate of 1 Hz, with an operating voltage of about 1.8–3.6 V. The RF transmitter consumes the most current, more than 1 mA, and a typical battery capacity of 220–600 mAh can sustain normal operation for about 5–10 years [1].

Research on self-propelled TPMS combining mechanical kinetic-energy-harvesting techniques has been ongoing for more than a decade [1, 5]. The available kinetic energy at the wheel mainly consists of rotational kinetic energy of the tire and cyclic strain kinetic energy. Therefore, the development of a self-propelled TPMS by converting the two available energies has been a hot research topic in the field of energy harvesting and faces great challenges and opportunities. Depending on the conversion mechanism, the generally reported energy-conversion methods for self-propelled TPMS mainly include electromagnetic [6–12], electrostatic [13–15], triboelectric electric [16–18], piezoelectric [19–34], or multi-mechanism hybrid methods [35]. Among these research methods, piezoelectric methods are widely used for kinetic energy harvesters due to their high power density and simple construction [36–43]. The current piezoelectric strain energy harvester can achieve a maximum output power of 6.5 mW [19, 20] with a matching resistance of 42 kΩ at 10 km/h. The harvester can power an autonomously developed TPMS for pressure sensing and data transmission every 2.5 seconds at 60 km/h, but due to overload or unexpected shocks, the attached tire, however, the PZT attached to the tread wall is easily damaged due to overload or accidental impact, which may cause some problems, such as air leakage and tire imbalance, and so far, the common commercial TPMS on the market is difficult to power effectively.

Figure 6.3 shows a three-dimensional model of the presented piezoelectric rotary motion energy harvester with the proposed self-propelled TPMS designed to be installed in the positions shown in Figures 6.3b and c. The radius of the designed hemispherical teeth on the gear is equal to the compression displacement provided to the piezoelectric buckled bridge, which is sufficient to induce the buckled bridge

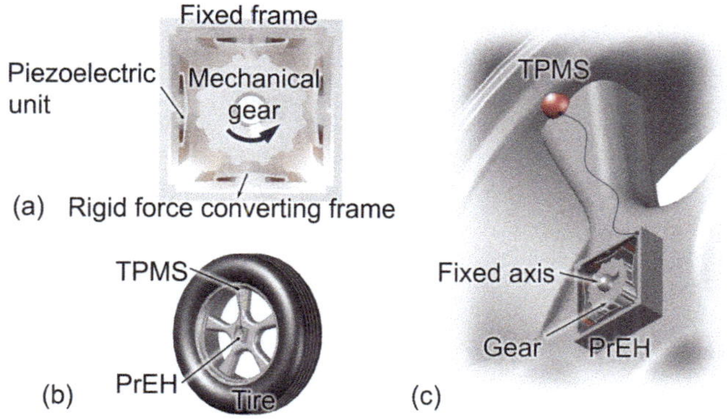

Figure 6.3 Schematic illustration for the working principle of the presented buckled-bridges-based piezoelectric rotational kinetic energy harvester (PrEH). (a) Three-dimensional model of the assembled PrEH. (b) The mounting position of the presented batteryless TPMS. (c) The enlarged image. Source: From Yi et al. [1] / with permission of IEEE.

to remain in a high output state. According to the study of the buckled-bridge-type vibration energy harvester [44, 45], when there is a relative motion between the mechanical gear and the fixed frame, each piezoelectric unit will be compressed by the hemispherical teeth of the gear through the rigid force conversion mechanism.

Figure 6.4 shows the circuit to evaluate the practical availability of the output electrical energy of the buckled-bridge-type piezoelectric rotary motion energy harvester. Among all, a 330 μF capacitor was used to store the converted electrical

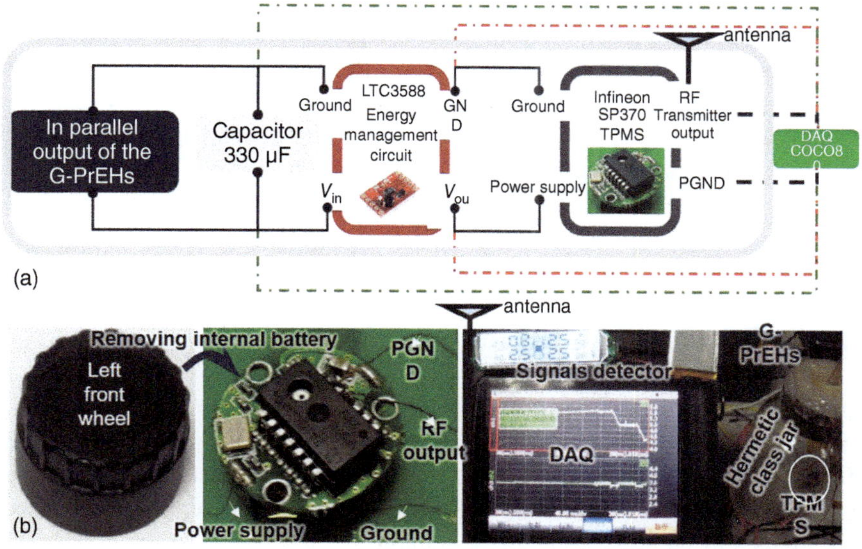

Figure 6.4 Connected circuit for the powering TPMS by the PrEH. (a) The diagram of power supply connection circuit for batteryless TPMS. (b) A commercial outward attaching TPMS removing the battery and the photograph of the power supply connection circuit. Source: Yi et al. [1] / with permission of IEEE.

Table 6.1 The comparison between the presented PEH and other reported PEHs for the batteryless TPMS.

Volume/mm^3	Excited condition	Output performance	Powering the TPMS	References
196	80 rpm	6.5 mW	60 km/h TI = 1 s	Makki et al. [20]
6	80 km/h	78 μW	No	Keck et al. [21]
—	10 km/h	10 μW	60 km/h TI > 10 s	Roundy et al. [46]
20	80.5 km/h	47 μW	No	
—	70 km/h	30 μW	No	Schaijk et al. [26]
300	60 km/h	380 μJ/cycle	No	Lee et al. [27]
120	215 rpm	36 μW	No	Wu et al. [28]
23	5–9.2 Hz	12 μW	No	Zhang et al. [30]
32	20 km/h	0.38 μJ/cycle	No	Eshghi et al. [31]
72	8.3 Hz	8.9 mW	~60 km/h TI = 0 s	Yi et al. [1]

TI: Transmission interval time.

energy, and then a commercial energy management module (LTC-3588, Linear Technologies, Inc., United States) was used to convert the capacitor output to a stable DC output of 3.3 V. Finally, the stable output was connected to a commercial TPMS module with the battery removed. The TPMS module contains an SP370 integrated circuit chip with a microcontroller unit, temperature sensor, barometric pressure sensor, and acceleration sensor inside. The energy flow throughout the process is monitored in real-time by a dynamic data collector, and Figure 6.4 shows a photograph of the connection status of the components in the experiment.

When the applied rotation frequency is not less than 8.3 Hz, this commercial TPMS can be driven by this piezoelectric energy harvester in real time to achieve a self-propelled TPMS. the output performance of this energy harvester is listed in Table 6.1 with some reported studies of energy harvesters used to achieve de-battery of TPMS, illustrating the superior characteristics of the proposed buckled-bridge-type piezoelectric rotary motion energy harvester.

6.4 Self-Powered 5G NB-IoT System

As fifth-generation communication and cloud technology are gradually maturing, numerous wireless electronic devices have access to cloud server in the new era of artificial intelligence (AIoT), which allow the users to better acquire and analyze big data [47–50]. It is predicted by IHS Markit that the number of smart device connections will be approximately 125 billion by 2030 [51]. Currently, wireless monitoring system integrating different types of sensors and electronic devices with communication capacity is widely distributed in various application scenarios ranging from smart city, intelligent agriculture, healthcare to environmental detection [52–62]. However, the power supply of these systems mainly relies on batteries. Considering their abundant distribution in remote area or even poor accessible environments, such as forests or farmlands, conventional batteries always fail to support long-term operation of these system due to their limited lifespan and requirement for replacement or recharging frequently [37, 63–65]. Besides, the battery can result in environmental pollution [66]. Therefore, these significant challenges indicate that a sustainable and maintenance-free remote monitoring system is highly desired [67–70].

Researchers have focused on looking for renewable energies in the ambient environment, such as solar, water, and wind. Particularly, wind is considered to be one of significant green energy due to its abundant and ubiquitous existence in remote and harsh environments. However, the traditional harvesters of wind energy have disadvantages of large volume, high cost, and complex installation, which limit the application of self-powered systems. To this end, great efforts have focused on miniaturized harvesters to convert mechanical energy to electrical power based on different mechanisms, such as electromagnetic [71, 72], piezoelectric [1, 73–76], triboelectric [54, 77–79], and hybrid approaches [80–85]. Compared with individual mechanisms, the hybrid harvesters in a single device have been substantiated that can harvest the same kind of mechanical energy simultaneously to improve the energy-conversion efficiency [86]. Nevertheless, due to the abrasions of material or structure, which induced heat dissipation and device failure [87–89], especially for the long-time high-speed rotation movement, implementing hybridized energy harvesters more efficiently and reliably is still a challenge.

Self-powered wireless monitoring systems are vital for the Internet of Thing (IoT) applications in remote areas. Although there is still a large gap between the output power of energy harvesters and power consumption of systems, real-time operation is not necessary in most cases and harvesters can charge up a capacitor with enough energy to power them. Self-powered wireless systems based on Bluetooth (1–20 m) [90, 91], Wi-Fi (<100 m) [92–94], and Zigbee (<300 m) [95, 96] are intensively reported. However, these short-range communication technologies cannot meet the requirement of long-range wireless transmission in remote areas. Although 3G, 4G, and other solutions based on cellular networks can provide broad coverage, they consume too much electric energy so energy harvesters have pretty difficulties in powering them [97].

To overcome aforementioned obstacles, low-power wide-area network (LPWAN) technologies have been proposed and are becoming a promising alternative to realize sustainable IoT due to the distinct features of long coverage and low power consumption [98]. To trade off these two factors, one emerging technology is narrowband IoT(NB-IoT)defined by the Third Generation Partnership Project (3GPP) that enhanced signal-to-noise ratio by focusing signals in a narrow bandwidth [99, 100]. 3GPP 5G technology with the NB-IoT included has been approved as 5G standard and NB-IoT becomes a crucial part of 5G massive Machine Type Communication. Furthermore, 3GPP R16 standards enable each NB-IoT terminal to access 5G core network, which will provide an important opportunity for NB-IoT ecology. It can thus be expected that, with the introduction of NB-IoT, self-sustained remote monitoring system will be achieved toward practical applications.

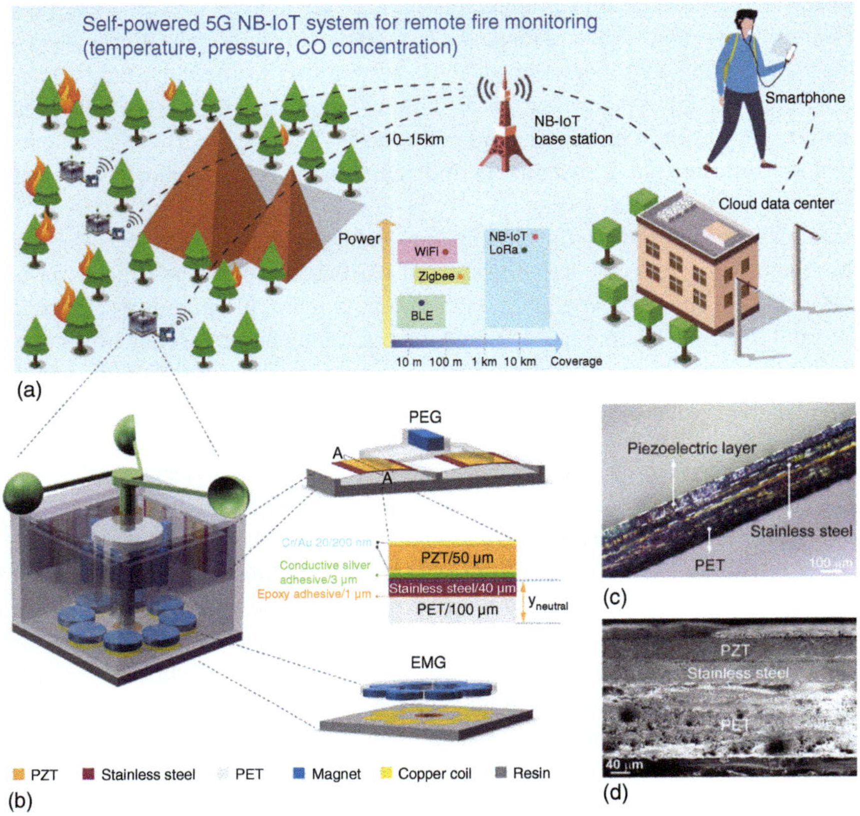

Figure 6.5 The self-powered 5G NB-IoT system driven by the MN-EH for remote fire monitoring. (a) Conceptual diagram of the self-powered remote wireless fire monitoring system. (b) Schematic illustration of the MN-EH including PEG and EMG. The Inset shows the position (A-A′) of the cross-section for buckled piezoelectric bridge. (c), (d) The Cross-sectional microscope image (c) and SEM image (d) with different functional layers. Source: From Hu et al. [101] / with permission of Elsevier.

Figure 6.5 depicts an application scenario of the self-powered remote wireless fire monitoring system, the architecture of which mainly consists of the MN-EH, physical sensors, communication module, NB-IoT base station, cloud data center, and mobile terminal. The electrical power is generated by the MN-EH and stored in a supercapacitor. When the fire happens, the NB-IoT system allows the sensor data to be remotely transmitted to the cloud server through NB-IoT base station and the mobile phone can acquire the real-time sensing information, which contributes to early warning for decreasing the loss of natural resources and human-living environment. NB-IoT provides connectivity to massive IoT devices that require low power and long-range communication. The performance comparison of some common wireless communication technologies is included in Table 6.1.

Figure 6.5b shows a durable magnetic-assisted noncontact energy harvester (MN-EH) scavenging wind energy to release a self-powered 5G NB-IoT system for remote fire monitoring. The hybrid MN-EH consisting of piezoelectric harvester

Table 6.2 Performance comparison of several common wireless communication technologies.

	Bluetooth	Wi-Fi	Zigbee	Lora	NB-IoT
Type		Wireless Local Area Network (WLAN)		Lower Power Wide Area Network (LPWAN)	
				Noncellular	Cellular
Spectrum	2.4 GHz IEEE 802.15.1	2.4 GHz IEEE 802.11	2.4 GHz IEEE 802.15.4	Unlicensed 433/868/915 MHz	Licensed LTE in-bands, GSM standalone
Bandwidth	1 MHz	20 MHz	2 MHz	125/250/500 kHz	180 kHz
Data rate	1 Mbps	150 Mbps	170 kbps	<250 kbps	<50 kbps
Coverage area	1–20 m	100 m	50–300 m	<10 km 154 dB	<15 km 164 dB

Source: [98, 102].

and electromagnetic harvester (EMG), delivers the capability to operate over a broad range of frequency for different wind speeds. The piezoelectric harvester is developed by using thinned-piezoelectric thick- film-based composite structure as shown in Figure 6.5. Figure 6.5 shows the image, which clearly illustrates the detailed multilayer structure, obtained by scanning electron microscopy. With the aid of the magnetic coupling mechanism, the noncontact mechanical movement can protect the device from degradation and abrasion to improve the operation lifespan. The rotational peak torque decreases by 93% through optimizing the magnetic arrangement. Furthermore, the MN-EH is demonstrated for powering the programmable NB-IoT system that can transmit the temperature and CO concentration to cloud server in the practical application. To our best knowledge, this represents the first demonstration of a wind-driven battery-less fully self-powered 5G NB-IoT system for remote fire monitoring (Tables 6.2 and 6.3).

6.5 Summary

Flexible buckled piezoelectric vibration energy-harvesting and conversion mechanism under low-frequency-forced-vibration excitation conditions are explored through the modification of the free-vibration-type flexible piezoelectric energy-harvesting and conversion model, and the high dependence characteristic of its output power on the applied frequency is found. It is proposed that the conversion of rotation frequency is carried out through a rotating gear to convert the low-frequency rotational motion into higher-frequency forced-vibration and reduce the frequency dependence.

TPMS have been included in the strict regulatory framework for automobiles in many countries, but commercial TPMS must rely on batteries with a limited life span to supply energy, thus posing some safety risks and environmental concerns. Notably, energy harvesting methods are expected to enable self-propelled TPMS, and

Table 6.3 Review of the proposed rotational energy harvesters for IoT applications.

Reference	[84]	[4]	[5]	[6]	[101]
Energy source	Rotation	Rotation	Rotation	Wind	Wind
Mechanism	TENG, EMG	PENG, TENG	TENG, EMG	TENG, EMG	PEG, EMG
Structure	Stator and rotator, contact	Stator and rotator, contact	Stator and rotator, contact	Stator and rotator, contact	Stator and rotator, noncontact
Volume	Cylinder ——	Cylinder ——	Cylinder ——	Cuboid $150 \times 100 \times 67\ mm^3$	Cuboid $65 \times 65 \times 50\ mm^3$
Average output power	——	100 rpm 10.88 mW	1200 rpm TENG: 2.13 mW EMG: 6.048 mW (Peak power)	9 m/s TENG: 0.7 mW EMG: 37 mW (Peak power)	8 m/s PEG: 1.4 mW EMG: 23.3 mW
Average power density	1000 rpm TENG: 438.9 mW/kg EMG: 181.9 mW/kg (Peak power density)	$6.04\ mW/cm^2$	——	TENG: $0.69\ \mu W/cm^3$ EMG: $36.82\ \mu W/cm^3$ (Peak power density)	PEG: $6.63\ \mu W/cm^3$ EMG: $110.3\ \mu W/cm^3$
Application of device	Environmental monitoring	Portable electronics	Disaster monitoring	Environmental monitoring	Remote fire monitoring
Demonstration of the sensor	Humidity and temperature sensor	Multi-functional meter, calculator	Temperature and acceleration sensor	Humidity and temperature sensor	CO sensor
Demonstration of the wireless system	No	RF-wireless	RF-wireless	Bluetooth	NB-IoT

low-frequency energy harvesting for self-propelled TPMS has wide implications in smart tires, energy harvesting, and the Internet of Things. In this chapter, we introduced a flexible buckled-based piezoelectric energy-conversion mechanism [1] to realize the kinetic energy harvesting of automotive wheel rotation and study its application in the power supply of commercial TPMS. The details are as follows:

This chapter also introduced a wind-driven self-powered 5G NB-IoT system [101] for remote fire monitoring and warning. The rotational MN-EH device, which hybridizes piezoelectric and electromagnetic harvesters can efficiently convert wind energy into electrical power. More importantly, the MN-EH has demonstrated to power a programmable battery-less 5G NB-IoT system that can acquire temperature and CO concentration. This research [101] provides an attractive strategy for realizing a self-powered remote monitoring system and exhibits great prospects in the applications of smart city, intelligent agriculture, and environmental monitoring in remote areas.

References

1 Yi, Z.R., Yang, B., Zhang, W.M. et al. (2021). Batteryless tire pressure real-time monitoring system driven by an ultralow frequency piezoelectric rotational energy harvester. *IEEE Transactions on Industrial Electronics* 68 (4): 3192–3201.

2 Reddy, J.N. (2003). *Mechanics of Laminated Composite Plates and Shells: Theory and Analysis*, 858. CRC Press.

3 Cottone, F., Gammaitoni, L., Vocca, H. et al. (2012). Piezoelectric buckled beams for random vibration energy harvesting. *Smart Materials and Structures* 21: 035021.

4 World Health Organization, Department of Violence, Injury Prevention, World Health Organization. Violence, Injury Prevention, & World Health Organization (2009). *Global Status Report on Road Safety: Time for Action*. World Health Organization.

5 Löhndorf, M., Kvisterøy, T., Westby, E. et al. (2007). Evaluation of energy harvesting concepts for tire pressure monitoring systems. *Proceedings of Power MEMS* 331–334.

6 Hatipoglu, G. and Urey, H. (2010). FR4-based electromagnetic energy harvester for wireless sensor nodes. *Smart Materials and Structures* 19: 015022.

7 Wang, Y.-J., Chen, C.-D., and Sung, C.-K. (2010). Design of a frequency-adjusting device for harvesting energy from a rotating wheel. *Sensors and Actuators A: Physical* 159: 196–203.

8 Kyoung Il, L., Byung Jik, L., Seong Hyun, K. et al. (2012). Energy harvesting by rotation of wheel for tire monitoring system. In: *2012 IEEE Sensors*, 4.

9 Lee, S. and Kim, D.-H. (2013). Durable and sustainable strap type electromagnetic harvester for tire pressure monitoring system. *Journal of Magnetics* 18: 473–480.

10 Wang, Q., Zhang, Y., Sun, N.X. et al. High power density energy harvester with high permeability magnetic material embedded in a rotating wheel.

In: *Proceedings of SPIE 8347, Nondestructive Characterization for Composite Materials, Aerospace Engineering, Civil Infrastructure, and Homeland Security 2012, SPIE Smart Structures and Materials + Nondestructive Evaluation and Health Monitoring, 2012, San Diego, California, United States*, 83470V, vol. 8347. https://doi.org/10.1117/12.916980.

11 Liu, H., Wang, S., Zhang, Y., and Wang, W. (2015). Study on the giant magnetostrictive vibration-power generation method for battery-less tire pressure monitoring system. *Proceedings of the Institution of Mechanical Engineers Part C: Journal of Mechanical Engineering Science* 229: 1639–1651.

12 Wang, Y.-J., Chen, C.-D., Lin, C.-C., and Yu, J.-H. (2015). A nonlinear suspended energy harvester for a tire pressure monitoring system. *Micromachines* 6: 312–327.

13 Westby, E.R. and Halvorsen, E. (2012). Design and modeling of a patterned-electret-based energy harvester for tire pressure monitoring systems. *IEEE/ASME Transactions on Mechatronics* 17: 995–1005.

14 Renaud, M., Fujita, T., Goedbloed, M. et al. (2015). Shock reliability analysis and improvement of MEMS electret-based vibration energy harvesters. *Journal of Micromechanics and Microengineering* 25: 104010.

15 Naito, Y. and Uenishi, K. (2019). Electrostatic MEMS vibration energy harvesters inside of tire treads. *Sensors* 19: 890.

16 Zhang, H.L. et al. (2014). Single-electrode-based rotating triboelectric nanogenerator for harvesting energy from tires. *ACS Nano* 8: 680–689.

17 Guo, T. et al. (2018). Compressible hexagonal-structured triboelectric nanogenerators for harvesting tire rotation energy. *Extreme Mechanics Letters* 18: 1–8.

18 Qian, J., Kim, D.-S., and Lee, D.-W. (2018). On-vehicle triboelectric nanogenerator enabled self-powered sensor for tire pressure monitoring. *Nano Energy* 49: 126–136.

19 Makki, N. and Pop-Iliev, R. (2011). Piezoelectric power generation for sensor applications: design of a battery-less wireless tire pressure sensor. In: *Proceedings of SPIE 8066, Smart Sensors, Actuators, and MEMS, SPIE Microtechnologies, 2011, Prague, Czech Republic*, V, 806618, vol. 8066. https://doi.org/10.1117/12.887112.

20 Makki, N. and Pop-Iliev, R. (2012). Battery-and wire-less tire pressure measurement systems (TPMS) sensor. *Microsystem Technologies* 18: 1201–1212.

21 Keck, M. A new approach of a piezoelectric vibration-based power generator to supply next generation tire sensor systems," SENSORS, 2007 IEEE, Atlanta, GA, USA, 28–31 October 2007, 2007, pp. 1299–1302, doi: https://doi.org/10.1109/ICSENS.2007.4388648.

22 Gu, L. and Livermore, C. (2010). Passive self-tuning energy harvester for extracting energy from rotational motion. *Applied Physics Letters* 97: 081904.

23 Singh, K.B., Bedekar, V., Taheri, S., and Priya, S. (2012). Piezoelectric vibration energy harvesting system with an adaptive frequency tuning mechanism for intelligent tires. *Mechatronics* 22: 970–988.

24 Zimmermann, T., Frey, A., Schreiter, M. et al. (2012) In MEMS-based piezoelectric energy harvesting modules for distributed automotive tire sensors, International Multi-Conference on Systems, Signals & Devices, IEEE: 2012; pp 1–4.

25 Suciu, C.V. and Koyanagi, K. (2013). Evaluation of the energy harvestable from an airless tire equipped with piezoelectric bimorphs on the lamellar spokes. *Journal of Telecommunications and Information Technology* 4: 79–84.

26 van Schaijk, R. et al. (2013). Smart sensors, actuators, and MEMS VI. In: *Proceedings of SPIE*, vol. 8763 (ed. U. Schmid, J. Aldavero and M. LeesterSchaedel), 876305.

27 Lee, J. and Choi, B. (2014). Development of a piezoelectric energy harvesting system for implementing wireless sensors on the tires. *Energy Conversion and Management* 78: 32–38.

28 Wu, X., Parmar, M., and Lee, D.-W. (2014). A seesaw-structured energy harvester with superwide bandwidth for TPMS application. *IEEE/ASME Transactions on Mechatronics* 19: 1514–1522.

29 Lee, J., Oh, J., Kim, H., and Choi, B. (2015). Strain-based piezoelectric energy harvesting for wireless sensor systems in a tire. *Journal of Intelligent Material Systems and Structures* 26: 1404–1416.

30 Zhang, Y., Zheng, R., Shimono, K. et al. (2016). Effectiveness testing of a piezoelectric energy harvester for an automobile wheel using stochastic resonance. *Sensors* 16: 1727.

31 Eshghi, A.T. et al. (2017). Design optimization under uncertainty and speed variability for a piezoelectric energy harvester powering a tire pressure monitoring sensor. *Smart Materials and Structures* 26: 105037.

32 Aliniagerdroudbari, H.; Esmaeeli, R.; Hashemi, S. R., et al. (2019) In A piezoelectric sandwich structure for harvesting energy from tire strain to power up intelligent tire sensors, 2019 IEEE Power and Energy Conference at Illinois (PECI), IEEE; pp 1–7.

33 Esmaeeli, R. et al. (2019). A rainbow piezoelectric energy harvesting system for intelligent tire monitoring applications. *Journal of Energy Resources Technology-Transactions of the ASME* 141: 062007.

34 Hu, Y.F. et al. (2011). A nanogenerator for energy harvesting from a rotating tire and its application as a self-powered pressure/speed sensor. *Advanced Materials* 23: 4068–4071.

35 Du, X. et al. (2018). Hybridized nanogenerators for harvesting vibrational energy by triboelectric-piezoelectric-electromagnetic effects. *Advanced Materials Technologies* 3: 1800019.

36 Bowen, C.R. and Arafa, M.H. (2015). Energy harvesting technologies for tire pressure monitoring systems. *Advanced Energy Materials* 5: 1401787.

37 Yang, Z., Zhou, S., Zu, J., and Inman, D. (2018). High-performance piezoelectric energy harvesters and their applications. *Joule* 2: 642–697.

38 Zhao, L. et al. (2019). A water-proof magnetically coupled piezoelectric-electromagnetic hybrid wind energy harvester. *Applied Energy* 239: 735–746.

39 Yi, Z. et al. (2017). High performance bimorph piezoelectric MEMS harvester via bulk PZT thick films on thin beryllium-bronze substrate. *Applied Physics Letters* 111: 013902.

40 Yi, Z., Hu, Y., Ji, B. et al. (2018). Broad bandwidth piezoelectric energy harvester by a flexible buckled bridge. *Applied Physics Letters* 113: 183901.

41 Liu, H., Zhong, J., Lee, C. et al. (2018). A comprehensive review on piezoelectric energy harvesting technology: materials, mechanisms, and applications. *Applied Physics Reviews* 5: 041306.

42 Tan, T., Yan, Z., and Hajj, M. (2016). Electromechanical decoupled model for cantilever-beam piezoelectric energy harvesters. *Applied Physics Letters* 109: 79–82.

43 Fu, H. and Yeatman, E.M. (2019). Rotational energy harvesting using bi-stability and frequency up-conversion for low-power sensing applications: theoretical modelling and experimental validation. *Mechanical Systems and Signal Processing* 125: 229–244.

44 Udani, J.P. and Arrieta, A.F. (2017). Sustaining high-energy orbits of bi-stable energy harvesters by attractor selection. *Applied Physics Letters* 111: 213901.

45 Fu, H., Sharif-Khodaei, Z., and Aliabadi, F. (2019). A bio-inspired host-parasite structure for broadband vibration energy harvesting from low-frequency random sources. *Applied Physics Letters* 114: 143901.

46 Roundy, S. and Tola, J. (2014). Energy harvester for rotating environments using offset pendulum and nonlinear dynamics. *Smart Materials and Structures* 23: 105004.

47 Zhang, X. et al. (2019). Two-dimensional MoS_2-enabled flexible rectenna for Wi-Fi-band wireless energy harvesting. *Nature* 566: 368–372.

48 Luo, J. et al. (2019). Flexible and durable wood-based triboelectric nanogenerators for self-powered sensing in athletic big data analytics. *Nature Communications* 10: 5147.

49 Chettri, L. and Bera, R. (2020). A comprehensive survey on internet of things (IoT) toward 5G wireless systems. *IEEE Internet of Things Journal* 7: 16–32.

50 Niu, S.M. et al. (2019). A wireless body area sensor network based on stretchable passive tags. *Nature Electronics* 2: 361–368.

51 Sharma, S.K. and Wang, X. (2020). Toward massive machine type communications in ultra-dense cellular IoT networks: current issues and machine learning-assisted solutions. *IEEE Communications Surveys & Tutorials* 22: 426–471.

52 Zhou, Z. et al. (2018). Wireless self-powered sensor networks driven by triboelectric nanogenerator for in-situ real time survey of environmental monitoring. *Nano Energy* 53: 501–507.

53 Pang, Y.K. et al. (2020). Multilayered cylindrical triboelectric nanogenerator to harvest kinetic energy of tree branches for monitoring environment condition and forest fire. *Advanced Functional Materials* 30: 8.

54 Xu, L.Y., Xu, L., Luo, J.J. et al. (2020). Hybrid all-in-one power source based on high-performance spherical triboelectric nanogenerators for harvesting environmental energy. *Advanced Energy Materials* 10 (36): 2001669.

55 Maurya, D. et al. (2020). 3D printed graphene-based self-powered strain sensors for smart tires in autonomous vehicles. *Nature Communications* 11: 5392.

56 Ma, L. et al. (2020). A machine-fabricated 3D honeycomb-structured flame-retardant triboelectric fabric for fire escape and rescue. *Advanced Materials* 32: e2003897.

57 Lee, B. et al. (2020). High-performance compliant thermoelectric generators with magnetically self-assembled soft heat conductors for self-powered wearable electronics. *Nature Communications* 11: 5948.

58 Shi, Q. et al. (2020). Deep learning enabled smart mats as a scalable floor monitoring system. *Nature Communications* 11: 4609.

59 Jin, T. et al. (2020). Triboelectric nanogenerator sensors for soft robotics aiming at digital twin applications. *Nature Communications* 11: 5381.

60 Zhu, M., Yi, Z., Yang, B. et al. (2021). Making use of nanoenergy from human – Nanogenerator and self-powered sensor enabled sustainable wireless IoT sensory systems. *Nano Today* 36: 101016.

61 Yi, Z., Xie, F., Tian, Y. et al. (2020). A battery- and leadless heart-worn pacemaker strategy. *Advanced Functional Materials* 30: 2000477.

62 Li, N. et al. (2019). Direct powering a real cardiac pacemaker by natural energy of a heartbeat. *ACS Nano* 13: 2822–2830.

63 Zhong, T. et al. (2019). A self-powered flexibly-arranged gas monitoring system with evaporating rainwater as fuel for building atmosphere big data. *Nano Energy* 60: 52–60.

64 Wang, L., He, T.Y.Y., Zhang, Z.X. et al. (2021). Self-sustained autonomous wireless sensing based on a hybridized TENG and PEG vibration mechanism. *Nano Energy* 80: 105555.

65 Shen, Q., Xie, X.K., Peng, M.F. et al. (2018). Self-powered vehicle emission testing system based on coupling of triboelectric and chemoresistive effects. *Advanced Functional Materials* 28 (10): 1703420.

66 Razmjoo, A. et al. (2021). A technical analysis investigating energy sustainability utilizing reliable renewable energy sources to reduce CO_2 emissions in a high potential area. *Renewable Energy* 164: 46–57.

67 Ryu, H., Yoon, H.J., and Kim, S.W. (2019). Hybrid energy harvesters: toward sustainable energy harvesting. *Advanced Materials* 31: e1802898.

68 Chen, X., Ren, Z., Han, M. et al. (2020). Hybrid energy cells based on triboelectric nanogenerator: from principle to system. *Nano Energy* 75.

69 Zhang, Q. et al. (2021). Shadow enhanced self-charging power system for wave and solar energy harvesting from the ocean. *Nature Communications* 12: 616.

70 Zou, Y. et al. (2019). A bionic stretchable nanogenerator for underwater sensing and energy harvesting. *Nature Communications* 10: 2695.

71 Zhang, Y.L. et al. (2020). Rotational electromagnetic energy harvester for human motion application at low frequency. *Applied Physics Letters* 116: 5.

72 Liu, H.C. et al. (2018). A non-resonant rotational electromagnetic energy harvester for low-frequency and irregular human motion. *Applied Physics Letters* 113: 5.

73 Kan, J. et al. (2016). Study on a piezo-windmill for energy harvesting. *Renewable Energy* 97: 210–217.

74 Rashidi, R., Summerville, N., and Nasri, M. (2019). Magnetically actuated piezoelectric-based rotational energy harvester with enhanced output in wide range of rotating speeds. *IEEE Transactions on Magnetics* 55: 1–8.

75 Yang, B., Yi, Z.R., Tang, G., and Liu, J.Q. (2019). A gullwing-structured piezoelectric rotational energy harvester for low frequency energy scavenging. *Applied Physics Letters* 115: 5.

76 Hwang, G.-T. et al. (2016). Self-powered wireless sensor node enabled by an aerosol-deposited PZT flexible energy harvester. *Advanced Energy Materials* 6.

77 Huang, L.-b. et al. (2016). Wind energy and blue energy harvesting based on magnetic-assisted noncontact triboelectric nanogenerator. *Nano Energy* 30: 36–42.

78 Zhu, G., Chen, J., Zhang, T. et al. (2014). Radial-arrayed rotary electrification for high performance triboelectric generator. *Nature Communications* 5: 3426.

79 Ronan Hinchet, H.-J.Y., Ryu, H., Kim, M.-k. et al. (2019). Transcutaneous ultrasound energy harvesting using capacitive triboelectric technology. *Science* 365: 491–494.

80 Zhao, C. et al. (2019). Hybrid piezo/triboelectric nanogenerator for highly efficient and stable rotation energy harvesting. *Nano Energy* 57: 440–449.

81 Qian, J. and Jing, X. (2018). Wind-driven hybridized triboelectric-electromagnetic nanogenerator and solar cell as a sustainable power unit for self-powered natural disaster monitoring sensor networks. *Nano Energy* 52: 78–87.

82 Zhang, B.B. et al. (2016). Rotating-disk-based hybridized electromagnetic-triboelectric nanogenerator for sustainably powering wireless traffic volume sensors. *ACS Nano* 10: 6241–6247.

83 Fan, X. et al. (2020). Triboelectric-electromagnetic hybrid nanogenerator driven by wind for self-powered wireless transmission in internet of things and self-powered wind speed sensor. *Nano Energy* 68.

84 Wang, P. et al. (2018). An ultra-low-friction triboelectric-electromagnetic hybrid nanogenerator for rotation energy harvesting and self-powered wind speed sensor. *ACS Nano* 12: 9433–9440.

85 Cao, R. et al. (2017). Rotating-sleeve triboelectric-electromagnetic hybrid nanogenerator for high efficiency of harvesting mechanical energy. *ACS Nano* 11: 8370–8378.

86 Zhao, Z. et al. (2020). Tumbler-shaped hybrid triboelectric nanogenerators for amphibious self-powered environmental monitoring. *Nano Energy* 76.

87 Wang, X., Wang, Z.L., and Yang, Y. (2016). Hybridized nanogenerator for simultaneously scavenging mechanical and thermal energies by electromagnetic-triboelectric-thermoelectric effects. *Nano Energy* 26: 164–171.

88 Zhao, L.-C. et al. (2019). A water-proof magnetically coupled piezoelectric-electromagnetic hybrid wind energy harvester. *Applied Energy* 239: 735–746.

89 Guo, H., Wen, Z., Zi, Y.L. et al. (2016). A water-proof triboelectric-electromagnetic hybrid generator for energy harvesting in harsh environments. *Advanced Energy Materials* 6 (6): 1501593.

90 Kim, C.Y. et al. (2021). Soft subdermal implant capable of wireless battery charging and programmable controls for applications in optogenetics. *Nature Communications* 12: 535.

91 Maharjan, P. et al. (2020). A fully functional universal self-chargeable power module for portable/wearable electronics and self-powered IoT applications. *Advanced Energy Materials* 10.

92 Xu, X. et al. (2018). A real-time wearable UV-radiation monitor based on a high-performance p-CuZnS/n-TiO_2 photodetector. *Advanced Materials* 30: e1803165.

93 Zhang, N. et al. (2020). Imperceptible sleep monitoring bedding for remote sleep healthcare and early disease diagnosis. *Nano Energy* 72.

94 Yichen Cai, J.S., Yang, C.-W., Wan, Y. et al. (2020). Mixed-dimensional MXene-hydrogel heterostructures for electronic skin sensors with ultrabroad working range. *Science Advances* 6: eabb5367.

95 Zhao, X. et al. (2017). Self-powered triboelectric nano vibration accelerometer based wireless sensor system for railway state health monitoring. *Nano Energy* 34: 549–555.

96 Xianjie, P., Guo, H., Chen, J. et al. (2017). Eye motion triggered self-powered mechnosensational communication system using triboelectric nanogenerator. *Science Advances* 6: eabb5367.

97 Zhang, X. et al. (2019). A low-power wide-area network information monitoring system by combining NB-IoT and LoRa. *IEEE Internet of Things Journal* 6: 590–598.

98 Qin, Z., Li, F.Y., Li, G.Y. et al. (2019). Low-power wide-area networks for sustainable IoT. *IEEE Wireless Communications* 26: 140–145.

99 Yang, W. et al. (2017). Narrowband wireless access for low-power massive internet of things: a bandwidth perspective. *IEEE Wireless Communications* 24: 138–145.

100 Beyene, Y.D. et al. (2017). NB-IoT technology overview and experience from cloud-ran implementation. *IEEE Wireless Communications* 24: 26–32.

101 Hu, G. et al. (2021). Self-powered 5G NB-IoT system for remote monitoring applications. *Nano Energy* 87: 106140.

102 Migabo, E.M., Djouani, K.D., and Kurien, A.M. (2020). The narrowband internet of things (NB-IoT) resources management performance state of art, challenges, and opportunities. *IEEE Access* 8: 97658–97675.

7

Fluid-Induced Piezoelectric Energy Harvesters

7.1 Introduction

Piezoelectric fluid vibration energy-harvesting technology can convert fluid kinetic energy, such as ocean energy, ocean current energy, tidal energy, river current energy, airflow energy, and wind energy, into electrical energy. Among them, the low-velocity constant flow environment in nature and industry has great value and significance for the research work of piezoelectric fluid vibration energy harvester because of its abundant source channels, constant environmental parameters, easy installation and arrangement, and the need to monitor the environmental conditions by wireless sensors.

Fluid-induced mechanical structures that produce regular coupling motion can be used as the source of piezoelectric fluid vibration energy harvester. Figure 7.1 depicts the detailed categories of fluid-induced motion, according to the form of motion it can be divided into two categories, which are rotary motion type and reciprocating vibration type [1, 2], reciprocating vibration type can be subdivided into resonance type and instability type, the former mainly contains rotational oscillation, vortex vibration, and jitter vibration, the latter mainly contains galloping and flutter [3]. According to the characteristics of various types of fluid-induced motion, different types of piezoelectric fluid vibration energy harvesters can be developed.

7.2 Types of Flow-Induced PEH

7.2.1 Rotating-Type PEH

Rotating structure [4] is a typical design solution for piezoelectric fluid vibration energy harvesters, inspired by windmills and turbines. The rotating piezoelectric fluid vibration energy harvester usually consists of a blade, a rotating disk, an actuator, and a piezoelectric element. It is based on the principle that when the fluid flow rate exceeds the minimum starting flow rate of the system, the fluid drives the blade and the rotating disk to rotate, and the actuator on the rotating disk excites the piezoelectric element to vibrate by magnetic force or toggling, thus realizing fluid-energy-harvesting and conversion. Depending on whether the actuator is

Flexible Piezoelectric Energy Harvesters and Sensors, First Edition. Bin Yang, Zhiran Yi, and Chengkuo Lee.

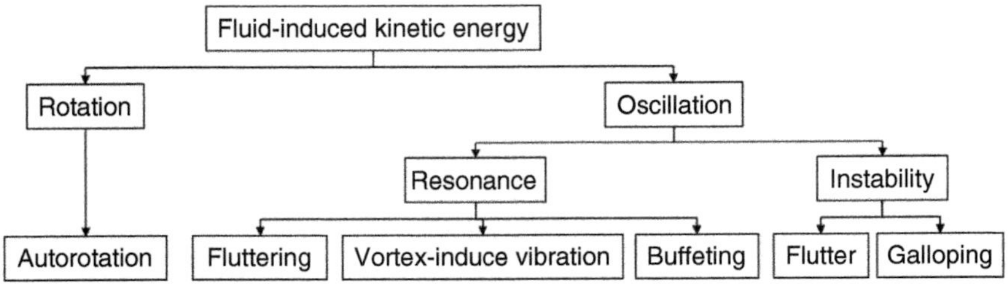

Figure 7.1 Classification of flow-induced motions.

in contact with the piezoelectric element or not, the rotating piezoelectric fluid vibration energy harvester can be divided into contact type [5] and noncontact type.

Impact-induced resonance was always used to enable effective excitation of the piezoelectric cantilevers' vibration modes and obtain optimum deformation for improving the output power of a rotational piezoelectric wind energy harvester [6], as shown in Figure 7.2a. In this device, the impact force is introduced by forming a piezoelectric bimorph cantilever polygon that is fixed at the circumference of the rotating fan's internal surface. Elastic balls are placed inside the polygon. When wind rotates the device, the balls strike the piezoelectric cantilevers, and thus electricity is generated by the piezoelectric effect. The impact point is carefully chosen to use the first bending mode as much as possible, and thus maximize the harvesting efficiency. The design enables each bimorph to be struck in a similar area and every bimorph is struck in that area at different moments. As a result, a relatively stable output frequency can be obtained. The output frequency can also be changed by choosing different bimorph dimensions, which will also make the device simpler and the costs lower. A prototype piezoelectric energy harvester (PEH) consisting of 12 piezoelectric cantilevers can provide the optimal DC output power 613 μW across the 20 kΩ resistor at a rotation speed of 200 r/min with an inscribed circle diameter of 31 mm [6].

Figure 7.2b shows a design of a magnetic force exciter for a small-scale windmill that vibrates a piezocomposite-generating element to convert wind energy into electrical energy [7]. A small-scale windmill was designed to be sensitive to low-speed wind in urban regions for the purpose of collecting wind energy. The magnetic force exciter consists of exciting magnets attached to the device's input rotor and a secondary magnet fixed at the tip of the piezocomposite-generating element. The piezocomposite-generating element is fixed to a clamp that can be adjusted to slide on the windmill's frame to change the gap between exciting and secondary magnets. Under an applied wind force, the input rotor rotates to create a magnetic force interaction that excites the piezocomposite-generating element. In a battery-charging test, the charging time for a 40 mAh battery was approximately three hours for natural wind in an urban region. Figure 7.2c shows a novel piezoelectric transducer [8], where the rotation of the blades results in large oscillations of piezoelectric beams. The piezoelectric bimorphs are made bistable by incorporation of repelling magnetic force. The magnetic force is due to interaction of permanent magnets at the tip of the beams with permanent magnets rotating

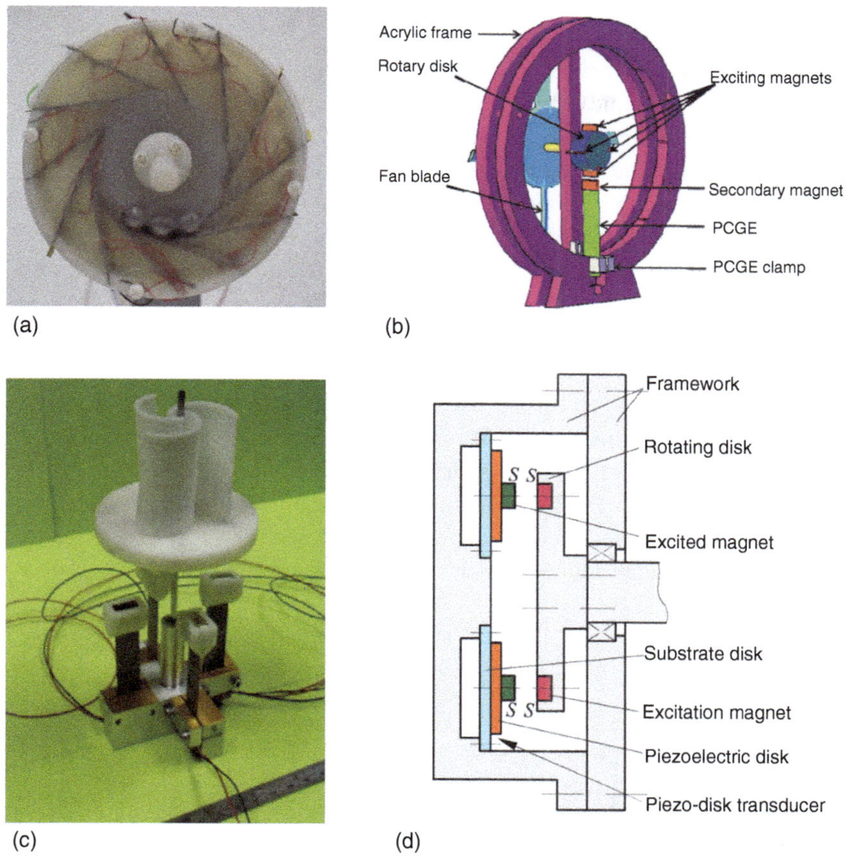

Figure 7.2 (a) Rotational piezoelectric wind energy harvesting using impact-induced resonance. Source: From Zhang et al. [6] / with permission of AIP Publishing, (b) a wind energy harvester using a piezo-composite-generating element. Source: Luong and Goo [7] / with permission of IOP Publishing, (c) a nonlinear piezoelectric wind turbine transducer. Source: From Karami et al. [8] / with permission of Elsevier, (d) a piezo-windmill energy harvester excited by rotary magnets. Source: [9] / with permission of Elsevier. PCGE, piezo-composite generating element.

with the blades. Since the magnetic force changes with blade rotation, the dynamics of the beams changes in time and the system is thus parametrically excited.

Figure 7.2d shows a piezo-disk energy harvester excited by rotary magnets to harvest energy from rotating structures [9]. The excitation force exerted on the piezo disk is generally periodic instead of harmonic. The wave-forms of amplitude-ratio/output-voltage in an excitation period perform multimodal damped oscillations. The maximal amplitude-ratio/output voltage is picked out and used to denote response performance. Under other parameters given, multiple optimal speeds of excited magnets can make the maximal amplitude-ratio/generated voltage to achieve the peak. With the increasing of piezo-disk stiffness denoted by radius and thickness, the optimal rotary speeds increase and the related voltage as well as voltage fluctuation decrease.

7.2.2 Flutter-Type PEH

Flutter is the self-excited vibration of a structure caused by aeroelastic or hydroelastic instabilities [10, 11]. A typical flutter piezoelectric fluid vibration energy harvester consists of a piezoelectric cantilever beam and a wing structure fixed at the end of the beam, where when the fluid flow velocity exceeds the "cut-in velocity," the wing structure is excited by the incoming flow and produces a coupled motion of lifting and sinking When the fluid velocity exceeds the "cut-in velocity," the wing structure is excited by the incoming flow and will produce a coupled motion of lifting and sinking and rocking, thus serving as a source of vibration for the piezoelectric cantilever.

Figure 7.3a shows a novel piezoelectric-energy-harvesting device driven by aeroelastic flutter vibrations of a simple pin-connected flap and beam. The system is subjected to a modal convergence flutter response above a critical wind speed and then oscillates in a limit cycle at higher wind speeds. Figure 7.3b shows a single vibration energy harvester integrated with an airfoil to concurrently harness energy from ambient vibrations and wind. The integrated device is shown to have superior performance under the combined loading when compared to utilizing two separate devices to harvest energy independently from the two available energy sources.

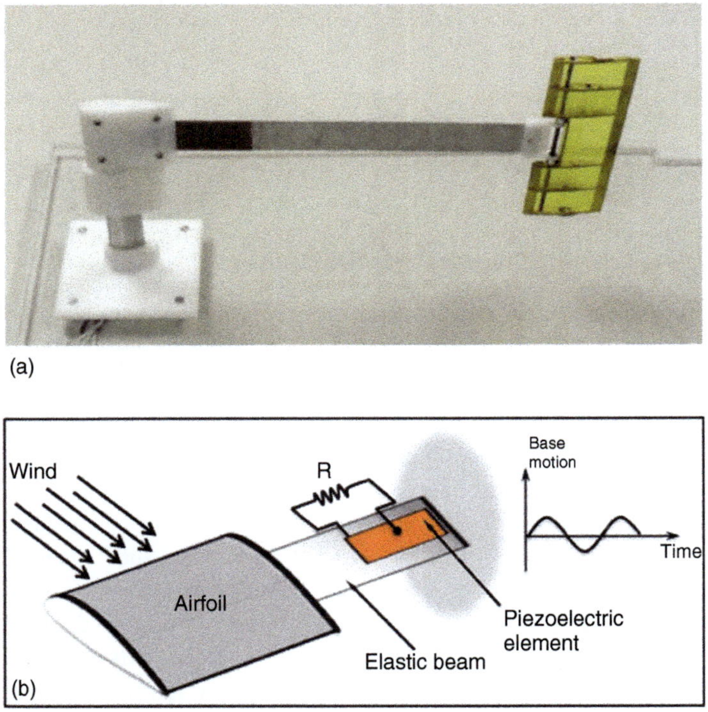

Figure 7.3 (a) An aeroelastic flutter energy harvester. Source: From Refs. [12, 13] / with permission of American Society of Mechanical Engineers (ASME), (b) a piezoelectric energy harvester excited by ambient vibrations and wind. Source: Bibo and Daqaq [14] / with permission of AIP Publishing.

Even below its flutter speed, the proposed device was able to provide 2.5 times the power obtained using two separate harvesters.

7.2.3 Galloping-Type PEH

Galloping is an unstable vibration of an elastomeric structure induced by the presence of a disturbance phenomenon inside the fluid [15], which occurs in the plane perpendicular to the incoming flow direction. A typical chirally vibrating piezoelectric fluid vibration energy harvester consists of a piezoelectric cantilever beam and a prism fixed at the end of the beam, with common prism cross-sectional shapes, such as triangular, square, and D-shaped. When the incoming flow velocity is greater than the tangential velocity, the prism is excited by the incoming flow and produces a single degree of freedom vibration perpendicular to the incoming flow direction, thus becoming the vibration source of the piezoelectric beam. According to the form of motion, the piezoelectric fluid vibration energy harvester can be divided into the translational vibration type and the torsional vibration type.

Figure 7.4a shows a model for harvesting energy from galloping oscillations of a bar with an equilateral triangle cross section attached to two cantilever beams [16, 17]. The energy is harvested by attaching piezoelectric sheets to cantilever beams holding the bar. The power levels that can be generated from these vibrations, and the variations of these levels with the load resistance and wind speed, are determined. Figure 7.4b shows a small-scale wind energy harvester by exploiting the galloping phenomenon of a bluff body attached to a piezoelectric cantilever [18].

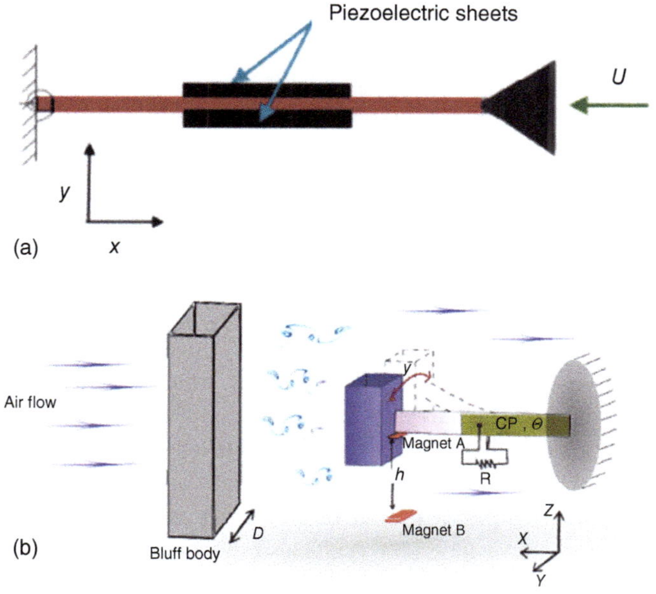

Figure 7.4 (a) A galloping-based piezoaeroelastic energy harvester. Source: Refs. [16, 17] / with permission of IOP Publishing, (b) a broadband bistable flow-energy harvester based on the wake-galloping phenomenon. Source: Alhadidi and Daqaq [19] / with permission of AIP Publishing. CP, composite patch.

Figure 7.4c shows a nonlinear bistable piezoelectric cantilever harvester to harvest wind energy [19]. It is demonstrated that the lock-in region of a wake-galloping flow-energy harvester can be improved by exploiting a bistable restoring force. A von Kármán vortex street is generated by placing a rectangular rod in the windward direction of the harvester and the voltage response of the harvester is evaluated as a function of the wind speed. It is shown that, compared to the linear design, bistability can be used to improve the steady-state bandwidth considerably. Figure 7.4d shows a harvester inspired by the trembling of aspen leaves in barely noticeable winds. The galloping energy harvester, a curved blade oriented perpendicular to the flow, can produce self-sustained oscillations at uncharacteristically low wind speeds. Dissimilar to many other galloping harvester designs, the flow is found to be attached at the rear surface of the blade when the blade is close to its zero displacement position, hence acting more closely to aerofoils rather than to conventionally used bluff bodies.

7.2.4 Vortex-Excited Vibration

Under certain conditions of flow velocity, when the uniform incoming flow in the flow field bypasses a certain obtuse structure, two asymmetric vortex trains with opposite rotational directions will be shed on both sides of it, and the asymmetric periodic flow field pressure distribution generated by the vortex trains will cause the elastic obtuse body itself or other elastomers in the wake domain to vibrate.

Two typical vortex-excited vibration piezoelectric fluid vibration energy harvesters are the de-vortex vibration type and the tail vortex vibration type. In the de-vortex oscillation type, the periodic vortex train shedding from the obtuse body generates a simple harmonic force on the obtuse body perpendicular to the incoming flow direction, which serves as the source of the piezoelectric beam; in the tail vortex oscillation type, the periodic vortex train shedding from the upstream obtuse body induces periodic changes in the flow field pressure on the upper and lower sides of the downstream oscillator, which also generates a simple harmonic force on the oscillator perpendicular to the incoming flow direction, which serves as the source of the piezoelectric beam. The vortex-excited vibrating piezoelectric fluid vibration energy harvester can only utilize the resonance effect in the self-locking region to maximize the energy conversion capability.

Figure 7.5a shows a PEH to concurrently harness energy from base excitations and vortex-induced vibrations [21]. The harvester consists of a multilayered piezoelectric cantilever beam with a circular cylinder tip mass attached to its free end that is placed in a uniform airflow and subjected to direct harmonic excitations. A significant improvement in the level of the harvested power can attain 150% compared to using two separate harvesters. The results also show that an increase in the base acceleration results in a reduction in the vortex-induced vibrations effects, an increase in the difference between the resonant excitation frequency and the pull-out frequency, and a significant effect associated with the quenching phenomenon. Figure 7.5b shows a flexible PEH [22], by using commercially available piezoelectric polymers, polyvinylidene fluoride (PVDF) to convert the

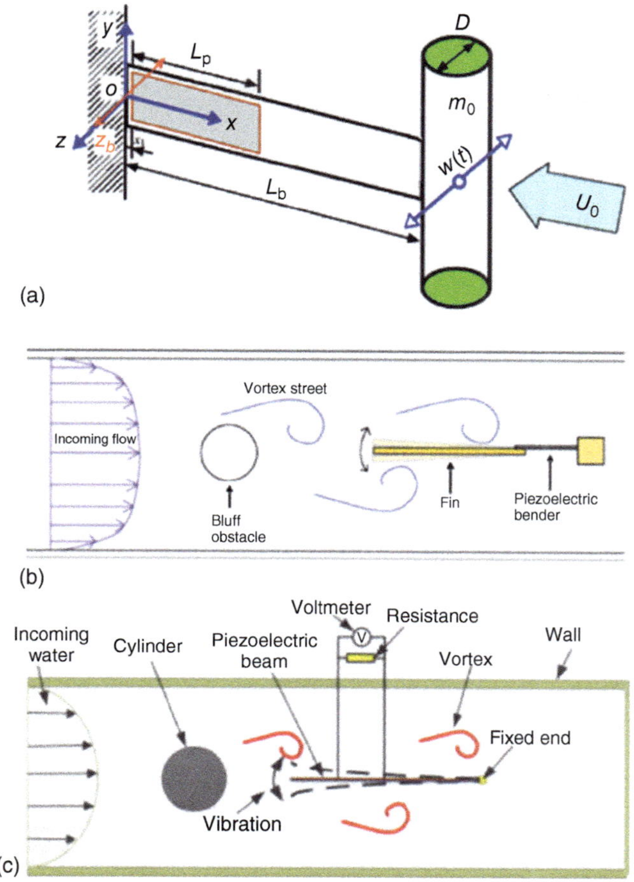

Figure 7.5 (a) Piezoelectric energy harvesting from concurrent-shedding-vortex-induced vibrations and base excitations. Source: Dai et al. [21] / with permission of Springer, (b) a wake vortex piezoelectric energy harvester under flow-induced vibration in heating, ventilation and air conditioning flows. Source: Weinstein et al. [23] / with permission of IOP Publishing, (c) a MFC piezoelectric energy harvester in the water vortex. Source: Shan et al. [24] / with permission of Elsevier. MFC, macro-fiber composite.

mechanical flow energy, available in oceans and rivers, to electrical power. The harvester makes use of the regular trail of traveling vortices behind a bluff body to strain the piezoelectric elements; the resulting undulating motion resembles that of a natural eel swimming. Figure 7.5c shows a cantilevered piezoelectric beam [23] excited in heating, ventilation, and air-conditioning flow. This excitation is amplified by the interactions between an aerodynamic fin attached at the end of the piezoelectric cantilever and the vortex shedding downstream from a bluff body placed in the airflow ahead of the fin/cantilever assembly. In a 15 cm diameter air duct, power generation of 200 μW for a flow speed of 2.5 m/s and power generation of 3 mW for a flow speed of 5 m/s was achieved. Figure 7.5d shows a macrofiber composite PEH [24], which is immersed in the water vortex field shedding from an upstream cylinder. It is found that the water velocity plays a stronger role in the output power than the cylinder diameter does. The output power of the PEH rises with the increase of the water velocity. The harvester achieved a maximum output power of 1.32 μW with a power density of 1.1 mW/m^2 at the water velocity of 0.5 m/s and a cylinder diameter of 30 mm.

7.3 Water Vortex-Shedding-Induced PEH

PEHs from fluid energy sources, particularly vortex-shedding-induced vibrations, have received considerable attention in recent years. However, there still exists a common problem with these harvesters – the current optimum position of a vortex-shedding-induced PEH located in the wake of a bluff body is commonly obtained through a series of experiments. Figure 7.6 shows a vortex-shedding-induced PEH [25]. An optimal ratio of distance between the bluff body and the proof mass to the diameter of bluff body was obtained about at the 0.63–1.87 under the water flow velocity from 0.25 to 0.75 m/s.

Figure 7.7 shows a further study on water-shedding-induced PEH. An E-shaped PEH, as shown in Figure 7.7a, is based on vortex-shedding-induced vibration for achieving broadband and enhanced energy capture from the liquid flow with low velocities. The PEH is realized by introducing two symmetrical vice piezoelectric beams to a traditional structure consisting of a drive sheet and a main piezoelectric beam as shown in comparison of Figure 7.7b,c. By changing the mass blocks on the sheet and vice beams, the first two-order resonance frequencies can be tuned to be close enough to obtain a wide bidirectional tunable operating bandwidth. Experimental results demonstrate that the proposed harvester can adapt to a wider fluid velocity spectrum and bring out higher output performances than the conventional PEH. Under the excitation of vortexes from the liquid flow with low velocities (0.15–0.7 m/s), the maximum increase in power, efficiency, and velocity

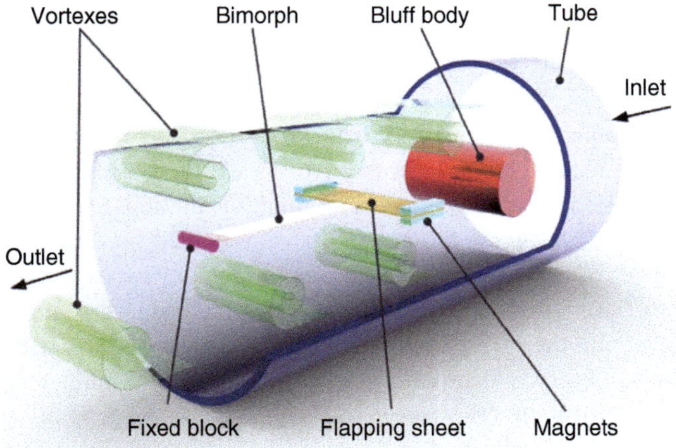

Figure 7.6 Schematic of the proposed vortex-shedding-induced piezoelectric energy harvester. Source: Hu et al. [25] / with permission of Elsevier.

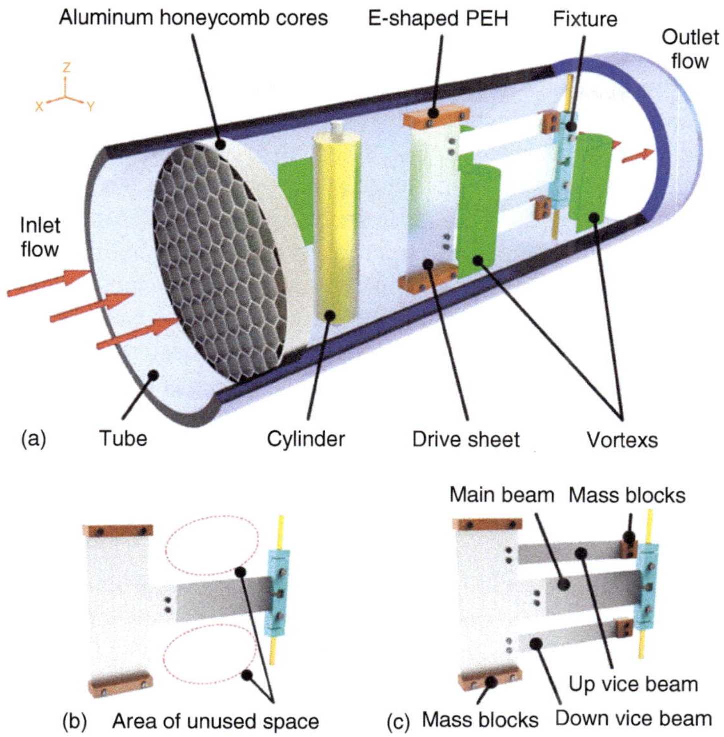

Figure 7.7 A broadband E-shaped piezoelectric energy harvester based on vortex-shedding-induced vibration from low-velocity liquid flow, schematic of (a) the proposed PEH system; (b) a conventional PEH; and (c) an E-shaped PEH. Source: Hu et al. [26] / AIP Publishing / CC BY 4.0.

spectrum over 20 μW can be 70%, 326%, and 60%, respectively, compared to its conventional counterpart.

7.4 Aeroacoustics-Driven Jet-Stream Wind Energy Harvester

Wind, as a common gas fluid, energy harvesting has attracted broad attention from researchers since wind energy is ubiquitous in remote and harsh environments [25, 27–39]. Despite the various transduction mechanisms, such as piezoelectric [40–42], electromagnetic [43], triboelectric [44, 45], and hybrid [46, 47], traditional wind energy harvesters, such as wind turbines, have disadvantages of large volumes and complex installation that limit the application to wireless sensor network (WSN). Miniaturization has become one of the emerging trends over the past few decades [48]. Currently, wind energy harvesters can be classified into two major kinds according to the moving parts – rotary type [27, 30, 49–51] and flow-induced vibrations type [52–57].

Aeroacoustics wind-energy-harvesting technologies started in the middle of last century. Wind can be utilized to induce an acoustic field with steady frequency and large excitation amplitude. This self-sustained oscillation can consequently apply to the conversion systems to power commercial electronic devices. Zou et al. studied a high-performance PEH with a jet-resonator system to harvest jet-stream wind energy [55]. The jet-resonator system was designed to create a dipole acoustic excitation source that is known as edge tones [58]. However, although the aeroacoustics principle has been proved experimentally to be capable of generating stimulus with steady frequency and large amplitude, the study of its fundamental mechanism was still not enough clear [55, 58–64] owing to its complexity. Researchers have mostly focused on the design of edge-tones generation systems, its potential of harvesting jet-stream wind energy was not fully revealed due to the limitation of manufacture and key functional materials.

Figure 7.8 shows the working mechanism of aeroacoustics-driven PEH that can be divided into three stages [65]. In stage 1, as shown in Figure 7.8a (i), the airflow converts into high-speed jet when it is forced through the nozzle. Then, high-speed jet will interact with the ambient slower fluid, causing resistance on the boundary of the high-speed jet. Therefore, it tends to peel back and forms a pair of vortices in opposite direction. The generated vortices subsequently impact on jet stream, exerting a lateral influence on it and producing a relatively weak sound. Then the jet stream and vortices impinge against the sharp edge, causing an increase in pressure at the tip of the edge and generating alternating vortices in opposite directions. The interaction of jet stream with edge subsequently creates an oscillation system, which acts as an acoustic dipole and propagates back to the nozzle region.

In stage 2, as a result of the feedback, the jet stream near nozzle is pushed upward and the one near edge is forced downward, forming an undulation, as shown in Figure 7.8a (ii). The undulation of stream further reinforces the vortices shedding developed at the nozzle and made it match with the oscillation at the edge. Audible edge tones develop as the vortices-shedding frequency near nozzle matches with the ones near the edge. Therefore, there is a critical jet velocity to generate audible pure edge tones. The feedback loop is believed to contribute to the self-sustained oscillation, which holds the pitch of edge tones relatively steady.

In stage 3, as shown in Figure 7.8a (iii), the edge tones travel through the quarter-wavelength resonator and reflect from the piezoelectric circular diaphragm unit, which is the closed-end of the resonator. Then standing wave is formed as a result of interference between waves reflected work and forth in the quarter-wavelength resonator. The standing wave not only applies to the piezoelectric circular diaphragm unit but also feedbacks to the edge and the flow field near nozzle, stabilizing the whole process of self-sustained oscillation. Normally, the frequency of edge tones without resonator is proportional to jet velocity until the mode transition. However, here the frequency of the edge tones is governed principally by the quarter-wavelength resonator. The jet velocity can be varied over a range without hugely affecting the frequency of edge tone because of the quarter-wavelength resonator, facilitating the energy harvesting under varied working situations. In addition, when coupling with the resonator, the amplitude of

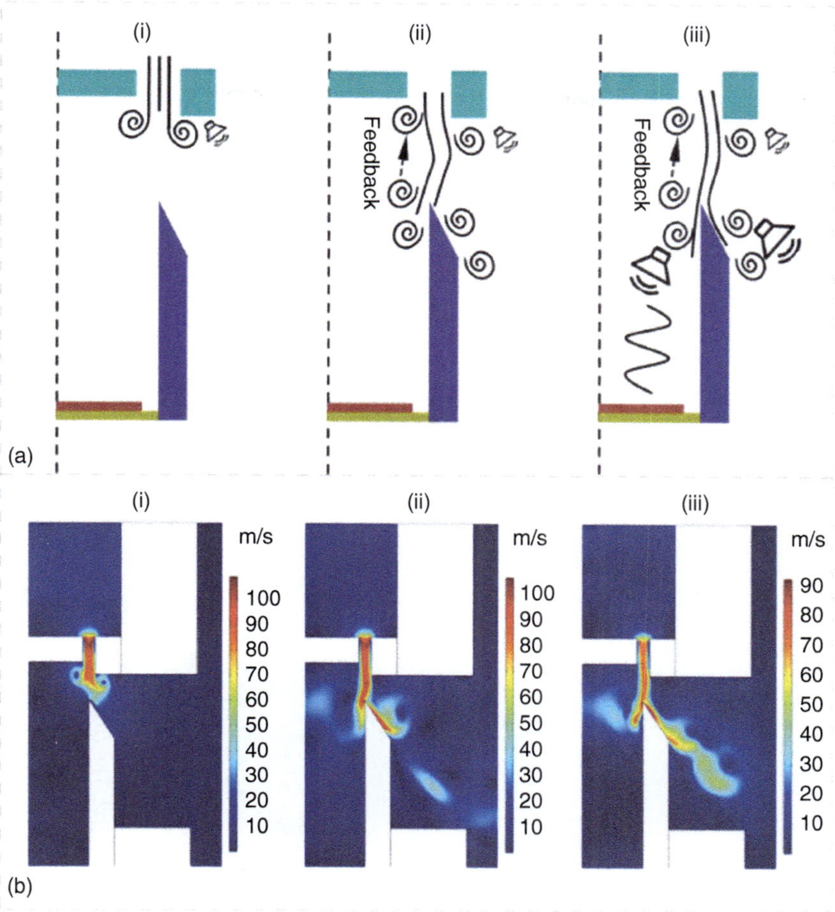

Figure 7.8 Schematic illustration of the working principle of the aeroacoustics-driven piezoelectric energy harvester. (a) Working mechanism of PEH in edge-tones generation (i–iii) and (b) simulated velocity distributions for PEH at three stages by COMSOL employing the finite element method (i–iii). Source: Huang et al. [65] / with permission of Elsevier.

the edge tones will be amplified, which enhances the electrical performance of the PEH. The results of computational fluid dynamics (CFD) simulation presented in Figure 7.8b are consistent with that of Figure 7.8a, consolidating the understanding of the working mechanism of PEH.

In 2021, we reported an aeroacoustics-driven PEH [65], whose design and key fabrication process of piezoelectric unit are shown in Figure 7.9a–c. The PEH is composed of jet-edge-resonator (JER) system and piezoelectric circular diaphragm unit. The piezoelectric circular diaphragm unit is primarily composed of metal substrate and piezoelectric layer. Specifically, the beryllium bronze foil is served as metal substrate for supplying uniform stress distribution to the piezoelectric layer, while the lead zirconate titanate (PZT) piezoelectric thick film (about 50-μm thick, as shown in Figure 7.9d–e) is used for converting mechanical vibration to electrical power.

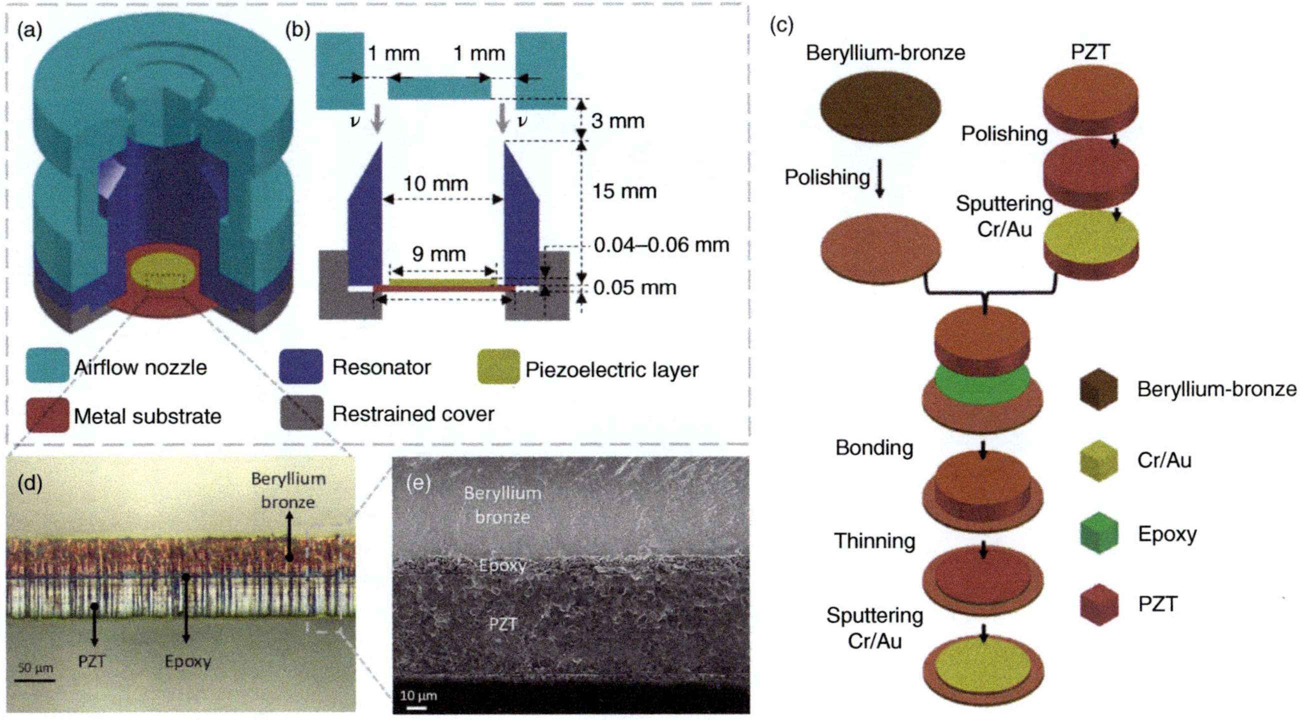

Figure 7.9 Schematic illustration (a) and the cross section (b) of the PEH including jet-edge resonator (JER) and piezoelectric circular diaphragm. (c) The detailed fabrication process of the piezoelectric circular diaphragm unit. The cross-sectional microscope image (d) and scanning electron microscopy (SEM) image (e) with different functional layers for the piezoelectric circular diaphragm unit. Source: (a–c) Huang et al. [65] / with permission of Elsevier, (d, e) From Huang et al. [65] / with permission of Elsevier.

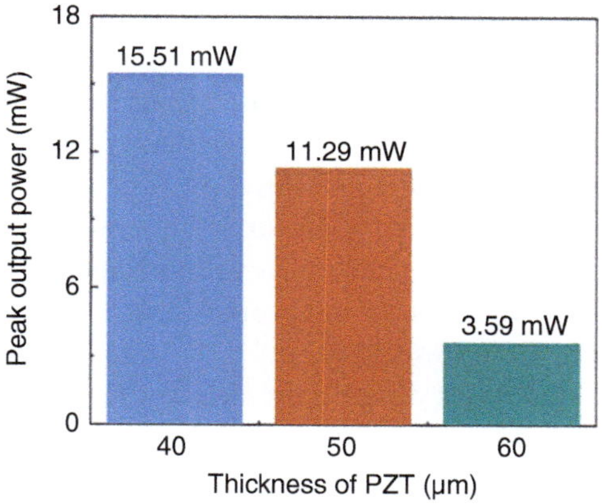

Figure 7.10 Output power of PEH with different thicknesses of PZT. Source: Huang et al. [65] / with permission of Elsevier.

At 79.6 m/s jet velocity, the PEH exhibits high closed-circuit current (4.7 mA) at 1.5 kΩ, open-circuit voltage (12.9 V), and large average output power (15.5 mW). Figure 7.10 shows the output power of PEH with different thicknesses of PZT. The output power was further validated by powering up different commercial electronic devices.

7.5 Summary

Rotating piezoelectric fluid vibration energy-harvester technology is richly accumulated, the number of piezoelectric elements can be set so that the excitation frequency can be adjusted to achieve resonance effect, and the circumferential array arrangement can significantly improve the system performance. However, there are disadvantages of this type of scheme – firstly, the system has a minimum start-up speed, when the flow rate is too low, the turntable cannot be started; secondly, the system structure is large and complex, not suitable for radial space-constrained occasions; finally, the system has a rotating disk optimal speed condition, when the flow rate is too high, the vibration response of piezoelectric elements cannot follow the rapid movement of the rotating disk, the system performance will be significantly reduced.

The flutter piezoelectric fluid vibration energy harvester has the advantages of early technical start, wing-like structure can be borrowed from the experience of aerospace field, simple and compact structure, response under supercritical conditions is not subject to initial conditions and external interference, and high output power in high-flow-velocity environment. The disadvantages are mainly in the following aspects – first, the system has 2 or 3 degrees of freedom of motion, with a wide range of parameters and a complex design process; furthermore, the airfoil-like structure transfers the kinetic energy from the lifting and sinking motion to the piezoelectric beam, while its pitching motion cannot transfer kinetic energy and thus leads to unavoidable energy loss; in addition, the system has a tangential velocity, and the structure cannot generate effective vibration at low-flow velocities.

Galloping-type piezoelectric fluid vibration energy harvester also has two distinctive advantages – the first one is the single-degree-of-freedom vibration system, which is easy to design theoretically and control parametrically; secondly, the vibration response of the system increases significantly with the increase of flow velocity, and the strong correlation between amplitude and flow velocity makes it particularly suitable for high-flow-velocity environments. The disadvantages of this type of energy harvester are also obvious – on the one hand, the amplitude of vibration requires more space for movement, which is inconvenient to be applied in narrow and small tubes with limited space; on the other hand, the system has a large cut-in speed, and cannot produce effective vibration when the flow velocity is below its threshold value.

The vortex vibration piezoelectric fluid vibration energy harvester has attracted a lot of attention in the field of fluid power generation because it can work with the classical "Carmen vortex street" phenomenon, which has the following advantages: First, the system has a single degree of freedom, the design environment is friendly, and the control method is simple; second, the system has no cut-in speed, and can enhance the energy conversion through the resonance effect. Secondly, the system has no cut-in speed and can enhance energy conversion through resonance effect, which is especially suitable for low-flow velocity applications. Thirdly, the structure is simple and compact, requiring little space for movement, and can be arranged in transverse or longitudinal arrays, which is easy to install in space-constrained situations, as well as easy to integrate with the sensing units in the monitoring system. The main problem of this type of system is that it can only achieve resonance and maximize energy-harvesting capability in the self-locking region, which results in a narrow frequency band and is only suitable for constant flow environments; in addition, the output performance in high-flow-velocity environments has a certain gap compared to the previous types of systems.

In terms of size, the overall structure of the rotating system and the blunt-body amplitude of the chirp system are both large and therefore not suitable for applications where space requirements are limited; however, the chirp and vortex vibration systems are relatively simple and compact, making them more suitable for applications in small spaces and easy to miniaturize. From the viewpoint of flow velocity range and output performance, flutter and chirp systems have high output performance at high flow velocities due to the divergence of response curves; rotary systems can perform better at lower flow velocities to meet the minimum start-up speed; vortex vibration systems have no cut-in speed and can use vortex resonance to enhance vibration response at low velocities, so they have better output performance at low flow velocities. From the design and control point of view, the galloping and vortex vibration systems are single-degree-of-freedom vibration systems with environment-friendly design and convenient parameter control.

References

1 Rostami, A.B. and Armandei, M. (2017). Renewable energy harvesting by vortex-induced motions: review and benchmarking of technologies. *Renewable and Sustainable Energy Reviews* 70: 193–214.

2 Abdelkefi, A. (2016). Aeroelastic energy harvesting: a review. *International Journal of Engineering Science* 100: 112–135.

3 Jung, H.-J. and Lee, S.-W. (2011). The experimental validation of a new energy harvesting system based on the wake galloping phenomenon. *Smart Materials and Structures* 20: 055022.

4 Zhou, G., Li, Z., Zhu, Z. et al. (2019). A new piezoelectric bimorph energy harvester based on the vortex-induced-vibration applied in rotational machinery. *IEEE/ASME Transactions on Mechatronics* 24: 700–709.

5 Janphuang, P., Lockhart, R., Isarakorn, D. et al. (2015). Harvesting energy from a rotating gear using an AFM-like MEMS piezoelectric frequency up-converting energy harvester. *Journal of Microelectromechanical Systems* 24: 742–754.

6 Zhang, J., Fang, Z., Shu, C. et al. (2017). A rotational piezoelectric energy harvester for efficient wind energy harvesting. *Sensors and Actuators A: Physical* 262: 123–129.

7 Luong, H.T. and Goo, N.S. (2012). Use of a magnetic force exciter to vibrate a piezocomposite generating element in a small-scale windmill. *Smart Materials and Structures* 21: 025017.

8 Karami, M.A., Farmer, J.R., and Inman, D.J. (2013). Parametrically excited nonlinear piezoelectric compact wind turbine. *Renewable Energy* 50: 977–987.

9 Kan, J., Fu, J., Wang, S. et al. (2017). Study on a piezo-disk energy harvester excited by rotary magnets. *Energy* 122: 62–69.

10 Perez, M., Boisseau, S., Gasnier, P. et al. (2015). An electret-based aeroelastic flutter energy harvester. *Smart Materials and Structures* 24: 035004.

11 Dias, J. A., De Marqui, C., and Erturk, A. Dimensionless analysis and scaling of a hybrid 3DOF airfoil-based piezoelectric-inductive aeroelastic energy harvester, AIAA 2013-1795. *54th AIAA/ASME/ASCE/AHS/ASC Structures, Structural Dynamics, and Materials Conference*, 8–11 April 2013. Boston, Massachusetts. https://doi.org/10.2514/6.2013-1795.

12 Bryant, M. and Garcia, E. (2011). Modeling and testing of a novel aeroelastic flutter energy harvester. *ASME Journal of Vibration and Acoustics* 133 (1).

13 Bryant, M., Wolff, E., and Garcia, E. (2011). Aeroelastic flutter energy harvester design: the sensitivity of the driving instability to system parameters. *Smart Materials and Structures* 20 (12), 125017.

14 Bibo, A. and Daqaq, M.F. (2013). Investigation of concurrent energy harvesting from ambient vibrations and wind using a single piezoelectric generator. *Applied Physics Letters* 102 (24): 243904.

15 Lan, C., Tang, L., Hu, G., and Qin, W. (2019). Dynamics and performance of a two degree-of-freedom galloping-based piezoelectric energy harvester. *Smart Materials and Structures* 28: 045018.

16 Abdelkefi, A., Yan, Z., and Hajj, M.R. (2013). Modeling and nonlinear analysis of piezoelectric energy harvesting from transverse galloping. *Smart Materials and Structures* 22: 025016.

17 Abdelkefi, A., Yan, Z., and Hajj, M.R. (2014). Performance analysis of galloping-based piezoaeroelastic energy harvesters with different cross-section geometries. *Journal of Intelligent Material Systems and Structures* 25: 246–256.

18 Tang, L., Zhao, L., Yang, Y., and Lefeuvre, E. (2015). Equivalent circuit representation and analysis of galloping-based wind energy harvesting. *IEEE/ASME Transactions on Mechatronics* 20: 834–844.

19 Alhadidi, A. and Daqaq, M. (2016). A broadband bi-stable flow energy harvester based on the wake-galloping phenomenon. *Applied Physics Letters* 109: 033904.

20 Tucker Harvey, S., Khovanov, I., and Denissenko, P. (2019). A galloping energy harvester with flow attachment. *Applied Physics Letters* 114: 104103.

21 Dai, H.L., Abdelkefi, A., and Wang, L. (2014). Piezoelectric energy harvesting from concurrent vortex-induced vibrations and base excitations. *Nonlinear Dynamics* 77: 967–981.

22 Taylor, G.W., Burns, J.R., Kammann, S. et al. (2001). The energy harvesting eel: a small subsurface ocean/river power generator. *IEEE Journal of Oceanic Engineering* 26: 539–547.

23 Weinstein, L.A., Cacan, M.R., So, P., and Wright, P. (2012). Vortex shedding induced energy harvesting from piezoelectric materials in heating, ventilation and air conditioning flows. *Smart Materials and Structures* 21: 045003.

24 Shan, X., Song, R., Liu, B., and Xie, T. (2015). Novel energy harvesting: a macro fiber composite piezoelectric energy harvester in the water vortex. *Ceramics International* 41: S763–S767.

25 Hu, Y., Yang, B., Chen, X. et al. (2018). Modeling and experimental study of a piezoelectric energy harvester from vortex shedding-induced vibration. *Energy Conversion and Management* 162: 145–158.

26 Hu, Y., Mou, F., Yang, B. et al. (2018). A broadband E-shaped piezoelectric energy harvester based on vortex-shedding induced vibration from low velocity liquid flow. *AIP Advances* 8 (12): 125214.

27 Fang, Y., Tang, T., Li, Y. et al. (2021). A high-performance triboelectric–electromagnetic hybrid wind energy harvester based on rotational tapered rollers aiming at outdoor IoT applications. *iScience* 24 (4): 102300.

28 Guo, Y., Chen, Y., Ma, J. et al. (2019). Harvesting wind energy: a hybridized design of pinwheel by coupling triboelectrification and electromagnetic induction effects. *Nano Energy* 60: 641–648.

29 Zhao, J., Mu, J., Cui, H. et al. (2021). Hybridized triboelectric–electromagnetic nanogenerator for wind energy harvesting to realize real-time power supply of sensor nodes. *Advanced Materials Technologies* 6 (4): 2001022.

30 Zhao, C., Zhang, Q., Zhang, W. et al. (2019). Hybrid piezo/triboelectric nanogenerator for highly efficient and stable rotation energy harvesting. *Nano Energy* 57: 440–449.

31 Phan, H., Shin, D.M., Jeon, S.H. et al. (2017). Aerodynamic and aeroelastic flutters driven triboelectric nanogenerators for harvesting broadband airflow energy. *Nano Energy* 33: 476–484.

32 Bi, M., Wu, Z., Wang, S. et al. (2020). Optimization of structural parameters for rotary freestanding-electret generators and wind energy harvesting. *Nano Energy* 75: 104968.

33 Zhao, Z., Zhang, Z., Xu, L. et al. (2020). Tumbler-shaped hybrid triboelectric nanogenerators for amphibious self-powered environmental monitoring. *Nano Energy* 76: 104960.

34 Wang, F., Wang, Z., Zhou, Y. et al. (2020). Windmill-inspired hybridized triboelectric nanogenerators integrated with power management circuit for harvesting wind and acoustic energy. *Nano Energy* 78: 105244.

35 Sun, W., Ding, Z., Qin, Z. et al. (2020). Wind energy harvesting based on fluttering double-flag type triboelectric nanogenerators. *Nano Energy* 70: 104526.

36 Bae, J., Lee, J., Kim, S.M. et al. (2014). Flutter-driven triboelectrification for harvesting wind energy. *Nature Communications* 5 (1): 1–9.

37 Wang, J., Gu, S., Zhang, C. et al. (2020). Hybrid wind energy scavenging by coupling vortex-induced vibrations and galloping. *Energy Conversion and Management* 213: 112835.

38 Zhang, Z., Wang, S., Kan, J. et al. (2020). A pneumatic piezoelectric vibration energy harvester based on the compressed air-transducer-structure interaction. *Energy Conversion and Management* 213: 112861.

39 Wang, J., Zhou, S., Zhang, Z., and Yurchenko, D. (2019). High-performance piezoelectric wind energy harvester with Y-shaped attachments. *Energy Conversion and Management* 181: 645–652.

40 Persano, L., Dagdeviren, C., Su, Y. et al. (2013). High performance piezoelectric devices based on aligned arrays of nanofibers of poly(vinylidenefluoride-*co*-trifluoroethylene). *Nature Communications* 4 (1, 10): 1.

41 Yang, Z., Zhou, S., Zu, J., and Inman, D. (2018). High-performance piezoelectric energy harvesters and their applications. *Joule* 2: 642–697.

42 Lu, L., Ding, W., Liu, J., and Yang, B. (2020). Flexible PVDF based piezoelectric nanogenerators. *Nano Energy* 78: 105251.

43 Carneiro, P., dos Santos, M.P.S., Rodrigues, A. et al. (2020). Electromagnetic energy harvesting using magnetic levitation architectures: a review. *Applied Energy* 260: 114191.

44 Wang, S., Lin, L., and Wang, Z.L. (2012). Nanoscale triboelectric-effect-enabled energy conversion for sustainably powering portable electronics. *Nano Letters* 12: 6339–6346.

45 Wang, Z.L. (2013). Triboelectric nanogenerators as new energy technology for self-powered systems and as active mechanical and chemical sensors. *ACS Nano* 7: 9533–9557.

46 Ryu, H., Yoon, H.-J., and Kim, S.-W. (2019). Hybrid energy harvesters: toward sustainable energy harvesting. *Advanced Materials* 31: 1802898.

47 Ahmad, M.M. and Khan, F.U. (2021). Review of vibration-based electromagnetic–piezoelectric hybrid energy harvesters. *International Journal of Energy Research* 45: 5058–5097.

48 Joselin Herbert, G.M., Iniyan, S., Sreevalsan, E., and Rajapandian, S. (2007). A review of wind energy technologies. *Renewable and Sustainable Energy Reviews* 11: 1117–1145.

49 Hu, G., Yi, Z., Lu, L. et al. (2021). Self-powered 5G NB-IoT system for remote monitoring applications. *Nano Energy* 87: 106140.

50 Qian, J. and Jing, X. (2018). Wind-driven hybridized triboelectric-electromagnetic nanogenerator and solar cell as a sustainable power unit for self-powered natural disaster monitoring sensor networks. *Nano Energy* 52: 78–87.

51 Chen, J., Tang, W., Han, K. et al. (2019). Bladeless-turbine-based triboelectric nanogenerator for fluid energy harvesting and self-powered fluid gauge. *Advanced Materials Technologies* 4 (3), 1800560.

52 Na, Y., Lee, M.-S., Lee, J.W., and Jeong, Y.H. (2020). Wind energy harvesting from a magnetically coupled piezoelectric bimorph cantilever array based on a dynamic magneto-piezo-elastic structure. *Applied Energy* 264: 114710.

53 Wang, J., Geng, L., Ding, L. et al. (2020). The state-of-the-art review on energy harvesting from flow-induced vibrations. *Applied Energy* 267: 114902.

54 Seyed-Aghazadeh, B., Samandari, H., and Dulac, S. (2020). Flow-induced vibration of inherently nonlinear structures with applications in energy harvesting. *Physics of Fluids* 32: 071701.

55 Zou, H., Chen, H., and Zhu, X. (2015). Piezoelectric energy harvesting from vibrations induced by jet-resonator system. *Mechatronics* 26: 29–35.

56 Tsujiura, Y., Suwa, E., Nishi, T. et al. (2017). Airflow energy harvester of piezoelectric thin-film bimorph using self-excited vibration. *Sensors and Actuators A: Physical* 261: 295–301.

57 Sun, D., Xu, Y., Chen, H. et al. (2012). A mean flow acoustic engine capable of wind energy harvesting. *Energy Conversion and Management* 63: 101–105.

58 Yu, Y., Sun, D., Wu, K. et al. (2011). CFD study on mean flow engine for wind power exploitation. *Energy Conversion and Management* 52: 2355–2359.

59 Li, Z., Li, J., and Chen, H. (2020). Experimental research on excitation condition and performance of airflow-induced acoustic piezoelectric generator. *Micromachines* 11: 913.

60 Brown, G.B. (1937). The mechanism of edge-tone production. *Proceedings of the Physical Society* 49: 508–521.

61 Brown, G.B. (1937). The vortex motion causing edge tones. *Proceedings of the Physical Society* 49: 493–507.

62 Iwagami, S., Tabata, R., Kobayashi, T. et al. (2021). Numerical study on edge tone with compressible direct numerical simulation: sound intensity and jet motion. *International Journal of Aeroacoustics* 20 (3-4): 283–316.

63 Nagy, P.T., Szabó, A., and Paál, G. (2021). A feedback model of the edge tone, using the adjoint Orr–Sommerfeld equation. *Journal of Fluid Mechanics* 915.

64 Jordan, P., Jaunet, V., Towne, A. et al. (2018). Jet–flap interaction tones. *Journal of Fluid Mechanics* 853: 333–358.

65 Huang, Y., Lu, L., Yi, Z. et al. (2021). Aeroacoustics-driven jet-stream wind energy harvester induced by jet-edge-resonator. *Nano Energy* 89: 106441.

8

Wearable Flexible Piezoelectric Energy Harvesters

8.1 Introduction

For decades, miniaturization of electronics and advances in materials are paving the way for prosperous development of wearable and implantable devices. Human is experiencing the great conveniences which are brought by diverse electronics, including smartwatch, glasses, shoes, glove, pacemaker, and necklace. With the aid of 5G communication, artificial intelligence, and edge computing, the highly digitized human will undergo a seamless involvement in the whole network or internet of things. Some of them possess the functions of the mobile console for processing and displaying the data. Meanwhile, various sensors with physical, optical, or chemical mechanisms form another major group in wearable or implantable systems. Hence, our physical activities, vital sign, physiobiochemical indicators can be continuously monitored for health inspection or performing the manipulations, besides, the environmental variations are also able to be detected for further utilization. However, the current commercialized wearable and implantable devices usually rely on the battery-based power supply, such as piezoresistive, optical, and capacitive sensors, and various actuators, which are inevitably encountered some constraints, including the long-term sustainability and the further shrinkage of device size.

With the rise of microelectronic devices, flexible PEHs are considered suitable alternatives for microbattery power. In recent years, with the rapid development of microtechnology, various low-power devices have brought new challenges to their power supply components [1]. For example, batteries are mainly used as power supply units for traditional miniature medical devices because of their advantages of high-power supply density and stability. However, they cannot meet the requirements of increasingly complex application environments due to frequent replacement of the battery. If the energy in the environment can be collected, such as conversing the mechanical energy in the body movement into electrical energy for continuous power supply, and form a self-powered system with wearable/implantable medical devices, a series of problems, such as pollution and replacement problems caused by chemical batteries can be solved. Therefore, it is highly desired for developing a sustainable clean energy harvester such as self-powered piezoelectric nanogenerator.

Flexible Piezoelectric Energy Harvesters and Sensors, First Edition. Bin Yang, Zhiran Yi, and Chengkuo Lee.

Wearable devices, such as heart rate detectors and smart bracelets that help to record human health data are closely connected with human-computer interaction according to the corresponding needs of consumer groups. Attributing to being able to connect with the network, their multifunction and synchronous information can be achieved. Therefore, it is important to design wearable self-powered devices, and there have been a number of studies on energy harvesting devices for wearable devices [2–4]. The key to structural design of wearable energy harvester is to collect the energy caused by vibration based on the principle of not disturbing the regular movement of human body and not causing a burden. Referring to the law of the body's own movement and power supply of portable mobile devices, the development of wearable piezoelectric energy harvester tends to the miniaturization of structure, low threshold, low-power consumption, and high-conversion efficiency of circuits.

8.2 Fiber Structure PEH

Although some breakthrough results in wearable sensors have been achieved to some extent by materials selection and structural design, it still remains challenging for two-dimensional sensors to effectively fit the irregular and deformable skin surface of our bodies [5, 6]. Besides, it is highly desired for wearable electronics to be breathable or comfortable in application, which is not possible for a film-type device. The above problems inevitably retard the advancement of wearable devices and need to be solved urgently.

Some inspirations have been found from the ancient textile technology to overcome the difficulties above [7–9]. If flexible piezoelectric materials are integrated onto the fiber substrate or making the piezoelectric devices into fiber shape, the resulting fiber devices can be woven into a textile that can simultaneously detect various physiological motion signals and collect energy generated by human motion [10–14]. Compared to two-dimensional sensors, these piezoelectric fibers can withstand a variety of complex and severe deformations, such as bending and knotting, while providing a comfortable and effective contact interface with human skin. The excellent flexibility and breathability ensure the wearer does not feel uncomfortable during long-term use, which greatly booms the application prospects.

Recently, some efforts were devoted to developing flexible 1D polyvinylidene fluoride (PVDF) piezoelectric fiber for more adaptation in wearable and stitchable applications. In 2020, a simple strategy for fabricating the piezoelectric microfiber with a novel core–sheath structure was proposed by directly electrospinning the poly(vinylidene fluoride-trifluoroethylene) (P[VDF-TrFE]) onto the flexible conductive wire [15]. They were successfully woven into a textile for assembling an intelligent fabric (Figure 8.1a), which was further fixed at wrist, foot, elbow, and knee, respectively (Figure 8.1b). Excellent flexibility and skin cooperation ensured that it had great potential for not only medical or sports apparel when employed as tactile and motion sensors but also micro energy generation in miniature flexible electronics industry for powering various electronic devices. The flexible stitchable

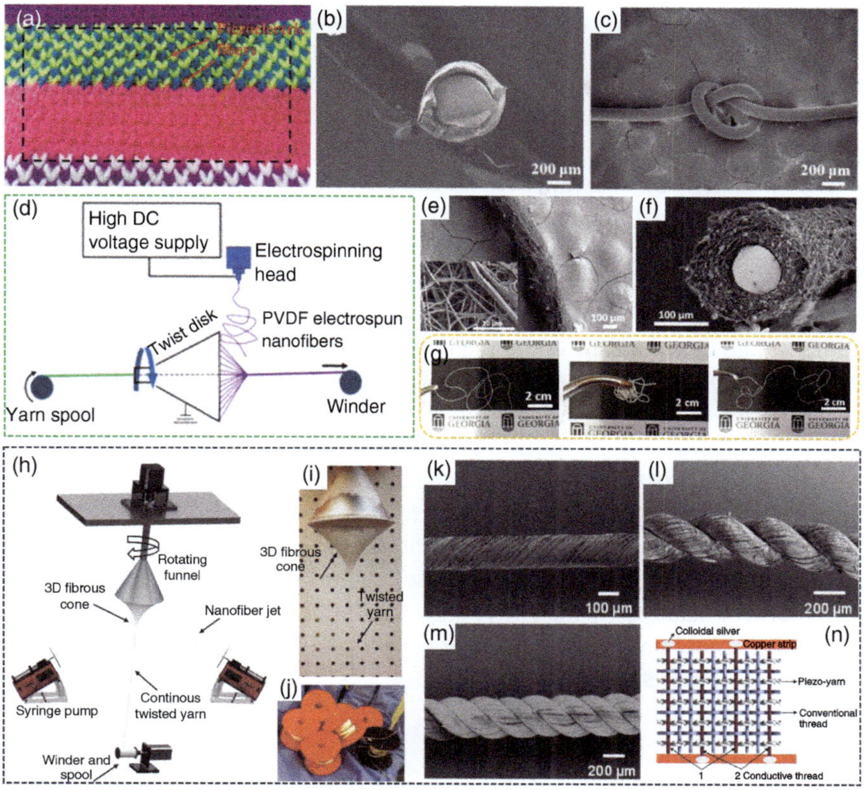

Figure 8.1 (a) Photograph of the smart fibers integrated with the textile. Cross-sectional SEM image (b) of Cu/P(VDF-TrFE) fiber. (c) SEM image of flexible Cu/P(VDF-TrFE) fiber in knotted structure. (d) Schematic for deposition of PVDF nanofibers around conductive core yarn by electrospinning. Longitude (e) and cross-section (f) SEM images of electrospun yarn. (g) Photographic images of EF yarn before, during, and after folding and twisting. (h) Schematic of the apparatus for manufacturing continuous twisted yarns. (i) Three-dimensional fibrous cone and twisted yarn from this cone. (j) Twisted yarns collected on spools. SEM images of P(VDF-TrFE) single yarn (k), two-ply yarn (l), and three-ply yarn (m). (n) Piezofabric with conductive threads indicated. Source: (a–c) From Lu et al. [15] / with permission of Elsevier, (d) Gao et al. [16] / with permission of IOP Publishing, (e–g) From Gao et al. [16] / with permission of IOP Publishing / with permission of CC BY, (h–n) From Yang et al. [14] / with permission of American Chemical Society.

fiber can bear various complexes, even severe deformations, such as bending and knotting, enabling itself to be woven into textile and further generate power and response when subjected to external deformation, which is of significance for the portable and wearable electronics that are expected to be breathable and comfortable. (Figure 8.1b, c).

In 2018, Gao et al. developed a new customized electrospinning device for directly integrating a layer of electrospun PVDF nanofibers around a silver-deposited nylon filament acting as an inner electrode (Figure 8.1d) [16]. Benefitted from the continuous fabrication process, the ultralong and flexible PVDF piezoelectric fiber with novel core–sheath structure can be obtained easily (Figure 8.1e and f). Toward the wearable application where large deflections are involved, the as-prepared fiber possesses superior flexibility and undergoes severe folding and twisting without any damage (Figure 8.1g). The advantage of core–sheath structure is that the PVDF layer was sandwiched between the inner core and outer electrode, making full use of piezoelectric materials to maximize energy collection in practical. Likewise, another structure and fabricated process for manufacturing 1D electrospun P(VDF-TrFE) stitchable yarn was proposed by Yang et al. [14]. Different from core–sheath structure, the inner core electrode was abandoned in the preparation, as shown in schematic Figure 8.1h–j. The process was divided into two steps. Firstly, the electrospun network was collected at the large end of the rotation funnel; secondly, the continuous twisted yarn was pulled out of the 3D fibrous cone and wrapped around the spool fixed on the winder. The flexible P(VDF-TrFE) yarns with different twist numbers, such as single yarn (Figure 8.1k), two-ply yarns (Figure 8.1l), and three-ply yarns (Figure 8.1m), can be prepared according to different requirements. When weaved into fabrics, the pair of conductive electrodes was designed according to the indication in Figure 8.1n. However, although the strength is greatly improved after being twisted, the preparation of these left–right electrodes inevitably has some problems, such as inefficient energy harvesting. Therefore, when fibrous piezoelectric devices are prepared, careful consideration should be taken into the placement of electrodes.

8.3 Film Structure PEH

Piezoelectric wearable energy harvester is developed rapidly due to its potential for the sustainable power supply with the emergence of wearable electronics, which is promising to get rid of the battery. Human body includes abundant kinetic energy, such as hand motion [17–20], wrist motion [21–26], knee-joint motion [27–32], foot motion [33–38], clothes motion, motion of the backpack, and joint rotation [17, 39–41]. The different piezoelectric energy harvesters, as shown in Figure 8.3, have been investigated by using these kinetic energies in human body.

8.3.1 Hand Motion

Portable handheld piezoelectric energy harvesters are driven by human handy motion vibration. Therefore, the excitation condition is with low frequency, large acceleration, and unabiding. Figure 8.2a shows a piezoelectric flexible energy harvester employing a metal ball to impact two sidewalls that act as the bases for two unimorph piezoelectric cantilevers [18, 19]. Each impact causes the free vibration of the piezoelectric cantilevers and this strategy offers mechanical frequency up-conversion. The fabricated harvester generated maximum 175 μW average power at the exciting frequency of 4.96 Hz and acceleration amplitude of 2 g. Additionally, part of researchers found the output energy waveforms contained information about human activities as detected by some motion sensors. Therefore, it also could be used as a low-power consumption device for human activity recognition [20].

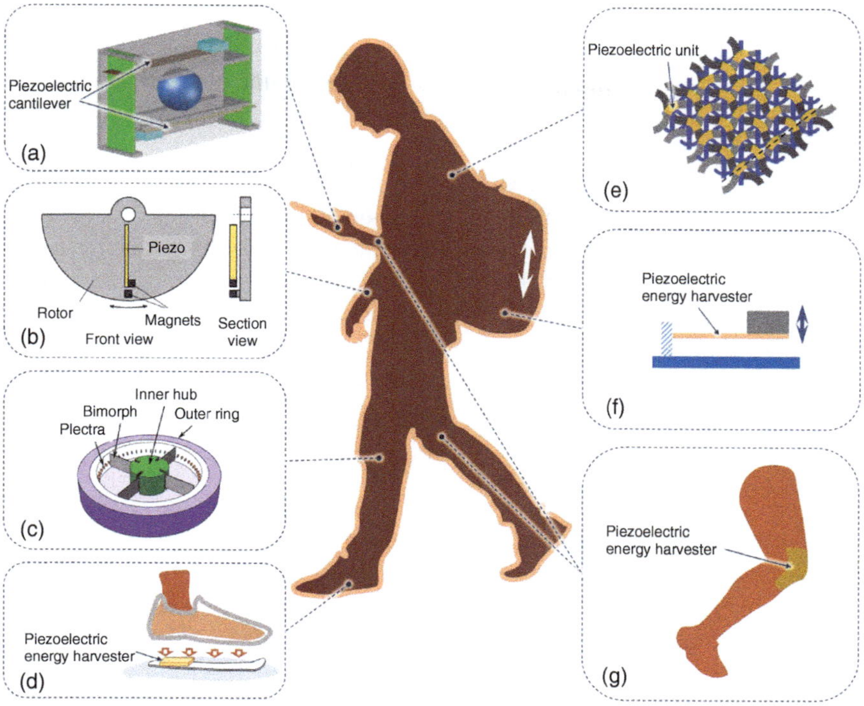

Figure 8.2 Piezoelectric wearable energy harvester. Energy from (a) hand motion, (b) wrist motion, (c) knee-joint motion, (d) foot pressure, (e) clothes motion, (f) motion of the backpack, and (g) joint rotation.

8.3.2 Wrist and Elbow Motion

Watch-like piezoelectric energy harvesters have been proposed to harvest the kinetic energy of the wrist motion. With the wrist motion, an eccentric proof mass embedded in the watch-like harvesters will induce free vibration of the piezoelectric structure (Figure 8.2b). This design with a frequency up-conversion feature is widely used to scavenge low-frequency vibrations [21]. Given the lack of electricity generation at rest, an external actuating approach through a magnetic reluctance coupling is presented for the watch-like piezoelectric rotational motion harvester [22]. With this strategy, an internal battery could be recharged during prolonged periods of inactivity. Lastly, the harvester with a functional volume of $1.85\,cm^3$ can generate $100\,\mu W$ power at the optimal driving frequency is 25 Hz [22]. To provide an estimation of maximum power output from wrist motion in different activities, Xue et al. presented a generalized 3D rotational harvester model [23]. Experimental results of the one piezoelectric beam prototype demonstrated an average power output of $3.8\,\mu W$ in pseudo-walking motion. These researchers further improved this harvester to ten piezoelectric beams, under the sinusoidal excitation of 25° and 0.8 seconds approximately $40\,\mu W$ was obtained [25]. Besides, some researchers directly utilized the cantilever piezoelectric energy harvester to scavenge the kinetic energy of the wrist motion. A smartwatch-sized harvester based on single piezoelectric cantilever performed average output powers of about 50 and 20 μW from the hand-arm and head motions, respectively [24]. The same type of harvester could generate about 100–250 μJ under the different hand movements [26]. The harvested energy from jogging for 30 minutes can supply the tracker with up to 37 sampling cycles.

In 2012, Wang group first proposed the concept of "human-motion-based energy harvesting" and designed a simple but practical and durable strategy of flexible ZnO NWs/PVDF plane/fiber piezoelectric nanogenerator for converting low-frequency (<1 Hz) mechanical activity into electricity at various amounts of deformation [42]. Uniquely, this generator can be smoothly attached to the elbow (Figure 8.3i), and the voltage and current signals were simultaneously obtained when the wearer folded and released his/her elbow at a low frequency below 1 Hz, indicating the feasibility of supplying portable electronics (Figure 8.3ii–iv).

8.3.3 Knee-Joint Energy Harvesting

Piezoelectric energy harvesters were proposed to be worn on the knee-joint as shown in Figure 8.2c. The harvesters relied on the plucking technique to achieve frequency up-conversion [27]. During a plucking action, the piezoelectric cantilevers were deflected by the plectrums. When released due to loss of contact, the

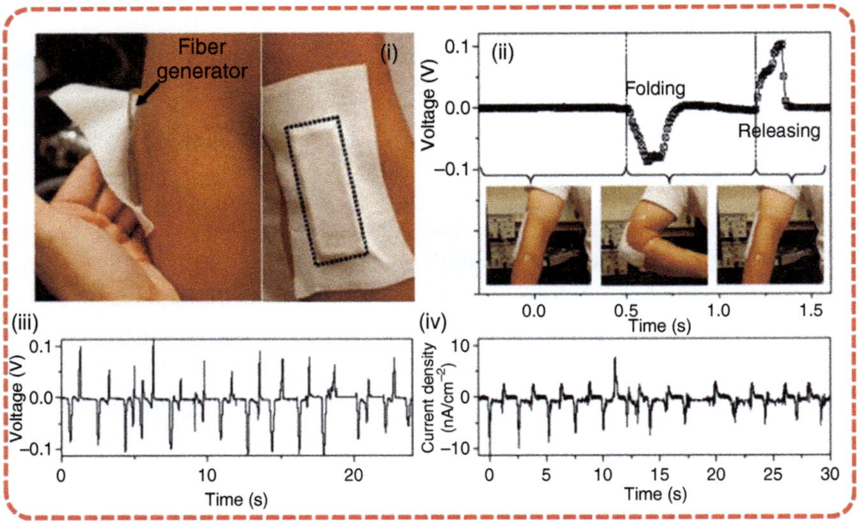

Figure 8.3 Piezoelectric wearable energy harvester. (a) Electrical measurements of our hybrid nanogenerator attached to a human arm. (i) Optical images of the hybrid-fiber device attached to an elbow. The hybrid device was attached to a plane PS substrate and covered with PDMS to experience uniform stress during the measurements. The device in the PDMS/PS supporting polymers was attached to the elbow. (ii) Open-circuit voltage output of a device depending on the folding and releasing of the elbow. A voltage output appeared in the event of the folding and releasing action of the elbow. (iii) Open-circuit voltage outputs of a fiber device during multiple folding-releasing events of the elbow. (iv) Closed-circuit current-density output of a device during multiple folding-releasing events of the elbow. Source: From Lee et al. [42] / with permission of John Wiley & Sons.

cantilever would be free to vibrate at its resonant frequency, generating electrical energy. A prototype, featuring four PZT-5H bimorphs, was built and studied in a knee simulator that reproduces the gait of a human subject. The average power output of the prototype was 2.06 ± 0.3 mW [28, 29, 31]. For further optimization, contactless magnetic plucking was presented. Record levels of rectified electrical power of over 50 and 70 mW per walking and running steps, respectively, were obtained [30]. Furthermore, as the walking speed increased from 3 to 7 km/h, power output of a magnetically plucked wearable knee-joint energy harvester could increase from 1.9 ± 0.12 to 4.5 ± 0.35 mW [32], which was sufficient for powering a wireless sensing system.

8.3.4 Foot Motion

The development of the piezoelectric wearable energy harvesters started early from harvesting the foot motion kinetic energy, as shown in Figure 8.2d. So far, the piezoelectric foot motion energy harvesters have been investigated as the most potential wearable piezoelectric energy-harvesting approach. At the end of the last century, the piezoelectric foot motion energy harvester was firstly presented by using a PVDF stave placed at the ball of the foot and a PZT Thunder at the heel [33]. This stave included 8-layer stacks of 28-μm PVDF sheets with the electrodes connected in parallel. The Thunder is a strip of composite PZT bonded to a curved piece of spring steel. The maximum output powers are seen to approach 20 mW for the PVDF stave and 80 mW for the PZT Thunder. The average net energy transfers of roughly 1 mJ/step for the PVDF and 2 mJ/step for the PZT Thunder. To enhance energy-conversion efficiency, a PVDF harvester based on a specially designed wavy structure is presented. The wavy structure can increase the average effective area of the piezoelectric film strain. The harvester provides an average output power of 1 mW during a walk at a frequency of roughly 1 Hz [34]. Recently, a heel harvester has been reported [36] to efficiently convert kinematic walking energy into electricity by utilizing the lever mechanism to gain maximum deformation of multilayer piezoelectric patches. The heel harvester uses a multilayered levered piezoelectric mechanism associated with a nonlinear mechanical-synchronized switching on inductor circuit interface. It tactfully switches on when the foot contacts the ground and switches off when the foot is lifted. The design takes full advantage of the user's weight and amplifies footstep displacement. Experimental results have shown an average effective output power of 6.13 mW under a harmonic excitation of 2.3 Hz (mimicking fast walking speed: 6.2 km/h for men and 5.55 km/h for women). Additionally, a sandwiched force amplification frame was used to transmit and amplify the vertical heel-strike force to the inner piezoelectric stack deployed in the horizontal direction [37]. The footwear harvester with six stacks performs 9 mW/shoe average output power at the walking speeds of 3.0 mph (4.8 km/h).

The design of all electrospinning procedures prevents the existence of rigid structure which decreases flexibility and enhances mechanical durability and energy harvesting. As shown in Figure 8.4, the as-prepared PVDF-based nonwoven nanogenerator can generate electrical energy when being struck by hand,

Table 8.1 Comparison of the reported piezoelectric energy harvesters on the excitations from walking or running.

References	Materials	Volume of the used piezoelectric materials (mm^3)	Harvesting mode	Excitation	Output average power
[27]	PZT	31.8 × 12.7 × 0.25	d_{31}	Single plucking action	2 μJ
[28, 29, 31]	PZT	31.8 × 12.7 × 0.25 × 4	d_{31}	Walking	2.06 ± 0.3 mW
[30]	PZT	27 × 6.5 × 0.25 × 16	d_{31}	Walking	50 mW/walking step 70 mW/running step
[32]	PZT	31.8 × 12.7 × 0.25 × 4	d_{31}	walking from 3 to 7 km/h	1.9 ± 0.12 to 4.5 ± 0.35 mW
[33]	PVDF+PZT	PVDF: 65 × 44.8 PZT: 25 × 38.1	d_{31}	Walking	1 mJ/step [PVDF] 2 mJ/step [PZT]
[34]	PVDF	40 × 0.3	d_{31}	Walking	1 mJ/step
[35]	PZT	76.2 × 31.75 × 2.28	d_{31}	Walking	11.55 μJ/step
[36]	PZT	46 × 20.8 × 0.3048	d_{31}	Walking	2.67 mJ/step
[37]	PZT	49 × 30 × 6	d_{33}	Walking	9 mW/shoe

walking/running, and arm swinging, respectively. Besides, increasing the force or frequency of human activities on the device could directly improve the electrical output. Karan et al. fabricated the PVDF/AlO-rGO wearable nanogenerator by using steel-woven fabric as the electrode, overcoming the stress mismatch between the electrode and functional piezoelectric layer [43]. A summary comparison of the representative walkingand running energy harvesters is proved in Table 8.1.

8.3.5 Clothes Motion

The ideal way to develop the wearable piezoelectric energy harvesters is to build on the existed wearable objects, such as clothes, gloves, shoes, or others. Among these wearable objects, clothes have the most potential due to their broad space and abundant kinetic energy resources. Fiber is the base of the clothes. Therefore, recently there were many researches focusing on the flexible piezoelectric fibers, as shown in Figure 8.2e. Depending on the coupled piezoelectric and semiconducting properties of ZnO, Bai et al. [44] developed a two-dimensional woven harvester, which was

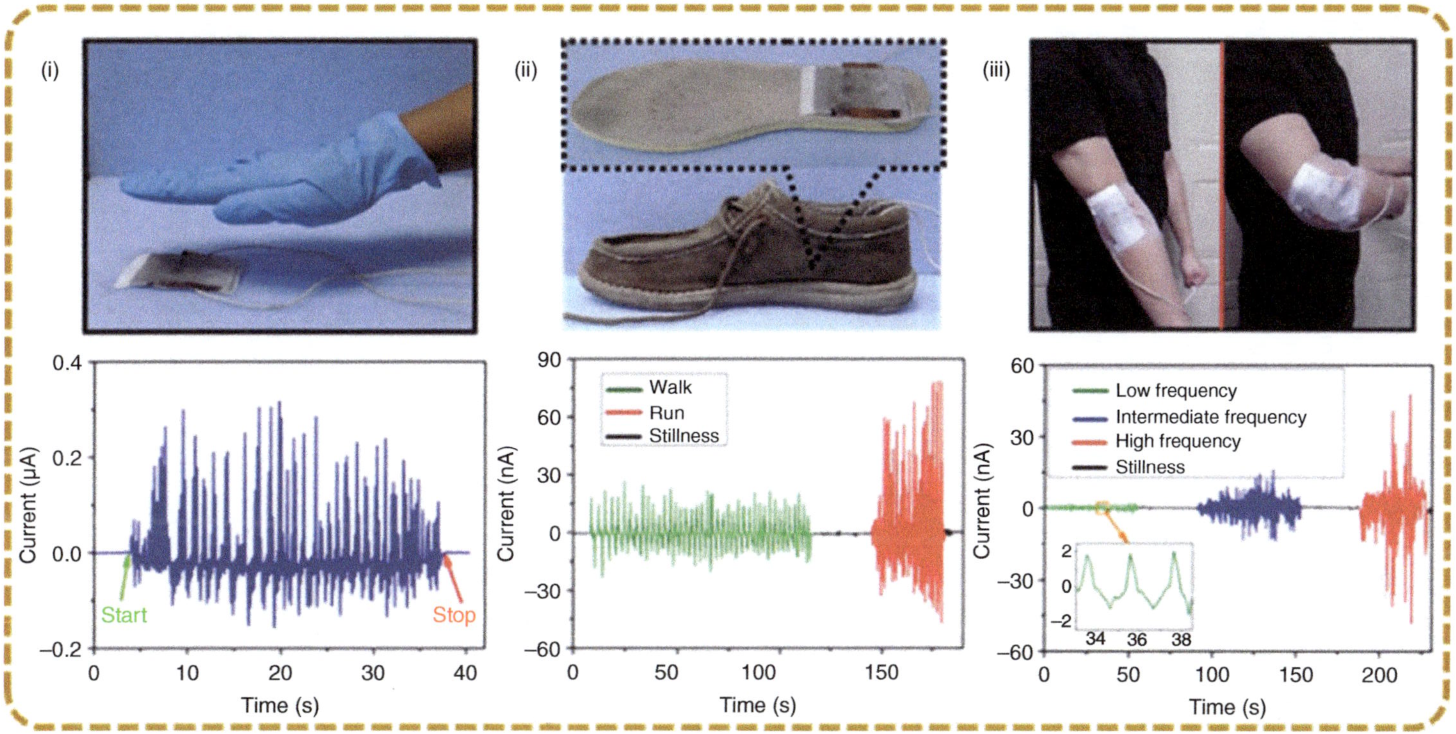

Figure 8.4 Piezoelectric wearable energy harvester. (i) Photograph of the NG being struck by hand and the corresponding output current. (ii) Photograph of the NG integrated into an insole and the output currents generated by walking and running. (iii) Photograph of the device fixed to an elbow and the output currents generated by arm swinging at low frequency, intermediate frequency, and high frequency.

composed of two kinds of fibers grown with ZnO nanowires crossing with each other. The generated open-circuit voltage and short-circuit current of the harvester were 3 mV and 17 pA, respectively. To improve the output performance, Zeng et al. [45] presented an all-fiber wearable electric power harvester based on a PVDF–$NaNbO_3$ nanofiber nonwoven fabric and an elastic conducting knitted fabric electrodes. The harvester could produce a peak open-circuit voltage of 3.4 V and a peak current of 4.4 mA in cyclic compression tests at the exciting frequency of 1 Hz and a maximum pressure of 0.2 MPa. Given the nonuniform deformation distribution in this piezoelectric nanofiber nonwoven fabric harvester, Sone et al. [46] proposed a wearable piezoelectric energy harvester, which was composed of warp and weft threads made of yarn strings and piezoelectric film straps, respectively. Maximum power densities of 81 and 125 $\mu W/cm^2$ were obtained from the device using warp-thread diameters of 5 and 3 mm, respectively. To improve the flexibility of the piezoelectric fibers, Sim et al. [47] developed a highly flexible two-ply piezoelectric yarn that consists of an electrospun PVDF-TrFE mat. Using the plying structure, the longitudinal tension force was effectively converted into a transverse compression force. The piezoelectric yarn showed a piezoelectric performance of 0.7 V by compression in the radial direction and 0.55 V by tension in the longitudinal direction. To provide more energy harvesting solutions, the possibility of flexible and durable piezoelectric generators based on simple procedures and low-cost strategies has been explored. Zhang et al. [48] report a flexible hybrid piezoelectric fiber-based two-dimensional fabric harvester. The hybrid piezoelectric fiber comprised aligned $BaTiO_3$ nanowires and PVC polymer. By attaching the fabric harvester to an elbow pad that was bent by human arms, the harvester can generate 1.9 V output voltage and 24 nA output current. Additionally, a novel yarn was explored as a wearable piezoelectric harvester consisting of a layer of electrospun PVDF nanofibers integrated with PVDF film around a silver-deposited nylon filament acting as an inner electrode [16, 49]. The resultant unique yarn produced an average peak voltage of 0.52 V, peak current of 18.76 nA under a cyclic compression of 0.02 MPa at 1.85 Hz.

8.3.6 Backpack

Backpack is an existed wearable object carried by a strap on humans back. Human motions can induce oscillation. An inertial energy harvester based on a piezoelectric cantilever is investigated for harvesting kinetic energy from a backpack [50], as shown in Figure 8.2f. Vibration levels in the backpack during walking or jogging were tested and analyzed. When the walking speed is 2–6 km/h, the amplitude of the acceleration is around 0.7–2.2g for the accelerator placed on the horizontal frame and the vibration level is higher in the frequency range of 2–12.7 Hz. The harvester can achieve a power output of 43.64 μW with a single piezoelectric element at the speed of 4 km/h.

8.3.7 Joint Rotation

Given the joint rotation is a big deformation motion, flexible piezoelectric materials are preferred to be used in the investigation of the piezoelectric wearable energy harvesters. PVDF as a common flexible piezoelectric material has been employed in the flexible piezoelectric energy harvester to scavenge the kinetic energy in human body. For instance, a reported pure PVDF-based piezoelectric energy harvester was placed on the back of the knee through an elastic cotton leotard, which allowed obtaining an output power of 1.45 μW during walking [39]. To improve the output power, Yang et al. presented a flexible energy harvester that consists of a polyvinylidene fluoride film attached to a curved substrate in a shell shape for harvesting energy from human motion [40]. The fabric, worn on the elbow joints and fingers, generates a high-output power of 0.21 mW in spite of slow and irregular motion. Due to the intrinsic limitation at the low piezoelectric constant of the PVDF, P(VDF-TrFE) piezoelectric material was developed for wearable piezoelectric energy harvesters. Solution-derived P(VDF-TrFE) piezoelectric thin films on cellulose paper substrates were prepared as flexible power harvesters [41]. P(VDF-TrFE) piezoelectric film-based paper energy harvesters were directly attached to the back of a human hand, and they generated a maximum output open-circuit voltage of 0.4 V at low-bending frequencies of 0.25 Hz. However, the amount of power harvested from conventional flexible piezoelectric energy harvesters is still insufficient for achieving energy autonomy; hence, they are only used for limited applications. Developing high-performance piezoelectric ceramics in flexible piezoelectric energy harvesters has been an effective approach to enhancing the energy extraction from the joint rotation. A flexible piezoelectric energy harvester based on PZT material with a total thickness of 170 μm is utilized [51], which provides sufficient flexibility for attachment to clothes or human skin. The harvester generated an open-circuit voltage of 165 V and a short-circuit current of 1.5 μA through mechanical bending motion on a linear stage.

Piezoelectric wearable energy harvesters from the joint rotation generally have some common features, such as being flexible and conformable for skin or clothes. However, the developments of the pure flexible piezoelectric materials, flexible piezoelectric co-polymer, or some flexible composites based on the piezoelectric ceramics are urgent to call for a breakthrough in the piezoelectric coupling coefficient. More innovative designs or some advanced fabrication technologies are needed to employ for improving the insufficient output power to meet the power consumption of some commercial wearable electronics.

8.4 Summary

In terms of energy harvesters as a power supply, the development of flexible piezoelectric energy harvesters greatly relies on the improvements of flexible piezoelectric materials or advanced micromachining processes. As to wearable PEHs on human body, the main development tendency includes high-output power, flexibility, and no irritation to the skin for practical feasibility. On the basis of abundant available kinetic energy in human body these energies, the reported researches on wearable PEHs focus on several main body parts, such as hand, wrist, and joint, as well as the wearable objects, such as clothes, backpack, glove, and shoes.

References

1 Mokhtari, F., Spinks, G.M., Fay, C. et al. (2020). Wearable electronic textiles from nanostructured piezoelectric fibers. *Advanced Materials Technologies* 5: 15.

2 Fuh, Y.K., Ye, J.C., Chen, P.C. et al. (2015). Hybrid energy harvester consisting of piezoelectric fibers with largely enhanced 20 V for wearable and muscle-driven applications. *ACS Applied Materials & Interfaces* 7: 16923–16931.

3 Maity, K. and Mandal, D. (2018). All-organic high-performance piezoelectric nanogenerator with multilayer assembled electrospun nanofiber mats for self-powered multifunctional sensors. *ACS Applied Materials & Interfaces* 10: 18257–18269.

4 Zheng, H.W., Zi, Y., He, X. et al. (2018). Concurrent harvesting of ambient energy by hybrid nanogenerators for wearable self-powered systems and active remote sensing. *ACS Applied Materials & Interfaces* 10: 14708–14715.

5 Sun, H., Zhang, Y., Zhang, J. et al. (2017). Energy harvesting and storage in 1D devices. *Nature Reviews Materials* 2: 12.

6 Wang, L., Wang, L., Zhang, Y. et al. (2018). Weaving sensing fibers into electrochemical fabric for real-time health monitoring. *Advanced Functional Materials* 28: 8.

7 Service, R.F. (2003). Technology – electronic textiles charge ahead. *Science* 301: 909–911.

8 Zeng, W., Shu, L., Li, Q. et al. (2014). Fiber-based wearable electronics: a review of materials,
fabrication, devices, and applications. *Advanced Materials* 26: 5310–5336.

9 Lu, L.J., Hu, Y.J., and Dai, K. (2017). The advance of fiber-shaped lithium ion batteries. *Materials Today Chemistry* 5: 24–33.

10 Dong, K., Peng, X., and Wang, Z.L. (2020). Fiber/fabric-based piezoelectric and triboelectric nanogenerators for flexible/stretchable and wearable electronics and artificial intelligence. *Advanced Materials* 32: 43.

11 Lu, X., Qu, H., and Skorobogatiy, M. (2017). Piezoelectric micro-and nanostructured fibers fabricated from thermoplastic nanocomposites using a fiber drawing technique: comparative study and potential applications. *ACS Nano* 11: 2103–2114.

12 Sim, H.J., Choi, C., Lee, C.J. et al. (2015). Flexible, stretchable and weavable piezoelectric fiber. *Advanced Engineering Materials* 17: 1270–1275.

13 Lu, L.J., Zhou, Y., Pan, J. et al. (2019). Design of helically double-leveled gaps for stretchable fiber strain sensor with ultralow detection limit, broad sensing range, and high repeatability. *ACS Applied Materials & Interfaces* 11: 4345–4352.

14 Yang, E.L., Xu, Z., Chur, L.K. et al. (2017). Nanofibrous smart fabrics from twisted yarns of electrospun piezopolymer. *ACS Applied Materials & Interfaces* 9: 24220–24229.

15 Lu, L., Yang, B., Zhai, Y., and Liu, J. (2020). Electrospinning core-sheath piezoelectric microfibers for self-powered stitchable sensor. *Nano Energy* 76: 104966.

16 Gao, H., Minh, P.T., Wang, H. et al. (2018). High-performance flexible yarn for wearable piezoelectric nanogenerators. *Smart Materials and Structures* 27: 11.

17 Cheng, Q., Peng, Z., Lin, J., Li, S., and Wang, F. (2015) Energy harvesting from human motion for wearable devices. In *10th International Conference on Nano/Micro Engineered and Molecular Systems International Conference on Nano-Micro Engineered and Molecular Systems*. pp. 409–412. IEEE.

18 Halim, M.A., Cho, H.O., and Park, J.Y. (2015). A handy motion driven, frequency up-converting piezoelectric energy harvester using flexible base for wearable sensors applications. *IEEE Sensors* 1–4.

19 Halim, M.A. and Park, J.Y. (2018). Piezoelectric energy harvester using impact-driven flexible side-walls for human-limb motion. *Microsystem Technologies-Micro-and Nanosystems-Information Storage and Processing Systems* 24: 2099–2107.

20 Khalifa, S., Lan, G., Hassan, M. et al. (2018). HARKE: human activity recognition from kinetic energy harvesting data in wearable devices. *IEEE Transactions on Mobile Computing* 17: 1353–1368.

21 Pillatsch, P., Yeatman, E.M., and Holmes, A.S (2013) A wearable piezoelectric rotational energy harvester. In *2013 IEEE International Conference on Body Sensor Networks*. pp. 1–6. IEEE

22 Pillatsch, P., Wright, P.K., Yeatman, E.M., Holmes, A.S., (2015) A wireless charging mechanism for a rotational human motion energy harvester. In *2015 IEEE 12th International Conference on Wearable and Implantable Body Sensor Networks* pp. 1–5. IEEE.

23 Xue, T., Ma, X., Rahn, C., Roundy, S. (2014) in *14th International Conference on Micro and Nanotechnology for Power Generation and Energy Conversion Applications* Vol. 557 *Journal of Physics Conference Series*. IOP.

24 Bai, Y., Tofel, P., Hadas, Z. et al. (2018). Investigation of a cantilever structured piezoelectric energy harvester used for wearable devices with random vibration input. *Mechanical Systems and Signal Processing* 106: 303–318.

25 Xue, T., Yeo, H.G., Trolier-McKinstry, S., and Roundy, S. (2018). Wearable inertial energy harvester with sputtered bimorph lead zirconate titanate (PZT) thin-film beams. *Smart Materials and Structures* 27: 085026.

26 Manjarres, J., Dasuki, K., Parody, H. *et al.* (2019) Monitoring system for kinetic energy harvesting in a mobile platform. In *2019 IEEE International Symposium on Circuits and Systems (ISCAS)* (pp. 1–5). IEEE.

27 Pozzi, M. and Zhu, M. (2011). Plucked piezoelectric bimorphs for knee-joint energy harvesting: modelling and experimental validation. *Smart Materials and Structures* 20: 055007.

28 Pozzi, M., Aung, M.S.H., Zhu, M. et al. (2012). The pizzicato knee-joint energy harvester: characterization with biomechanical data and the effect of backpack load. *Smart Materials and Structures* 21: 075023.

29 Pozzi, M. and Zhu, M. (2012). Characterization of a rotary piezoelectric energy harvester based on plucking excitation for knee-joint wearable applications. *Smart Materials and Structures* 21: 055004.

30 Pozzi, M. (2016). Magnetic plucking of piezoelectric bimorphs for a wearable energy harvester. *Smart Materials and Structures* 25: 045008.

31 Kuang, Y. and Zhu, M. (2016). Characterisation of a knee-joint energy harvester powering a wireless communication sensing node. *Smart Materials and Structures* 25: 055013.

32 Kuang, Y., Ruan, T., Chew, Z.J., and Zhu, M. (2017). Energy harvesting during human walking to power a wireless sensor node. *Sensors and Actuators A: Physical* 254: 69–77.

33 Kymissis, J., Kendall, C., Paradiso, J., Gershenfeld, N. (1998) Parasitic power harvesting in shoes. In *Digest of papers. Second international symposium on wearable computers (Cat. No. 98EX215)*. pp. 132–139. IEEE.

34 Zhao, J. and You, Z. (2014). A shoe-embedded piezoelectric energy harvester for wearable sensors. *Sensors* 14: 12497–12510.

35 Kalantarian, H. and Sarrafzadeh, M. (2016). Pedometers without batteries: an energy harvesting shoe. *IEEE Sensors Journal* 16: 8314–8321.

36 Hua, R., Liu, H., Yang, H. et al. (2018). A nonlinear interface integrated lever mechanism for piezoelectric footstep energy harvesting. *Applied Physics Letters* 113: 053902.

37 Qian, F., Xu, T.-B., and Zuo, L. (2018). Design, optimization, modeling and testing of a piezoelectric footwear energy harvester. *Energy Conversion and Management* 171: 1352–1364.

38 Singh, S., Gupta, V.K., and Mukherjee, S. (2018). Piezoelectric based energy harvester embedded in shoe for wearable electronics. *Materials Physics and Mechanics* 37: 159–167.

39 Proto, A., Fida, B., Bernabucci, I. *et al.* (2016) Wearable PVDF transducer for biomechanical energy harvesting and gait cycle detection. In *2016 IEEE EMBS Conference on Biomedical Engineering and Sciences (IECBES)*. pp. 62–66. IEEE.

40 Yang, B. and Yun, K.-S. (2012). Piezoelectric shell structures as wearable energy harvesters for effective power generation at low-frequency movement. *Sensors and Actuators A: Physical* 188: 427–433.

41 Won, S.S., Sheldon, M., Mostovych, N. et al. (2015). Piezoelectric poly(vinylidene fluoride trifluoroethylene) thin film-based power generators using paper substrates for wearable device applications. *Applied Physics Letters* 107: 202901.

42 Lee, M., Chen, C.Y., Wang, S. et al. (2012). A hybrid piezoelectric structure for wearable nanogenerators. *Advanced Materials* 24: 1759–1764.

43 Karan, S.K., Bera, R., Paria, S. et al. (2016). An approach to design highly durable piezoelectric nanogenerator based on self-poled PVDF/AlO-rGO flexible nanocomposite with high power density and energy conversion efficiency. *Advanced Energy Materials* 6: 12.

44 Bai, S., Zhang, L., Xu, Q. et al. (2013). Two dimensional woven nanogenerator. *Nano Energy* 2: 749–753.

45 Zeng, W., Tao, X.M., Chen, S. et al. (2013). Highly durable all-fiber nanogenerator for mechanical energy harvesting. *Energy & Environmental Science* 6: 2631–2638.

46 Song, S. and Yun, K.-S. (2015). Design and characterization of scalable woven piezoelectric energy harvester for wearable applications. *Smart Materials and Structures* 24: 045008.

47 Sim, H.J., Choi, C., Lee, C.J. et al. (2015). Flexible two-ply piezoelectric yarn energy harvester. *Current Nanoscience* 11: 539–544.

48 Zhang, M., Gao, T., Wang, J. et al. (2015). A hybrid fibers based wearable fabric piezoelectric nanogenerator for energy harvesting application. *Nano Energy* 13: 298–305.

49 Gao, H., Asheghali, D., Yadavalli, N.S. et al. (2019). Fabrication of core-sheath nanoyarn via touchspinning and its application in wearable piezoelectric nanogenerator. *Journal of the Textile Institute* 111 (6): 906–915.

50 Chen, F., He, M., Wang, S. et al. (2019). Study of an inertial piezoelectric energy harvester from a backpack. *Ferroelectrics* 550: 233–243.

51 Khan, M.B., Kim, D.H., Han, J.H. et al. (2019). Performance improvement of flexible piezoelectric energy harvester for irregular human motion with energy extraction enhancement circuit. *Nano Energy* 58: 211–219.

9

Implantable Piezoelectric Energy Harvesters

9.1 Introduction

Cardiac pacemaker, as a life-supporting biomedical device, has been widely used in clinic. Currently, millions of lives rely on implantable medical devices, such as cardiac pacemakers, to maintain normal vital signs [1]. Over the past half-century, significant advances in pacemaker technology have been made, such as internet-based remote monitoring. However, there is a lack of innovative development in the basic system paradigm of an extravascular pulse generator connected to leads traversing the venous system to contact myocardial tissue. Moreover, the basic powering method remains highly dependent on batteries [2–4]. Many pacemaker-related complications [2, 3, 5], such as infection, pneumothorax, lead failure, and invalid pacing, are related to this basic construct, particularly the power supply and leads. So, periodic battery replacement surgery is inevitable to ensure sufficient power supply, which greatly increases the health risks of patients with heightened morbidity and potential mortality [6]. Therefore, developing smart, multiple functions, leadless, and batteryless features have become the focal points of future cardiac pacemakers [3, 5, 7–9].

Great efforts have been made in implantable bioenergy harvesting to develop batteryless medical devices [3, 6, 10–22], especially cardiac pacemakers. One of the most promising approaches is to directly convert bioenergy, mainly including biomechanical energy, biochemical energy, and biothermal energy of human body, in vivo into electrical energy to avoid the dependence on environmental factors or other requirements outside the body in regular operations. In terms of biomechanical energy, it is rich distribution of energy in body, such as heartbeats, muscle stretching, and other motions of tissues or organs [3, 6, 11, 12, 14, 16, 20, 23]. The mechanisms of biomechanical energy conversion include electromagnetic induction [7, 24–26], triboelectric [10–12, 27], electrostatic [28], and piezoelectric [3, 5, 6, 14, 20, 29–42] approaches. As for biochemical energy mainly focusing on biofuel [18, 19, 43–58], it is the most fundamental energy in body without biocompatible concerns but is challenging for converting into electrical energy. Regarding biothermal energy [59, 60], it is the best stable available energy in vivo but has low efficiency due to small temperature gradient in human body.

Flexible Piezoelectric Energy Harvesters and Sensors, First Edition. Bin Yang, Zhiran Yi, and Chengkuo Lee.

Bioenergy mainly includes biomechanical energy, biochemical energy, and biothermal energy of human body. Researches on high-efficiency bioenergy conversion in vivo are also to be carried out around the three energies. The in vivo performance of these reported implantable biomechanical energy-harvesting technologies for cardiac pacemaker is summarized. Natural mechanical motions of organs can be converted into electrical energy for powering implantable devices [3, 6, 7, 10–12, 14, 20, 29–32]. The mechanisms of mechanical energy conversion include electromagnetic induction [7, 8, 24, 25], triboelectric [10–12, 27, 61, 62], electrostatic [28], and piezoelectric [14, 20, 29–32] methods.. Among them, piezoelectric approach is more promising for implantable batteryless medical devices owing to its high-power density, high-output stability, and device flexibility [3, 6, 14, 31, 33–36, 63, 64].

9.2 ZnO Nanowires-Based IPEH

ZnO nanowires-based PEH [65] was presented by Prof. Wang in 2007 for the first time and demonstrated its large potential solution for powering nanodevices and nanosystems. On account of cellular level biocompatibility and biosafety of ZnO nanowires [66], the technology achieved the first in vivo biomechanical energy harvesting was carried out in 2010 by using a flexible piezoelectric energy harvester based on ZnO nanowires [29] (shown in Figure 9.1), which was fixed at the diaphragm and heart of mice and converted the mechanical energy of such organ motions into electrical energy. This work, as a crucial step toward implantable self-powered nanosystems, demonstrated the in vivo application of a ZnO nanowires-based piezoelectric energy harvester for harvesting biomechanical energy inside a live animal and has shown a feasible approach to scavenge the biomechanical energy inside the body, such as heartbeat, blood flow, muscle stretching, or even irregular vibration.

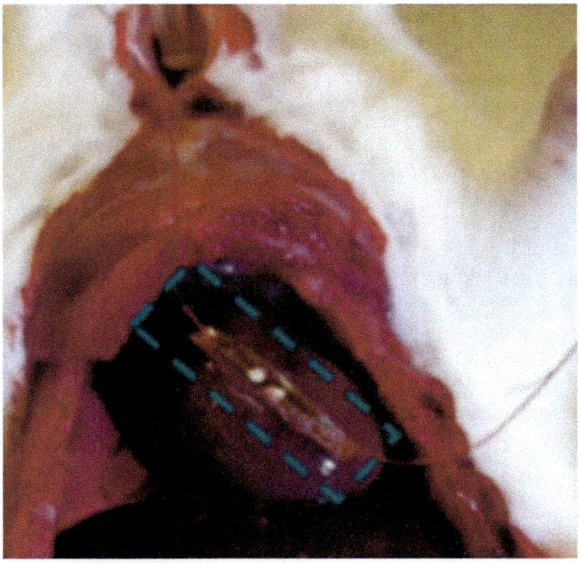

Figure 9.1 A ZnO nanowires-based single-wire generator attached to a live rat's heart. Source: From Li et.al. [29] / with permission of John Wiley & Sons.

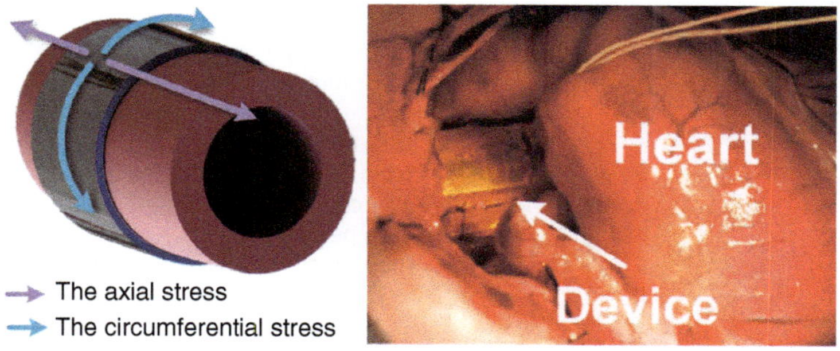

Figure 9.2 The implanted PVDF-based piezoelectric energy harvester wrapping around the ascending aorta. Source: From Cheng et al. [14] / with permission of Elsevier.

9.3 PVDF-Based IPEH

Polyvinylidene fluoride (PVDF), as an excellent flexible piezoelectric material [30, 67], has intrinsic advantages in harvesting mechanical energy in vivo; however, the low electromechanical coupling coefficient of PVDF makes the output of PVDF-based implantable piezoelectric energy harvesters perform generally, especially the output current. A PVDF-based piezoelectric energy harvester [14, 35], as shown in Figure 9.2, was developed by Peking University and Second Military Medical University in 2015. This research evaluated the feasibility and efficacy of generating electric power by utilizing the pulsating energy of ascending aorta with a flexible and implantable piezoelectric energy harvester by implanting the device into wrap around the ascending aorta of a porcine. The V_{max} and I_{max} of the presented IPEH were 1.5 V and 300 nA under the heart rate of 120 bpm and the blood pressure of 160/105 mmHg. The instantaneous output power can reach 30 nW with a long-lasting duration of 700 ms and 77.8% duty ratio. A sponge-like mesoporous PVDF mesoporous PVDF-based nanogenerators [32] was implanted into harvesting the gentle motion energy of rodent muscle to demonstrate its capacity in good biocompatibility and long-term in vivo operation. However, the low-energy-conversion efficiency always becomes the major barrier to the applications of implantable energy harvesting. Obtaining PVDF composite materials or built structures with the ability of enhancement may be possible to develop an available PVDF implantable energy harvester.

9.4 Ceramic-Type IPEH

In 2014, Rogers' research team at Northwestern University made the first attempt to prepare a flexible d_{31}-mode piezoelectric energy harvester using piezoelectric ceramic PZT film-transfer technology to harvest cardiac pulsation kinetic energy to solve the problem of sustainable power supply for pacemakers, with the output current of 0.1 μA [1, 3], which is a great improvement compared to ZnO piezoelectric energy harvester. The solution to the pacemaker power supply problem through

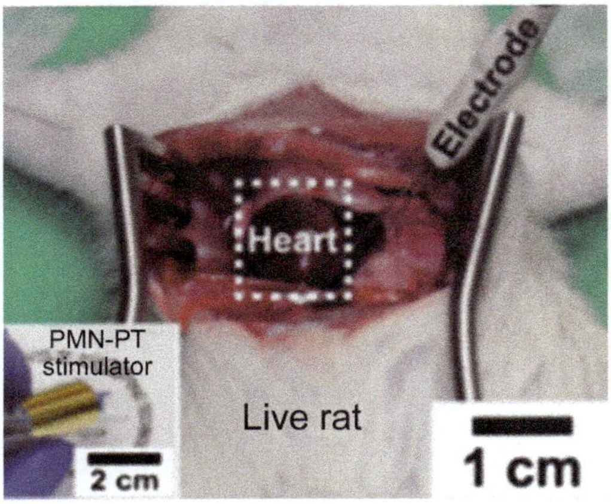

Figure 9.3 A photograph of the medical experiment on a living rat for stimulating the heart. The inset is a snapshot of flexible PMN-PT thin-film stimulator being used to stimulate the living heart of rat. Source: From Hwang et.al. [20] / with permission of John Wiley & Sons.

a multilayer integration scheme was also proposed, but no relevant validation experiments were conducted, and the output performance of the monolithic device was still too low compared to the pacemaker power supply requirement.

Professor Lee's research team at the Korea Advanced Institute of Technology used a single-crystal PMN-PT-based d_{31}-mode piezoelectric energy harvester (Figure 9.3) and proposed the idea of feeding a pacemaker with energy in this energy harvester, and the final validation experiments were also performed in vitro to achieve electrical stimulation of the mouse heart, for which the energy harvester has sufficient energy after implantation. Further studies and validation are needed for this energy harvester to have sufficient energy-conversion efficiency after implantation.

In contrast, high electromechanical coefficient piezoelectric ceramics or piezoelectric single crystals have attracted much attention. 2017 research team of Prof. Lee from Korea Advanced Institute of Technology addressed the previous research on piezoelectric energy harvester The researcher used PMN-PZT single-crystal piezoelectric ceramic material with high electromechanical coupling coefficient to prepare flexible d_{33}-mode piezoelectric energy harvester (Figure 9.4), and a large-animal pig with very similar heart size to human body was chosen to conduct the related experiments. The researchers used the reported implantation method

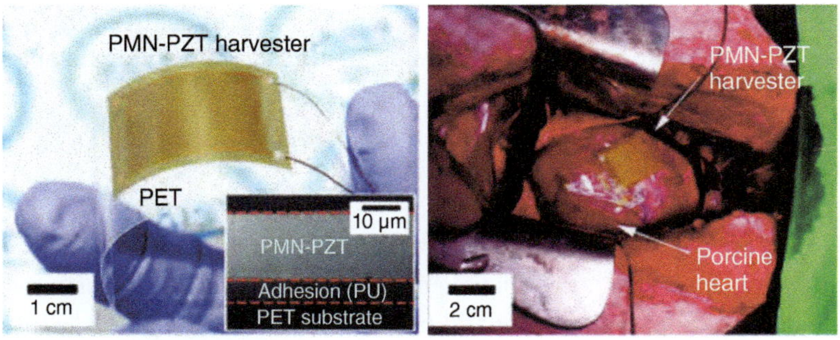

Figure 9.4 A photograph of the device affixed to a porcine heart between the right ventricle and left ventricular apex. Source: From Kim et.al. [6] / with permission of John Wiley & Sons.

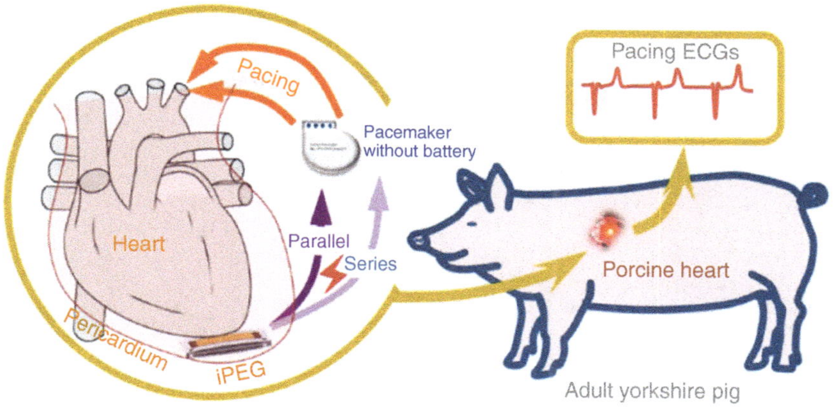

Figure 9.5 Direct Powering a Real Cardiac Pacemaker by Natural Energy of a Heartbeat. Source: From Li et.al. [5] / with permission of American Chemical Society.

and stitched the device to the heart surface to capture the kinetic energy during the heartbeat, achieving a peak open-circuit voltage output of nearly 17.8 V and a short-circuit current output of 1.75 μA, which is a 17.5-fold performance improvement compared to previous studies, but it still cannot meet the power consumption for normal pacemaker operation, so further studies are needed.

Recently, Prof. Yang et al. [5] from Shanghai Jiao Tong University have presented an integrated strategy for direct powering a modern and full-function cardiac pacemaker using a capsular piezoelectric energy harvester (Figure 9.5). The implanted harvester could generate a high-output current of 15 μA in vivo over state-of-the-art performance, which demonstrated the self-powered cardiac pacemaker strategy by implantable piezoelectric harvesting technology. For practical application, this self-powered pacing strategy faced an obstacle about how to set up the pacing leads. Consequently, a heart-worn pacemaker strategy with the batteryless powering and leadless pacing features was further explored by these researchers [68]. The batteryless feature was attained via heart-extrusion energy scavenging through a micromachined piezoelectric thick-film harvester, which obtained a record in vivo output current of 30 μA (Figure 9.6). The exocardial pacing method demonstrated by this batteryless pacemaker avoided a device or leads placed within the cardiac chambers.

Acoustic energy transfer is also a promising energy-harvesting technology candidate for implantable biomedical devices. However, it does not show competitive strength for enabling self-powered implantable biomedical devices due to two issues – large size of bulk piezoelectric ultrasound harvesters and output power fluctuation with transferred distance due to standing waves. Prof. Lee et al. from National University of Singapore report a MEMS-based broadband piezoelectric ultrasonic energy harvester to enable self-powered implantable biomedical devices (Figure 9.7) [64]. The piezoelectric ultrasonic energy harvester is a microfabricated PZT diaphragm array and has wide operation bandwidth. By adjusting frequency of the input ultrasound wave within the operation bandwidth, standing wave effect

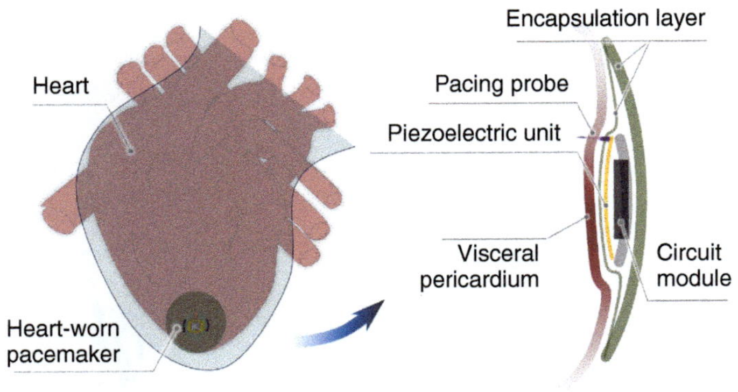

Figure 9.6 A heart-worn batteryless and leadless pacemaker strategy. Source: Yi et al. [68] / with permission of John Wiley & Sons.

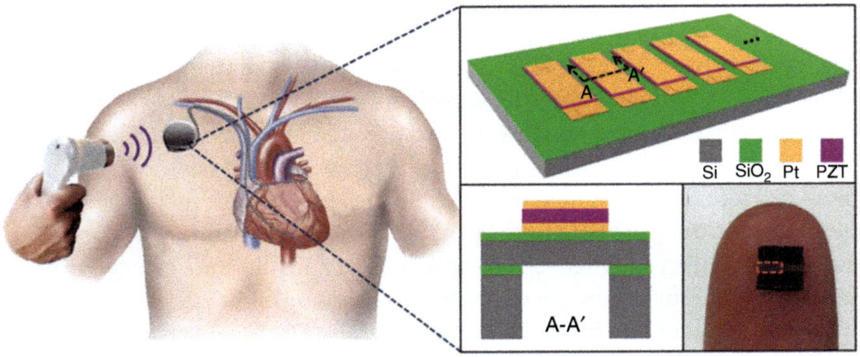

Figure 9.7 Acoustic energy transfer for implantable biomedical devices. Source: Shi et al. [64]/Springer Nature/CC BY 4.0.

can be minimized for any given distance. At 1 cm distance, power density can be increased from 0.59 to 3.75 $\mu W/cm^2$ at input ultrasound intensity of 1 mW/cm^2 when frequency changes from 250 to 240 kHz. The proposed piezoelectric ultrasonic energy harvester shows great potential to power implanted biomedical device chips.

Additionally, a very small-volume piezoelectric MEMS energy-harvesting strategy was reported for a leadless pacemaker [63, 69]. Spiral-shaped piezoelectric harvesters are introduced for their flexibility and large incoming mechanical energy. Nontrivial optimal electrodes placement and best spiral design obtained power of 3 $\mu J/cm^3$ per heartbeat and transduction efficiency of 0.0057 at 1.5 Hz [69]. Unfortunately, the harvested energy was still not sufficient for the power consumption of commercial cardiac pacemakers [70].

9.5 Summary

In vivo biomechanical energy conversion was first achieved by a flexible piezoelectric harvester based on zinc oxide nanowire film [29]. Flexible polyvinylidene difluoride was used to develop an implantable piezoelectric harvester [14, 35, 36]. Subsequently, a piezoelectric ceramic (lead zirconate titanate)-based flexible harvester could realize an in vivo output current of ~0.1 μA in large-animal models [3]. This provided promising evidence that the piezoelectric ceramic method can yield significant electrical power from the motion of internal organs. To further improve its performance, single-crystal piezoelectric materials with higher mechanical coupling coefficients were used to scavenge heartbeat kinetic energy [3, 6, 20, 37–40]. However, commercial cardiac pacemakers are still difficult to drive using implantable piezoelectric harvesters.

Recently, a capsular piezoelectric energy harvester [5] has been presented for direct powering a modern and full-function cardiac pacemaker. The implanted harvester could generate a high-output current of 15 μA in vivo over state-of-the-art performance. For practical application, this self-powered pacing strategy faced an obstacle about how to set up the pacing leads. Consequently, a heart-worn pacemaker strategy with the batteryless powering and leadless pacing features was further explored [68] in 2020. The batteryless feature was attained via heart-extrusion energy scavenging through a micromachined piezoelectric thick-film harvester, which obtained a record in vivo output current of 30 μA. The exocardial pacing method demonstrated by this batteryless pacemaker avoided a device or leads placed within the cardiac chambers. Taken together, sufficient power supply is still a critical obstacle for cardiac pacemakers by implantable piezoelectric harvesters, which are associated with materials, fabrication, structural design, implantation location, and conversion mode [5, 41, 42].

References

1 Dagdeviren, C. (2016). The future of bionic dynamos. *Science* 354: 1109.

2 Mulpuru, S.K., Madhavan, M., McLeod, C.J. et al. (2017). Cardiac pacemakers: function, troubleshooting, and management: part 1 of a 2-part series. *Journal of the American College of Cardiology* 69: 189–210.

3 Dagdeviren, C., Yang, B.D., Su, Y. et al. (2014). Conformal piezoelectric energy harvesting and storage from motions of the heart, lung, and diaphragm. *Proceedings of the National academy of Sciences of the United States of America* 111: 1927–1932.

4 Li, W., Xu, X., Liu, C. et al. (2017). Ultralight and binder-free all-solid-state flexible supercapacitors for powering wearable strain sensors. *Advanced Functional Materials* 27: 1702738.

5 Li, N., Yi, Z., Ma, Y. et al. (2019). Direct powering a real cardiac pacemaker by natural energy of a heartbeat. *ACS Nano* 13: 2822–2830.

6 Kim, D.H., Shin, H.J., Lee, H. et al. (2017). In vivo self-powered wireless transmission using biocompatible flexible energy harvesters. *Advanced Functional Materials* 27: 1700341.

7 Zurbuchen, A., Pfenniger, A., Stahel, A. et al. (2013). Energy harvesting from the beating heart by a mass imbalance oscillation generator. *Annals of Biomedical Engineering* 41: 131–141.

8 Zurbuchen, A., Haeberlin, A., Bereuter, L. et al. (2018). Endocardial energy harvesting by electromagnetic induction. *IEEE Transactions on Biomedical Engineering* 65: 424–430.

9 Choi, Y.S., Yin, R.T., Pfenniger, A. et al. (2021). Fully implantable and bioresorbable cardiac pacemakers without leads or batteries. *Nature Biotechnology* 39: 1228–1238.

10 Zheng, Q., Shi, B., Fan, F. et al. (2014). In vivo powering of pacemaker by breathing-driven implanted triboelectric nanogenerator. *Advanced Materials* 26: 5851–5856.

11 Zheng, Q., Zhang, H., Shi, B. et al. (2016). In vivo self-powered wireless cardiac monitoring via implantable triboelectric nanogenerator. *ACS Nano* 10: 6510–6518.

12 Ma, Y., Zheng, Q., Liu, Y. et al. (2016). Self-powered, one-stop, and multifunctional implantable triboelectric active sensor for real-time biomedical monitoring. *Nano Letters* 16: 6042–6051.

13 Song, K., Han, J.H., Lim, T. et al. (2016). Subdermal flexible solar cell arrays for powering medical electronic implants. *Advanced Functional Materials* 5: 1572–1580.

14 Cheng, X., Xue, X., Ma, Y. et al. (2016). Implantable and self-powered blood pressure monitoring based on a piezoelectric thinfilm: simulated, in vitro and in vivo studies. *Nano Energy* 22: 453–460.

15 Jeong, C.K., Park, K.I., Son, J.H. et al. (2014). Self-powered fully-flexible light-emitting system enabled by flexible energy harvester. *Energy & Environmental Science* 7: 4035–4043.

16 Hwang, G.-T., Kim, Y., Lee, J.H. et al. (2015). Self-powered deep brain stimulation via a flexible PIMNT energy harvester. *Energy & Environmental Science* 8: 2677–2684.

17 Kim, B.-Y., Lee, W.H., Hwang, H.G. et al. (2016). Resistive switching memory integrated with nanogenerator for self-powered bioimplantable devices. *Advanced Functional Materials* 26: 5211–5221.

18 Halamkova, L., Halámek, J., Bocharova, V. et al. (2012). Implanted biofuel cell operating in a living snail. *Journal of the American Chemical Society* 134: 5040–5043.

19 Mercier, P.P., Lysaght, A.C., Bandyopadhyay, S. et al. (2012). Energy extraction from the biologic battery in the inner ear. *Nature Biotechnology* 30: 1240–1243.

20 Hwang, G.T., Park, H., Lee, J.H. et al. (2014). Self-powered cardiac pacemaker enabled by flexible single crystalline PMN-PT piezoelectric energy harvester. *Advanced Materials* 26: 4880–4887.

21 Zhao, J.W., Ghannam, R., Htet, K.O. et al. (2020). Self-powered implantable medical devices: photovoltaic energy harvesting review. *Advanced Healthcare Materials* 9: 2000779.

22 Shi, B.J., Li, Z., and Fan, Y.B. (2018). Implantable energy-harvesting devices. *Advanced Materials* 30: 1801511.

23 Xu, S., Qin, Y., Xu, C. et al. (2010). Self-powered nanowire devices. *Nature Nanotechnology* 5: 366–373.

24 Zurbuchen, A., Haeberlin, A., Bereuter, L. et al. (2017). The Swiss approach for a heartbeat-driven lead- and batteryless pacemaker. *Heart Rhythm* 14: 294–299.

25 Goto, H., Sugiura, T., Harada, Y., and Kazui, T. (1999). Feasibility of using the automatic generating system for quartz watches as a leadless pacemaker power source. *Medical & Biological Engineering & Computing* 37: 377–380.

26 Roberts, P., Stanley, G., and Morgan, J.M. (2008). Harvesting the energy of cardiac motion to power a pacemaker. *Circulation* 118: S679–S680.

27 Ouyang, H., Liu, Z., Li, N. et al. (2019). Symbiotic cardiac pacemaker. *Nature Communications* 10: 1821.

28 Basset, P., Galayko, D., Paracha, A.M. et al. (2009). A batch-fabricated and electret-free silicon electrostatic vibration energy harvester. *Journal of Micromechanics and Microengineering* 19: 115025.

29 Li, Z., Zhu, G., Yang, R. et al. (2010). Muscle-driven in vivo nanogenerator. *Advanced Materials* 22: 2534–2537.

30 Sun, C., Shi, J., Bayerl, D.J., and Wang, X. (2011). PVDF microbelts for harvesting energy from respiration. *Energy & Environmental Science* 4: 4508–4512.

31 Dagdeviren, C., Shi, Y., Joe, P. et al. (2015). Conformal piezoelectric systems for clinical and experimental characterization of soft tissue biomechanics. *Nature Materials* 14: 728–736.

32 Yu, Y., Sun, H., Orbay, H. et al. (2016). Biocompatibility and in vivo operation of implantable mesoporous PVDF-based nanogenerators. *Nano Energy* 27: 275–281.

33 Madhavan, M., Mulpuru, S.K., McLeod, C.J. et al. (2017). Advances and future directions in cardiac pacemakers: part 2 of a 2-part series. *Journal of the American College of Cardiology* 69: 211–235.

34 Roundy, S. and Wright, P.K. (2004). A piezoelectric vibration based generator for wireless electronics. *Smart Materials and Structures* 13: 1131–1142.

35 Zhang, H., Zhang, X.S., Cheng, X. et al. (2015). A flexible and implantable piezoelectric generator harvesting energy from the pulsation of ascending aorta: in vitro and in vivo studies. *Nano Energy* 12: 296–304.

36 Persano, L., Dagdeviren, C., Su, Y. et al. (2013). High performance piezoelectric devices based on aligned arrays of nanofibers of poly(vinylidenefluoride-co-trifluoroethylene). *Nature Communications* 4: 1633.

37 Park, K.I., Xu, S., Liu, Y. et al. (2010). Piezoelectric $BaTiO_3$ thin film nanogenerator on plastic substrates. *Nano Letters* 10: 4939–4943.

38 Park, K.I., Son, J.H., Hwang, G.T. et al. (2014). Highly-efficient, flexible piezoelectric PZT thin film nanogenerator on plastic substrates. *Advanced Materials* 26: 2514–2520.

39 Dagdeviren, C., Su, Y., Joe, P. et al. (2014). Conformable amplified lead zirconate titanate sensors with enhanced piezoelectric response for cutaneous pressure monitoring. *Nature Communications* 5: 4496.

40 Qi, Y., Kim, J., Nguyen, T.D. et al. (2011). Enhanced piezoelectricity and stretchability in energy harvesting devices fabricated from buckled PZT ribbons. *Nano Letters* 11: 1331–1336.

41 Dagdeviren, C., Li, Z., and Wang, Z.L. (2017). Energy harvesting from the animal/human body for self-powered electronics. *Annual Review of Biomedical Engineering* 19: 85–108.

42 Parvez Mahmud, M.A., Huda, N., Farjana, S.H. et al. (2018). Recent advances in nanogenerator-driven self-powered implantable biomedical devices. *Advanced Energy Materials* 8: 1701210.

43 Pan, C., Fang, Y., Wu, H. et al. (2010). Generating electricity from biofluid with a nanowire-based biofuel cell for self-powered nanodevices. *Advanced Materials* 22: 5388–5392.

44 Cinquin, P., Gondran, C., Giroud, F. et al. (2010). A glucose biofuel cell implanted in rats. *PLoS One* 5.

45 Hansen, B.J., Liu, Y., Yang, R.S., and Wang, Z.L. (2010). Hybrid nanogenerator for concurrently harvesting biomechanical and biochemical energy. *ACS Nano* 4: 3647–3652.

46 Miyake, T., Haneda, K., Nagai, N. et al. (2011). Enzymatic biofuel cells designed for direct power generation from biofluids in living organisms. *Energy & Environmental Science* 4: 5008–5012.

47 Rasmussen, M., Ritzmann, R.E., Lee, I. et al. (2012). An implantable biofuel cell for a live insect. *Journal of the American Chemical Society* 134: 1458–1460.

48 Schroder, U. (2012). From in vitro to in vivo-biofuel cells are maturing. *Angewandte Chemie, International Edition* 51: 7370–7372.

49 Szczupak, A., Halámek, J., Halámková, L. et al. (2012). Living battery - biofuel cells operating in vivo in clams. *Energy & Environmental Science* 5: 8891–8895.

50 Southcott, M., MacVittie, K., Halámek, J. et al. (2013). A pacemaker powered by an implantable biofuel cell operating under conditions mimicking the human blood circulatory system–battery not included. *Physical Chemistry Chemical Physics* 15: 6278–6283.

51 Katz, E. and MacVittie, K. (2013). Implanted biofuel cells operating in vivo – methods, applications and perspectives – feature article. *Energy & Environmental Science* 6: 2791.

52 Castorena-Gonzalez, J.A., Foote, C., MacVittie, K. et al. (2013). Biofuel cell operating in vivo in rat. *Electroanalysis* 25: 1579–1584.

53 MacVittie, K., Halámek, J., Halámková, L. et al. (2013). From "cyborg" lobsters to a pacemaker powered by implantable biofuel cells. *Energy & Environmental Science* 6: 81–86.

54 Sales, F., Iost, R.M., Martins, M.V.A. et al. (2013). An intravenous implantable glucose/dioxygen biofuel cell with modified flexible carbon fiber electrodes. *Lab on a Chip* 13: 468–474.

55 Zebda, A., Cosnier, S., Alcaraz, J.P. et al. (2013). Single glucose biofuel cells implanted in rats power electronic devices. *Scientific Reports* 3.

56 Holade, Y., MacVittie, K., Conlon, T. et al. (2014). Pacemaker activated by an abiotic biofuel cell operated in human serum solution. *Electroanalysis* 26: 2445–2457.

57 MacVittie, K., Conlon, T., and Katz, E. (2015). A wireless transmission system powered by an enzyme biofuel cell implanted in an orange. *Bioelectrochemistry* 106: 28–33.

58 Shoji, K., Akiyama, Y., Suzuki, M. et al. (2016). Biofuel cell backpacked insect and its application to wireless sensing. *Biosensors & Bioelectronics* 78: 390–395.

59 Hoang, D.C., Tan, Y.K., Chng, H.B., and Panda, S.K. (2009). Thermal energy harvesting from human warmth for wireless body area network in medical healthcare system. In: *International Conference on Power Electronics and Drive Systems*, 1277–1282. Taipei, Taiwan: IEEE.

60 Cadei, A., Dionisi, A., Sardini, E., and Serpelloni, M. (2014). Kinetic and thermal energy harvesters for implantable medical devices and biomedical autonomous sensors. *Measurement Science and Technology* 25: 012003.

61 Hinchet, R., Yoon, H.J., Ryu, H. et al. (2019). Transcutaneous ultrasound energy harvesting using capacitive triboelectric technology. *Science* 365: 491–494.

62 Fan, F.R., Tian, Z.Q., and Wang, Z.L. (2012). Flexible triboelectric generator! *Nano Energy* 1: 328–334.

63 Jackson, N., Olszewski, Z., O'Murchu, C., and Mathewson, A. (2018). Ultralow-frequency PiezoMEMS energy harvester using thin-film silicon and parylene substrates. *Journal of Micro/Nanolithography, MEMS, and MOEMS* 17 (1): 015005. https://doi.org/10.1117/1.JMM.17.1.015005.

64 Shi, Q., Wang, T., and Lee, C. (2016). MEMS based broadband piezoelectric ultrasonic energy harvester (PUEH) for enabling self-powered implantable biomedical devices. *Scientific Reports* 6: 24946.

65 Wang, X.D., Song, J.H., Liu, J., and Wang, Z.L. (2007). Direct-current nanogenerator driven by ultrasonic waves. *Science* 316: 102–105.

66 Li, Z., Yang, R., Yu, M. et al. (2008). Cellular level biocompatibility and biosafety of ZnO nanowires. *The Journal of Physical Chemistry C* 112: 20114–20117.

67 Han, M., Zhang, X.S., Meng, B. et al. (2013). r-Shaped hybrid nanogenerator with enhanced
piezoelectricity. *ACS Nano* 7: 8554–8560.

68 Yi, Z., Xie, F., Tian, Y. et al. (2020). A battery- and leadless heart-worn pacemaker strategy. *Advanced Functional Materials* 30 (25): 2000477.

69 Deterre, M., Lefeuvre, E., Zhu, Y. et al. (2014). Micro blood pressure energy harvester for intracardiac pacemaker. *Journal of Microelectromechanical Systems* 23: 651–660.

70 Jackson, N., Olszewski, O.Z., O'Murchu, C., and Mathewson, A. (2018). Ultralow-frequency PiezoMEMS energy harvester using thin-film silicon and parylene substrates. *Journal of Micro/Nanolithography, MEMS and MOEMS* 17 (1): 015005. https://doi.org/10.1117/1.JMM.17.1.015005.

10

Flexible Piezoelectric Sensors

10.1 Introduction

Piezoelectric effect has been widely developed for lots of common commercial physical sensors for the measurement of acceleration, acoustics, force, load, pressure, shock, strain, torque, and vibration. With the rapid development of wearable electronics and requirement of some special applications, demand for flexible piezoelectric sensors is gradually built to obtain more accurate information from human body or other irregular surfaces. Flexible piezoelectric sensing was investigated as a potential approach in healthcare monitoring [1, 2], measurement of complex surfaces [3], or intelligent recognition [4, 5]. Researches on the flexible piezoelectric sensors mainly focused on cutaneous pressure, force, strain, tension, ultrasonic sensors, sensorimotor enhancer, and accelerometer.

10.2 Working Principle

Flexible piezoelectric sensing is one developing technology of flexible electronics based on the traditional piezoelectric effect. The fundamental sensing principle can be described by the formula as follows:

$$\text{Sensitivity} = \frac{\mathrm{d}V_\mathrm{N}}{\mathrm{d}F} = \frac{\mathrm{d}}{\mathrm{d}F}\left(\frac{\Delta V}{V_\mathrm{s}}\right)$$

where V_N is the normalized voltage, F is the applied pressure, V_s is the saturation voltage, and ΔV is the relative change in output voltage. Since, the charge $Q = d_{ij} \cdot F$, where d_{ij} is piezoelectric constant, while the output voltage $V = d_{ij} \cdot F/C$, where C is capacitance of the piezoelectric sensor. The open-circuit voltage is relatively independent of frequency since the charge is proportional to the applied force, which is benefit for cutaneous activity sensing.

Flexible Piezoelectric Energy Harvesters and Sensors, First Edition. Bin Yang, Zhiran Yi, and Chengkuo Lee.

10.3 Flexible Piezoelectric Sensors

10.3.1 Pressure/Force Sensor

(1) PVDF

Wearable piezoelectric pressure/force sensors were investigated for monitoring the motion, tactile, behavior, respiration, arterial pulse, or heart rate. Polyvinylidene difluoride (PVDF), as a flexible piezoelectric material, was developed as a film sensor with a double-sided arch structure and thin-shell structure [6] for respiration and pulse monitoring in real time, respectively. Its flexible characteristics make the device comfortable to wear and meet the patient's daily monitoring needs. Given near-field electro-spinning is capable of precisely depositing highly aligned micro/nanofibers, which are potential to obtain high sensitivity. Fuh et al. developed [7] wavy-substrate self-powered sensors based on near-field electrospun PVDF. To further improve its performance, the piezoelectric ceramics or other materials were used to mix into PVDF. A piezoelectric pressure sensor based on electrospun PVDF/$BaTiO_3$ nanowire nanocomposite fibers [8] performs a pressure sensitivity of 0.017 kPa^{-1} in the range from 1 to 40 kPa and a response time of 290 ms. Due to the good clothing-integrated characteristics of the fibers, Huang et al. [9] fabricated flexible piezoelectric sensors based on PVDF/graphene composite coating on commercially available fabrics. The piezoelectric sensor exhibits a sensitivity of 34 V/N and shows low detecting threshold of 0.6 mN. To enhance the effective strain in the piezoelectric layer, You et al. [10] fabricated a 3D wave-shaped piezoelectric composite made of P(VDF-TrFE) piezoelectric film and high-elastic amorphous alloy ribbon, as shown in Figure 10.1. The wave-shaped structure produces pre-stretch in P(VDF-TrFE) film to increase the effective piezoelectric strain coefficient. An applied pressure could not only produce compressive stress in the film but also change the curvature and suffer stress along the length direction; therefore, it can generate a much higher voltage output. For further study, Xin et al. [11] found that the curvature of the multi-arch structure pressure sensor was one of the main factors influencing the voltage output performance of PVDF piezoelectric film. The proposed sensor with a curvature of 200 m^{-1} increased the detection signal amplitude of the measurements by 141% compared with the counterpart of planar type.

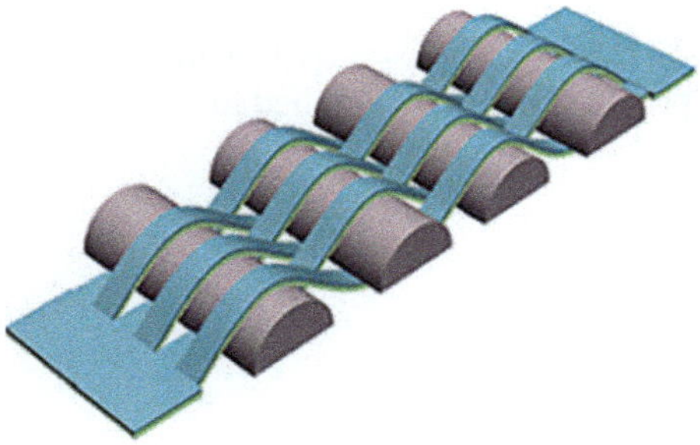

Figure 10.1 A 3D wave-shaped piezoelectric composite made of P(VDF-TrFE) piezoelectric film and high-elastic amorphous alloy ribbon. Source: From You et al. [10] / with permission of AIP Publishing.

(2) Piezoelectric ceramics
To further improve the sensitivity, the piezoelectric ceramics were employed for the wearable piezoelectric pressure sensor. Park et al. [12] demonstrated a piezoelectric pulse sensor to obtain a sensitivity of approximately 0.018 kPa^{-1}, response time of approximately 60 ms. Then, Yi et al. [13, 14] proposed a piezoelectric force sensor based on thinned bulk piezoelectric ceramics to monitor cutaneous activities in real time and recognize hand gestures. The sensitivity reached approximately 10 V/N and the response time was 11 ms. Inspired by the configuration of the human epidermis, Yan et al. [15] presented a flexible three-dimensional cellular sensor array based on lead zirconate titanate (PZT) via a one-step thermally induced phase separation method. The sensor holds a decent pressure sensitivity up to 0.19 V/kPa with a response time of less than 16 ms.

(2) Friendly piezoelectric materials
For exploiting an environment-friendly and flexible piezoelectric pressure sensor, Ghosh et al. presented a bioinspired piezoelectric pressure sensor (~0.8 V/kPa) from structurally stable fish gelatin nanofibers using large-area compatible electro-spinning technology [16]. Gao et al. [17] reported a lead-free flexible piezoelectric sensor with a sensitivity of approximately 0.55 V/kPa within the range of 5–85 kPa based on the composite with barium calcium zirconate titanate (BCZT) and PDMS (Table 10.1).

Table 10.1 Recent researches in the flexible piezoelectric pressure/force sensors.

References	Materials	Range	Sensitivity	Response time (ms)
[8]	PVDF/$BaTiO_3$ nanowire fibers	1–40 kPa	0.017 kPa^{-1}	290
[9]	PVDF/GO	>0.6 mN	34 V/N	
[12]	PZT	1–30 kPa	0.018 kPa^{-1}	60
[13, 14]	PZT	5–500 mN	10 V/N	11
[15]	Cellular structure with PZT		0.19 V/kPa^1	16
[18]	ZnO	>0.6 Pa >0.1 m/s^2 >57 dB <250 Hz	−6.8 kPa^{-1}	<5
[19]	PVDF/GO		35 nA/kPa below 2.45 kPa	
[20]	$PbTiO_3$ nanowires/ graphene		9.4×10^{-3} kPa^{-1}	5–7
[16]	Fish gelatin nanofibers		~0.8 V kPa^{-1}	
[17]	$(Ba_{0.85}Ca_{0.15})(Ti_{0.90}Zr_{0.10})O_3$ (BCZT) and PDMS	5–85 kPa	0.55 V/kPa	

Source: Zhu et al. [21] / with permission of Elsevier.

10.3.2 Strain Sensor

Flexible and stretchable electronics technologies have been attracting increasing attention owing to their potential applications in wearable human–machine interfaces [22, 23]. Cutaneous activities mainly come from motion in limbs, pulse waves in arteries, or disturbances associated with breathing. The subtle changes caused by cutaneous activities are expected to be measured via the sensors attached to the skin, which is valuable for in situ monitoring of vital signs in healthcare and recognizing different gestures in man–machine interaction [22, 24–28]. The conventional sensors are faced with some critical demerits, such as rigid structures and high-power consumption, which causes inconvenience in daily life for medical applications [29]. Many flexible pressure sensors are reported and based on various transduction mechanisms, including piezoresistive [30], capacitive [31], and optical types [32]. These sensors have detected physiological signals produced by human activities, but the power supply is inevitable for their operation, which limits their wearable application. Moreover, frequent replacement of power sources also severely restricts the development of implantable biomedical electronics in vivo environment [33]. An attractive and promising approach to solve these issues is to develop flexible sensors with conformal, highly sensitive, and self-powered properties [34]. Piezoelectric force sensors can directly generate electrical signals in response to mechanical strain for a self-powered sensor system [12]. However, low electromechanical coupling coefficient of the deposited PZT film [35] and the improper design of the mechanical neutral plane result in its poor performance in the arterial pulse monitoring.

To enhance wearability of the piezoelectric strain sensors, transparent and stretchable wearable mechanical sensors [36] were developed with a sensitivity of 0.12 mV/cm below 3 cm^{-1}. A significant enhancement of the piezoelectric wearable sensor was obtained by ultrathin ZnO p–n homojunction films [37] and a PVDF sensor [38]. Microelectromechanical system technologies and transfer printing processes are adopted for fabrication of the epidermal PVDF sensor. Based on a similar fabrication process, an ultrathin flexible piezoelectric sensor for monitoring eye fatigue was designed and fabricated [39], as shown in Figure 10.2. Then, Sato et al. developed fully printed flexible strain sensors by optimizing the concentration and viscosity of a ferroelectric polymer solution [40]. The strain rate sensitivity is

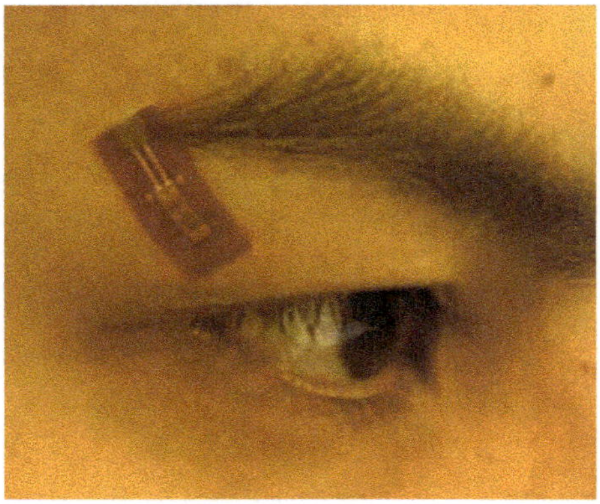

Figure 10.2 An ultrathin flexible piezoelectric sensor for monitoring eye fatigue. Source: From Lu et al. [39] / with permission of IOP Publishing.

about 8 V·s/m at the applying strains of 10–500 mm/s. Additionally, maintaining appropriate levels of food intake and developing regularity in eating habits are crucial to weight loss and the preservation of a healthy lifestyle [41–45]. Rapid weight gain during infancy increases the risk of obesity [46]. Moreover, awareness of eating habits is an important step toward portion control and weight loss. Food-intake monitoring system based around a wearable wireless-enabled necklace attracted researchers' attention.

Professor Yang and coworkers proposed a multifunctional piezoelectric force sensor with multilayer structure to monitor radial pulse signals and recognize hand gestures [13], as shown in Figure 10.3. The thinned bulk piezoelectric ceramic as the functional layer is utilized for the wearable force sensor to improve sensitivity. The bonded supporting metal foil provides more uniform stress distribution on the piezoelectric layer. They demonstrated a high sensitivity piezoelectric force sensor based on thinned bulk inorganic piezoelectric ceramic, which has a high electromechanical-coupling coefficient. Experimental studies provide characteristics of the sensor with sensitivity of ~10 V/N. The piezoelectric force sensor is successfully used for monitoring radial pulse signals in near-surface arteries in real time and recognizing different hand gestures. The monitored arterial pulse voltage amplitude is close to doubled improvement compared to recent reports based on the deposited PZT film. The arterial pulse force of the young man is about ~20 mN. Meanwhile, a wireless transmission system is designed and fabricated for acquiring in situ information of the sensor to achieve real-time monitoring. The sensor performs a promising application in healthcare or man–machine interaction.

10.3.3 Ultrasonic Sensor

Piezoelectric ultrasonic monitoring or detecting is a traditional wireless method widely used in many industries or medical applications. However, development of flexible piezoelectric ultrasonic devices is just beginning. Professor Xu and coworkers [47, 48] report a low-profile membrane-based stretchable ultrasonic probe, as shown in Figure 10.4, based on an array of thin and high-performance 1–3 piezoelectric composites. The resulting device has an electromechanical-coupling coefficient of 0.60, a high signal-to-noise ratio of 20.28 dB, a wide bandwidth of 47.11%, a negligible cross-talk level between adjacent elements of −70 dB, and a high spatial resolution of 610 mm at different depths. The "island-bridge" layout offers bi-axial stretchability of more than 50% with minimal impact on the transducer performance, which allows the device to work on nonplanar complex defects under flat, concave, and convex surfaces.

Meanwhile, Professor Yang and coworkers adopted a similar but simpler technology route to achieve a flexible piezoelectric ultrasonic device [49], as shown in Figure 10.5. The flexible piezoelectric ultrasonic energy harvester array was designed and fabricated by integrating a large number of piezoelectric active elements with multilayered flexible electrodes in an elastomer membrane. The developed flexible piezoelectric ultrasonic energy harvester can be driven by the ultrasonic wave to produce continuous voltage and current outputs on both planar and curved surfaces, reaching output signals of more than 2 V_{pp} and 4 μA, respectively. Then,

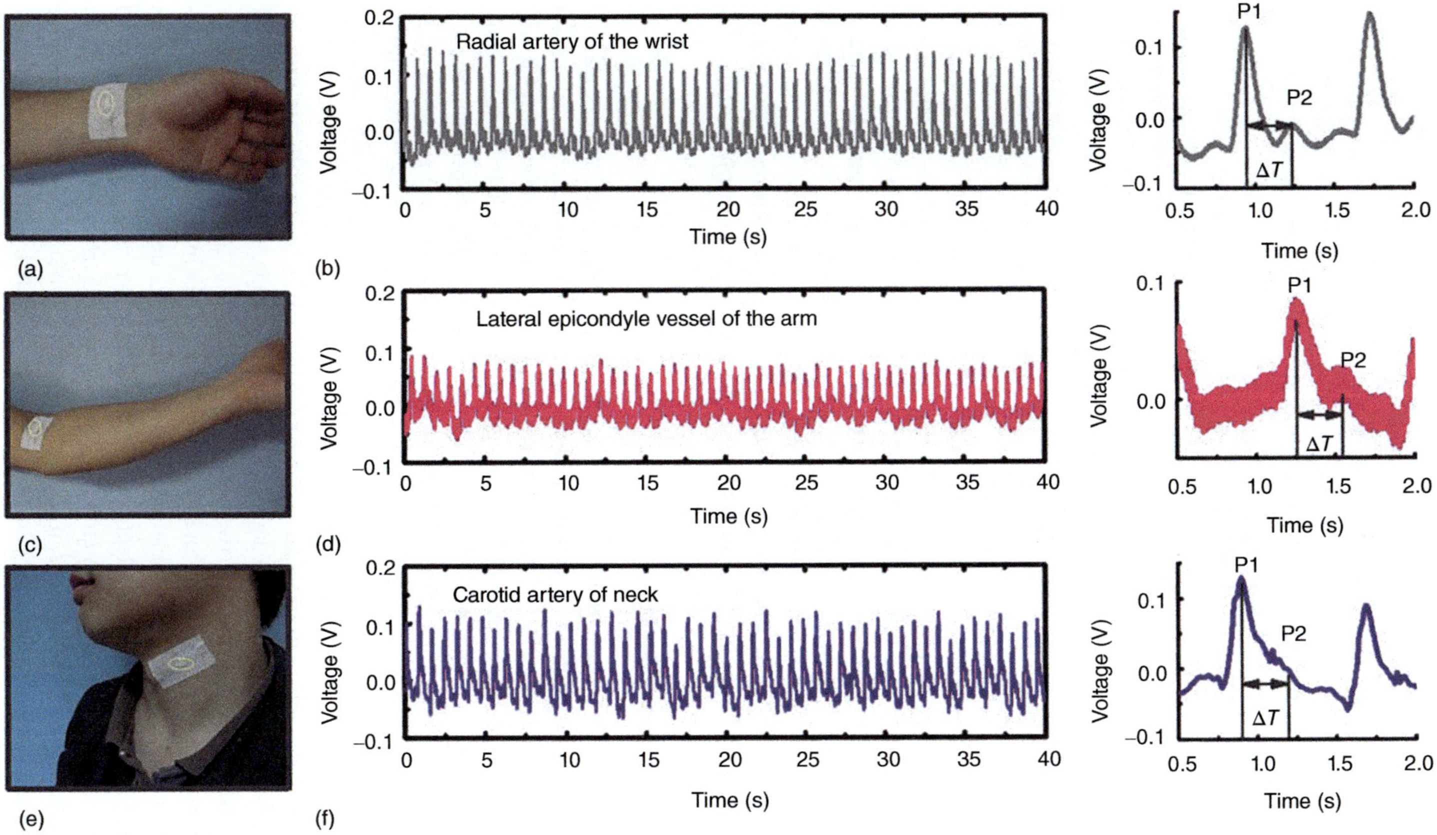

Figure 10.3 Measurement of atrial pulse via the piezoelectric sensor. Source: From Yi et al. [13] / with permission of IEEE. (a), (c), and (e) Photograph of a piezoelectric sensor attached to the wrist, the lateral epicondyle vessel, and the carotid artery of neck. (b), (d), and (f) Source: Yi et al. [13] / with permission of IEEE. The corresponding recorded radial artery pulse signals and the enlarged pulse signal.

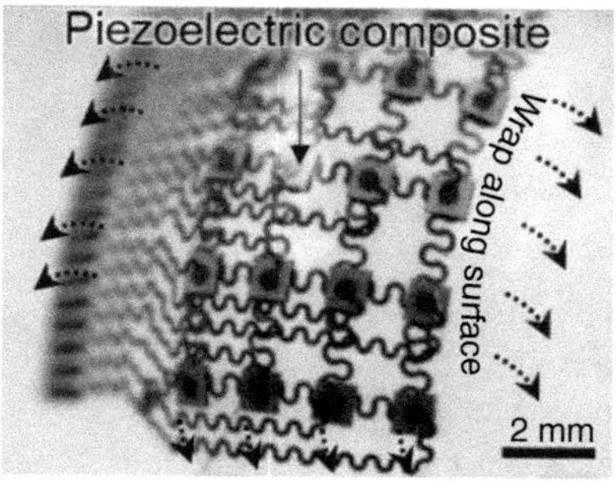

Figure 10.4 Low-profile membrane-based stretchable ultrasonic probe. Source: Wang et al. [48] / with permission of Springer Nature.

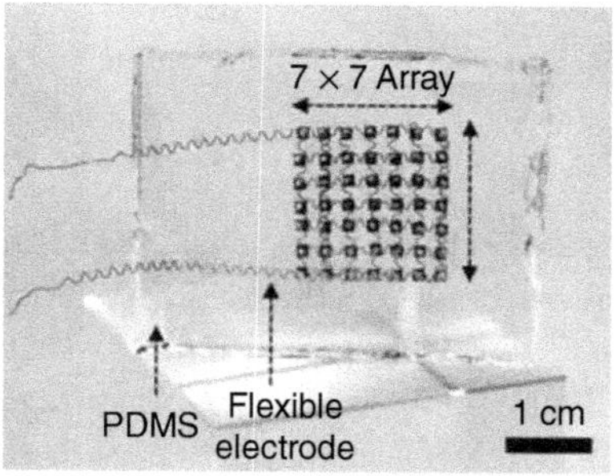

Figure 10.5 Flexible piezoelectric ultrasonic energy harvester array for bio-implantable wireless generator. Source: Jiang et al. [49] / with permission of Elsevier.

they also presented an electrical stimulation strategy using ultrasound-driven wireless energy-harvesting technology to convert acoustic energy to electricity through the piezoelectric effect [50]. The design, fabrication, and performance of a millimeter-scale flexible ultrasound patch that utilizes an environment-friendly lead-free piezocomposite are described. A modified dice-and-fill technique is used to manufacture the microstructure of the piezocomposite and to generate improved electrical and acoustic properties. The as-developed device can be attached to a complex surface and be driven by ultrasound to produce adjustable electrical outputs, reaching a maximum output power of 45 mW/cm^2. The considerable current signals (e.g. current > 72 μA and current density > 9.2 $nA/\mu m^2$) that are higher than the average thresholds of retinal stimulation are also obtained in the ex vivo experiment of an implanted environment, showing great potential to be integrated on implanted biomedical devices for electrical stimulation application.

Thrombosis and restenosis after vascular reconstruction procedures may cause complications, such as stroke, but a clinical means to continuously monitor vascular conditions is lacking. Conventional ultrasound probes are rigid, particularly for postoperative patients with fragile skin. Techniques based on photoplethysmography or thermal analysis provide only relative changes in flow volume and have a shallow detection depth. Professor Feng et al. introduced a flexible Doppler ultrasound device, as shown in Figure 10.6, for the continuous monitoring of the absolute velocity of blood flow in deeply embedded arteries based on the Doppler effect. The device is thin (1 mm), lightweight (0.75 g), and skin conforming. When the dual-beam Doppler method is used, the influence of the Doppler angle on the velocity measurement is avoided. Experimental studies on ultrasound phantoms and human subjects demonstrate accurate measurement of flow velocity. The

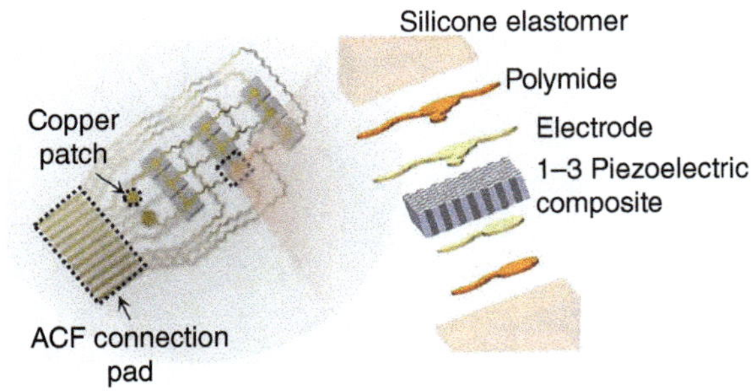

Figure 10.6 Flexible Doppler ultrasound device for the monitoring of blood flow velocity. Source: Wang et al. [51] / American Association for the Advancement of Science / CC BY-4.0. ACF, anisotropic conductive film.

wearable Doppler device has the potential to enhance the quality of care for patients after reconstruction surgery.

10.3.4 Wearable Piezoelectric Other Sensors

Yoshida et al. developed a tension sensor with a coaxial structure using a narrow slit ribbon made of a uniaxially stretched poly(L-lactic acid) (PLLA) film for application to a wearable device [52]. Construction of the tension sensor is shown as follows (Figure 10.7). Tinsel wire was used as the center conductor of the sensor. The tinsel wire consists of yarn of synthetic fibers arranged at the center, with a spirally wound rolled copper foil ribbon on the side surface. Next, slit ribbons obtained from a uniaxially oriented film of PLLA are wound helically on the side surface of the center conductor in the direction of a left-handed screw, at an angle of 45° to the central axis. The rolled copper foil is used as an outer conductor and covers the yarn without a gap. The prototype of the fabricated tension sensor has good flexibility since the sensor is in the form of a filament and consists of a highly flexible material. The piezoelectric line was found to have an output sensitivity per unit length of 14 pC/N mm. It was also found that the sensor maintained its room-temperature sensitivity up to 60 °C. Compared with an existing coaxial line sensor using PVDF, the sensor using PLLA does not exhibit pyroelectricity, meaning that no undesirable voltage is generated when in contact with body heat, which is a significant advantage of wearable sensors.

Hu et al. presented a chest-worn accelerometer with high sensitivity for continuous cardiorespiratory sound monitoring [53]. The accelerometer is based on an asymmetrically gapped cantilever, which is composed of a bottom mechanical layer and a top piezoelectric layer separated by a gap. A sensitivity of 86 V/g and a noise floor of 40 $ng/Hz^{1/2}$ are obtained. Preliminary tests for recording both cardiac and respiratory signals are carried out on human body and the new sensor exhibits better performance compared with a high-end electronic stethoscope.

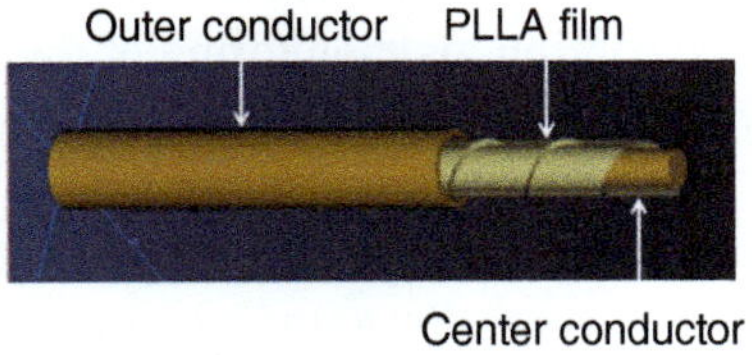

Figure 10.7 Structure of sensor having a coaxial structure. Source: Yoshida et al. [52] / with permission of IOP Publishing.

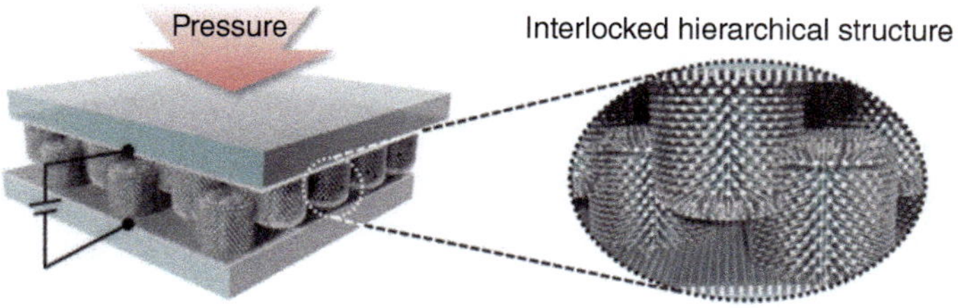

Figure 10.8 Bioinspired interlocked and hierarchical design of ZnO nanowire arrays for static and dynamic pressure-sensitive electronic skins. Source: From Ha et al. [18] / with permission of John Wiley & Sons.

10.4 Piezoelectric Static Sensing

Most piezoelectric-induced pressure sensors have challenges in measuring static signals. A combination mechanism based on the piezoresistive and piezoelectric modes was presented for detecting both static and dynamic tactile stimuli. A sensor based on the interlocked ZnO piezoelectric nanowires [18], as shown in Figure 10.8, performed a high-pressure sensitivity of $-6.8\ kPa^{-1}$ and an ultrafast response time of <5 ms in the piezoresistive mode and a high-frequency-dynamic stimuli ($\approx$250 Hz) in the piezoelectric mode. Then, the same idea is to develop fingerprint-like patterns and interlocked microstructures sensor with a sensitivity of 35 nA/kPa below 2.45 kPa [19, 54]. Additionally, a synergistic mechanism between strain-induced polarization charges in piezoelectric nanowires and the caused change of carrier scattering in graphene was developed to resolve this problem. Chen et al. [20] presented a pressure sensor with nanowires/graphene heterostructures based on this mechanism. This sensor performs the static pressures with a sensitivity of up to $9.4 \times 10^{-3}\ kPa^{-1}$ and a fast response time down to 5–7 ms.

10.5 Wireless Communication

Arterial pulse responses are closely associated with the body's health [55, 56]. Subtle cutaneous activities caused by arterial pulses can be captured by the attached sensors on the skin, which is valuable for healthcare monitoring [25]. At present, wearable sensors for healthcare monitoring mainly includes piezoelectric [13], piezoresistive [26, 57], triboelectric [58], optical [25], or hybrid [59–61] approaches. The piezoelectric approach with highly sensitive dynamic response characteristics and easy miniaturization has been widely deployed to obtain accurate arterial pulse signals [29, 62–64]. Considering the direct contact with the skin, wearable piezoelectric sensors are developing toward more flexible, wireless, and portable sensors. Commercially available arterial pulse sensors exist predominantly in the form of mechanically hard components with the leads for power supply or acquiring data. Developed wireless pulse sensors are ultimately limited by the dimensions and weight of the batteries for power supply [13]. Currently, near-field

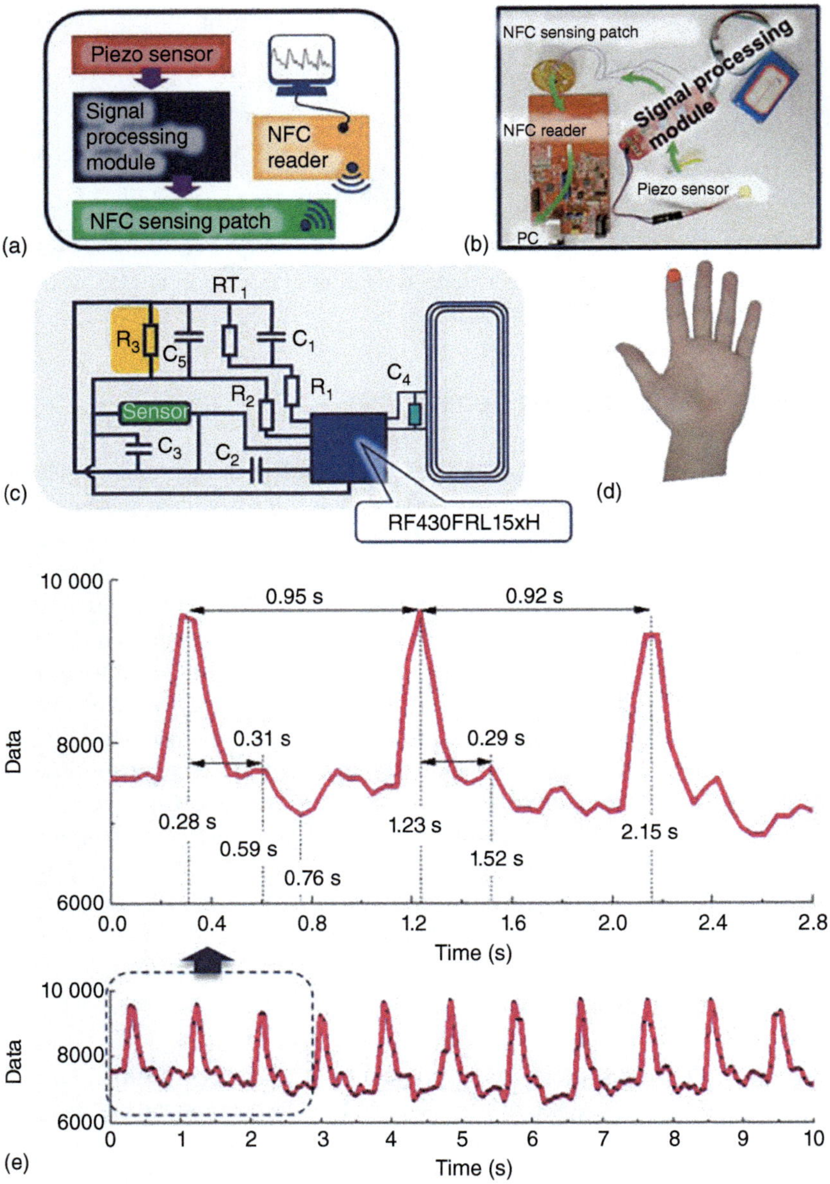

Figure 10.9 Feasibility evaluation of NFC-based piezoelectric arterial pulse waves monitoring. (a) Schematic illustration of the detailed signals transmission route. (b) Photograph of the verification circuit. (c) The portable NFC-based sensor monitoring circuit bulk diagram. (d) Photograph for the attached site of the piezoelectric force sensor. (e) The acquired arterial pulse waveforms of the index finger. Source: Yi et al. [14]/IEEE.

communication (NFC) technology has been exploited for wireless and battery-less wearable monitoring systems or medical devices due to its easy miniaturization and integration compared to Bluetooth, ZigBee, or Wi-Fi methods [65–69]. However, NFC sensors are generally used to complete a static measurement [70]. It is rare to be used for dynamic signal measurement. Therefore, exploiting NFC technology for wireless piezoelectric sensors to develop a more portable arterial pulse monitoring system is promising. Recently, Professor Yang et al. presented a wireless arterial pulse monitoring strategy by combining the piezoelectric force sensor and NFC technology, as shown in Figure 10.9.

The piezoelectric force sensor mainly consists of the thinned PZT thick-film functional layer and a flexible polymer substrate. It is characterized by different applied frequencies and different attached surfaces. Arterial pulse monitoring performances of single piezoelectric force sensor are characterized by the fingers and wrist. A commercial NFC sensing patch based on RF430FRL15xH chip is used after configuring the ADC with 20 Hz sampling rate in agreement with standard protocol. A bias and amplification processing are to avoid the negative or mismatched piezoelectric signals for ADC input. Figure 10.9a,b show schematic illustration of the detailed signals transmission route and the photograph of the practical verification circuit. The signals are received by an NFC reader module (RF430FRL152HEVM), which is available for PC applications. In experiment, the R3 of the NFC patch (Figure 10.9c) is substituted by the piezoelectric force sensor to create the wireless communication of the sensing signals. Finally, the feasibility of NFC technology-based wireless piezoelectric arterial pulse monitoring system is demonstrated by acquiring the arterial pulse signals of the finger. The piezoelectric is attached to the index finger (Figure 10.9d). Figure 10.9e displays the acquired arterial pulse waveforms.

10.6 Application in Body Sensor Network

With the rapid emergence of novel wearable electronics, piezoelectric wearable devices were widely investigated as body sensor networks for healthcare monitoring or motion recognition due to their low power consumption. A wearable sensorimotor enhancer intends to improve tactile sensitivity in human fingertips [71] by attaching a piezoelectric stack actuator to the radial side of the fingertip and stimulating tactile receptors by applying vibration. This approach was also used as a stimulator for the feedback of virtual reality. Ultralow power consumption could maintain their normal operations. For a human–computer interface glove, flexible PVDF sensors [23] were used to detect changes in the angles of the finger joints. Wireless module was attached to the wrist for portable wearing. The virtual hand could be controlled well by this interface glove. Similar research was reported by Volkinburg and Washington [22]. They fabricated a wearable gesture-based controller by using the PVDF sensors. The controller can repeatedly and accurately discern, in real time, between right- and left-hand gestures. In addition to these researches about motion recognition, healthcare monitoring also is a hot area. Due to the high permittivity, low-polarization response time, and outstanding durability of the ZnO-nanoneedle, the hybrid film performed the lowest detectable pressure of 4 Pa [72]. Furthermore, the reduced graphene oxide (rGO) electrode-based Bluetooth antenna was used to attain a high peak gain of 2.70 dBi, which offered a promising tool for wireless pressure monitoring in critical healthcare. Meier et al. [73] provided an energy-harvesting, shoe-mounted system for medical sensing using piezoelectric transducers for generating power. The sensing electronics were completely powered by the harvested energy from walking or running, generating 10–20 μJ of energy per step that is then consumed by capturing and storing the force sensor data. Park et al. [12] developed a self-powered piezoelectric pulse sensor for in vivo measurement of radial/carotid pulse signals in near-surface arteries. The sensor performed a sensitivity of $0.018\,kPa^{-1}$ and response time of 60 ms. Wireless transmission of detected arterial pressure signals to a smartphone demonstrated the possibility of self-powered and real-time pulse monitoring system. Yi et al. [14] proposed a wireless piezoelectric pulse sensor based on NFC technology, which greatly decreased the volume and enhanced its wearability. Maintaining appropriate levels of food intake and developing regularity in eating habits are crucial to weight loss and the preservation of a healthy lifestyle [42–44]. As a food-intake monitoring system based around a wearable wireless-enabled necklace, motion in the throat is captured and transmitted to a mobile application for processing and user guidance. Using Bayesian networks, average precision and recall of 90.17% and 88.9% were obtained to classify between chewable vitamins, saliva swallowing, medication capsules, speaking, and drinking water. Additionally, rapid weight gain during infancy increases the risk of obesity. To improve the precision of the piezoelectric jaw motion sensor [41], Farooq and Sazonov [46] trained two separate linear support vector machine classifiers for food intake and activity detection, which resulted in a weighted average of precision and recall of 99.85%. Unlike the previous state-of-the-art piezoelectric sensor-based system that employs spectrogram features, Hussain et al. [45] tried to fully exploit time-domain-based signals for optimal features. They demonstrated that the chewing sequence carried important information for food classification. They presented a system that yields an accuracy of 89.2% for food intake detection and 80.3% for food classification of over 17 food categories. Patient adherence is critical to the successful treatment of many diseases, as well as the effective assessment of treatment for research purposes. Kalantarian et al. [44] proposed a two-step system for detecting when a pill bottle is opened using commercial smart-bottle technologies, and when a pill is consumed using a custom-designed smart necklace using a piezoelectric sensor. These two mechanisms coupled with the mobile application can passively monitor adherence and inform caregivers of patient status. Kim et al. [54] presented a flexible wireless electronic skin sensor system achieved by an interlocked PVDF/rGO micro-dome array structure. A flexible system prototype was developed and experimentally verified to provide various wireless wearable sensing functions – including pulse wave, voice, chewing/swallowing, breathing, knee movements, and temperature – while their real-time sensed data are displayed on a smartphone.

References

1 Liu, Z.-D., Liu, J.K., Wen, B. et al. (2018). Cuffless blood pressure estimation using pressure pulse wave signals. *Sensors* 18: 4227.

2 Wang, T.-W. and Lin, S.-F. (2020). Wearable piezoelectric-based system for continuous beat-to-beat blood pressure measurement. *Sensors (Basel Switzerland)* 20: 851.

3 Xiong, W., Zhu, C., Guo, D. et al. (2021). Bio-inspired, intelligent flexible sensing skin for multifunctional flying perception. *Nano Energy* 90: 106550.

4 Wang, W., Cao, J., Yu, J. et al. (2019). Self-powered smart insole for monitoring human gait signals. *Sensors (Basel, Switzerland)* 19 (24): 5336.

5 Tahir, A.M., Chowdhury, M.E., Khandakar, A. et al. (2020). A systematic approach to the design and characterization of a smart insole for detecting vertical ground reaction force (vGRF) in gait analysis. *Sensors (Basel, Switzerland)* 20: 957.

6 Xin, Y., Qi, X., Qian, C. et al. (2014). A wearable respiration and pulse monitoring system based on PVDF piezoelectric film. *Integrated Ferroelectrics* 158: 43–51.

7 Fuh, Y.K., Wang, B.S., and Tsai, C.-Y. (2017). Self-powered pressure sensor with fully encapsulated 3D printed wavy substrate and highly-aligned piezoelectric fibers array. *Scientific Reports* 7 (1): 1–7.

8 Guo, W., Tan, C., Shi, K. et al. (2018). Wireless piezoelectric devices based on electrospun PVDF/$BaTiO_3$ NW nanocomposite fibers for human motion monitoring. *Nanoscale* 10: 17751–17760.

9 Huang, T., Yang, S., He, P. et al. (2018). Phase-separation-induced PVDF/graphene coating on fabrics toward flexible piezoelectric sensors. *ACS Applied Materials & Interfaces* 10: 30732–30740.

10 You, S., Shi, H., Wu, J. et al. (2016). A flexible, wave-shaped P(VDF-TrFE)/metglas piezoelectric composite for wearable applications. *Journal of Applied Physics* 120: 234103.

11 Xin, Y., Liu, T., Xu, Y. et al. (2019). Development of respiratory monitoring and actions recognition based on a pressure sensor with multi-arch structures. *Sensors and Actuators A: Physical* 296: 357–366.

12 Park, D.Y., Joe, D.J., Kim, D.H. et al. (2017). Self-powered real-time arterial pulse monitoring using ultrathin epidermal piezoelectric sensors. *Advanced Materials* 29: 1702308.

13 Yi, Z., Yang, H., Tian, Y. et al. (2018). Self-powered force sensor based on thinned bulk PZT for real-time cutaneous activities monitoring. *IEEE Electron Device Letters* 39: 1226–1229.

14 Yi, Z., Huang, J., Liu, Z. et al. (2019). Portable, wireless wearable piezoelectric arterial pulse monitoring system based on near-field communication approach. *IEEE Electron Device Letters* 41 (1): 183–186.

15 Yan, C., Deng, W., Jin, L. et al. (2018). Epidermis-inspired ultrathin 3D cellular sensor array for self-powered biomedical monitoring. *ACS Applied Materials & Interfaces* 10: 41070–41075.

16 Ghosh, S.K., Adhikary, P., Jana, S. et al. (2017). Electrospun gelatin nanofiber based self-powered bio-e-skin for health care monitoring. *Nano Energy* 36: 166–175.

17 Gao, X., Zheng, M., Yan, X. et al. (2019). The alignment of BCZT particles in PDMS boosts the sensitivity and cycling reliability of a flexible piezoelectric touch sensor. *Journal of Materials Chemistry C* 7: 961–967.

18 Ha, M., Lim, S., Park, J. et al. (2015). Bioinspired interlocked and hierarchical design of ZnO nanowire arrays for static and dynamic pressure-sensitive electronic skins. *Advanced Functional Materials* 25: 2841–2849.

19 Park, J., Kim, M., Lee, Y. et al. (2015). Fingertip skin-inspired microstructured ferroelectric skins discriminate static/dynamic pressure and temperature stimuli. *Science Advances* 1 (9): e1500661.

20 Chen, Z., Wang, Z., Li, X. et al. (2017). Flexible piezoelectric-induced pressure sensors for static measurements based on nanowires/graphene heterostructures. *ACS Nano* 11: 4507–4513.

21 Zhu, M., Yi, Z., Yang, B., and Lee, C. (2021). Making use of nanoenergy from human – nanogenerator and self-powered sensor enabled sustainable wireless IoT sensory systems. *Nano Today* 36: 101016.

22 Van Volkinburg, K. and Washington, G. (2017). Development of a wearable controller for gesture-recognition-based applications using polyvinylidene fluoride. *IEEE Transactions on Biomedical Circuits and Systems* 11: 900–909.

23 Cha, Y., Seo, J., Kim, J.-S., and Park, J.-M. (2017). Human–computer interface glove using flexible piezoelectric sensors. *Smart Materials and Structures* 26: 057002.

24 Tran Quang, T. and Lee, N.-E. (2016). Flexible and stretchable physical sensor integrated platforms for wearable human-activity monitoring and personal healthcare. *Advanced Materials* 28: 4338–4372.

25 Khan, Y., Ostfeld, A.E., Lochner, C.M. et al. (2016). Monitoring of vital signs with flexible and wearable medical devices. *Advanced Materials* 28: 4373–4395.

26 Pang, C., Koo, J.H., Nguyen, A. et al. (2015). Highly skin-conformal microhairy sensor for pulse signal amplification. *Advanced Materials* 27: 634–640.

27 Hu, W., Chen, S., Zhuo, B. et al. (2016). Highly sensitive and transparent strain sensor based on thin elastomer film. *IEEE Electron Device Letters* 37: 667–670.

28 Chen, Y., Lu, B., Chen, Y., and Feng, X. (2016). Biocompatible and ultra-flexible inorganic strain sensors attached to skin for long-term vital signs monitoring. *IEEE Electron Device Letters* 37: 496–499.

29 Dagdeviren, C., Su, Y., Joe, P. et al. (2014). Conformable amplified lead zirconate titanate sensors with enhanced piezoelectric response for cutaneous pressure monitoring. *Nature Communications* 5: 4496.

30 Pan, L., Chortos, A., Yu, G. et al. (2014). An ultra-sensitive resistive pressure sensor based on hollow-sphere microstructure induced elasticity in conducting polymer film. *Nature Communications* 5: 3002.

31 Boutry, C.M., Nguyen, A., Lawal, Q.O. et al. (2015). A sensitive and biodegradable pressure sensor array for cardiovascular monitoring. *Advanced Materials* 27: 6954–6961.

32 Yun, S., Park, S., Park, B. et al. (2014). Polymer-waveguide-based flexible tactile sensor array for dynamic response. *Advanced Materials* 26: 4474–4480.

33 Kim, D.H., Shin, H.J., Lee, H. et al. (2017). In vivo self-powered wireless transmission using biocompatible flexible energy harvesters. *Advanced Functional Materials* 27 (25), 1700341.

34 Jing, Q., Xie, Y., Zhu, G. et al. (2015). Self-powered thin-film motion vector sensor. *Nature Communications* 6: 8031.

35 Wang, L.P., Deng, K., Zou, L. et al. (2002). Microelectromechanical systems (MEMS) accelerometers using lead zirconate titanate thick films. *IEEE Electron Device Letters* 23: 182–184.

36 Lim, S., Son, D., Kim, J. et al. (2015). Transparent and stretchable interactive human machine interface based on patterned graphene heterostructures. *Advanced Functional Materials* 25: 375–383.

37 Pradel, K.C., Wu, W., Ding, Y., and Wang, Z.L. (2014). Solution-derived ZnO homojunction nanowire films on wearable substrates for energy conversion and self-powered gesture recognition. *Nano Letters* 14: 6897–6905.

38 Dong, W., Xiao, L., Hu, W. et al. (2017). Wearable human–machine interface based on PVDF piezoelectric sensor. *Transactions of the Institute of Measurement and Control* 39: 398–403.

39 Lu, C., Wu, S., Lu, B. et al. (2018). Ultrathin flexible piezoelectric sensors for monitoring eye fatigue. *Journal of Micromechanics and Microengineering* 28: 025010.

40 Sato, J., Sekine, T., Yi-Fei, W. et al. (2019). Ferroelectric polymer-based fully printed flexible strain rate sensors and their application for human motion capture. *Sensors and Actuators A: Physical* 295: 93–98.

41 Farooq, M., Chandler-Laney, P.C., Hernandez-Reif, M., and Sazonov, E. (2015). Monitoring of infant feeding behavior using a jaw motion sensor. *Journal of Healthcare Engineering* 6: 23–39.

42 Alshurafa, N., Kalantarian, H., Pourhomayoun, M. et al. (2015). Recognition of nutrition intake using time-frequency decomposition in a wearable necklace using a piezoelectric sensor. *IEEE Sensors Journal* 15: 3909–3916.

43 Kalantarian, H., Alshurafa, N., Le, T., and Sarrafzadeh, M. (2015). Monitoring eating habits using a piezoelectric sensor-based necklace. *Computers in Biology and Medicine* 58: 46–55.

44 Kalantarian, H., Motamed, B., Alshurafa, N., and Sarrafzadeh, M. (2016). A wearable sensor system for medication adherence prediction. *Artificial Intelligence in Medicine* 69: 43–52.

45 Hussain, G., Javed, K., Cho, J., and Yi, J. (2018). Food intake detection and classification using a necklace-type piezoelectric wearable sensor system. *IEICE Transactions on Information and Systems* E101D: 2795–2807.

46 Farooq, M. and Sazonov, E. (2016). A novel wearable device for food intake and physical activity recognition. *Sensors* 16: 1067.

47 Hu, H., Zhu, X., Wang, C. et al. (2018). Stretchable ultrasonic transducer arrays for three-dimensional imaging on complex surfaces. *Science Advances* 4: eaar3979.

48 Wang, C., Li, X., Hu, H. et al. (2018). Monitoring of the central blood pressure waveform via a conformal ultrasonic device. *Nature Biomedical Engineering* 2: 687–695.

49 Jiang, L., Yang, Y., Chen, R. et al. (2019). Flexible piezoelectric ultrasonic energy harvester array for bio-implantable wireless generator. *Nano Energy* 56: 216–224.

50 Jiang, L., Yang, Y., Chen, R. et al. (2019). Ultrasound-induced wireless energy harvesting for potential retinal electrical stimulation application. *Advanced Functional Materials* 29 (33), 1902522.

51 Wang, F., Jin, P., Feng, Y. et al. (2021). Flexible Doppler ultrasound device for the monitoring of blood flow velocity. *Science Advances* 7 (44): eabi9283.

52 Yoshida, M., Onishi, K., Tanimoto, K., and Nishikawa, S. (2017). Flexible tension sensor based on poly(L-lactic acid) film with coaxial structure. *Japanese Journal of Applied Physics* 56: 10PG02.

53 Hu, Y. and Xu, Y. (2012). An ultra-sensitive wearable accelerometer for continuous heart and lung sound monitoring. In: *2012 Annual International Conference of the IEEE Engineering in Medicine and Biology Society IEEE Engineering in Medicine and Biology Society Conference Proceedings*, 694–697. IEEE.

54 Kim, S.-W., Lee, Y., Park, J. et al. (2018). A triple-mode flexible E-skin sensor interface for multi-purpose wearable applications. *Sensors* 18: 78.

55 O'Rourke, M.F. (1971). The arterial pulse in health and disease. *American Heart Journal* 82: 687–702.

56 Hansoti, R.C. (1961). Arterial pulse in health and disease. *Indian Journal of Medical Sciences* 15: 987–990.

57 Amjadi, M., Kyung, K.-U., Park, I., and Sitti, M. (2016). Stretchable, skin-mountable, and wearable strain sensors and their potential applications: a review. *Advanced Functional Materials* 26: 1678–1698.

58 Shi, Q., Wang, H., Wang, T., and Lee, C. (2016). Self-powered liquid triboelectric microfluidic sensor for pressure sensing and finger motion monitoring applications. *Nano Energy* 30: 450–459.

59 Sun, C., Shi, Q., Hasan, D. et al. (2019). Self-powered multifunctional monitoring system using hybrid integrated triboelectric nanogenerators and piezoelectric microsensors. *Nano Energy* 58: 612–623.

60 Koh, K.H., Shi, Q., Cao, S. et al. (2019). A self-powered 3D activity inertial sensor using hybrid sensing mechanisms. *Nano Energy* 56: 651–661.

61 Zhu, M., Shi, Q., He, T. et al. (2019). Self-powered and self-functional cotton sock using piezoelectric and triboelectric hybrid mechanism for healthcare and sports monitoring. *ACS Nano* 13: 1940–1952.

62 Curry, E.J., Ke, K., Chorsi, M.T. et al. (2018). Biodegradable piezoelectric force sensor. *Proceedings of the National Academy of Sciences of the United States of America* 115: 909–914.

63 McLaughlin, J., McNeill, M., Braun, B., and McCormack, P.D. (2003). Piezoelectric sensor determination of arterial pulse wave velocity. *Physiological Measurement* 24: 693–702.

64 Ma, Y., Choi, J., Hourlier-Fargette, A. et al. (2018). Relation between blood pressure and pulse wave velocity for human arteries. *Proceedings of the National Academy of Sciences of the United States of America* 115: 11144–11149.

65 Kim, S.B., Lee, K.H., Raj, M.S. et al. (2018). Soft, skin-interfaced microfluidic systems with wireless, battery-free electronics for digital, real-time tracking of sweat loss and electrolyte composition. *Small* 14: e1802876.

66 Krishnan, S.R., Su, C.J., Xie, Z. et al. (2018). Wireless, battery-free epidermal electronics for continuous, quantitative multimodal thermal characterization of skin. *Small* 14: e1803192.

67 Heo, S.Y., Kim, J., Gutruf, P. et al. (2018). Wireless, battery-free, flexible, miniaturized dosimeters monitor exposure to solar radiation and to light for phototherapy. *Science Translational Medicine* 10: eaau1643.

68 Han, S., Kim, J., Won, S.M. et al. (2018). Battery-free, wireless sensors for full-body pressure and temperature mapping. *Science Translational Medicine* 10: eaan4950.

69 Rahimi, R., Brener, U., Chittiboyina, S. et al. (2018). Laser-enabled fabrication of flexible and transparent pH sensor with near-field communication for in-situ monitoring of wound infection. *Sensors and Actuators B: Chemical* 267: 198–207.

70 Chung, H.U., Kim, B.H., Lee, J.Y. et al. (2019). Binodal, wireless epidermal electronic systems with in-sensor analytics for neonatal intensive care. *Science* 363.

71 Kurita, Y., Shinohara, M., and Ueda, J. (2013). Wearable sensorimotor enhancer for fingertip based on stochastic resonance effect. *IEEE Transactions on Human-Machine Systems* 43: 333–337.

72 Shin, K.-Y., Lee, J.S., and Jang, J. (2016). Highly sensitive, wearable and wireless pressure sensor using free-standing ZnO nanoneedle/PVDF hybrid thin film for heart rate monitoring. *Nano Energy* 22: 95–104.

73 Meier, R., Kelly, N., Almog, O., and Chiang, P. (2014). A piezoelectric energy-harvesting shoe system for podiatric sensing. In: *2014 36th Annual International Conference of the IEEE Engineering in Medicine and Biology Society*, 622–625. IEEE.

11

Artificial Intelligence Algorithm for Flexible Sensors

11.1 Introduction

The advanced development in 5G and intelligence of things (IoT) technology provide cost-effective approaches for wireless network connectivity between various sensors and processors [1–3]. In this regard, it is possible for the IoT sensor system to real-time collect data and transmits the sensor information to the cloud server using big data and higher forms of artificial intelligence (AI) analysis, which could boost the performance of current flexible sensor in many applications, including personal healthcare, sports, smart home, and digital twins [4, 5]. The growing integration of AI with functional electronics has spawned a new development of intelligent systems. Therefore, artificial intelligence of things (AIoT) as a combination of AI and IoT has become the most advanced technology, which can realize an intelligent ecosystem in wide range of applications [6–10]. With the aid of AI, the intelligent flexible sensor is capable of detecting, analyzing, and making decisions, which will perform more complex and complete analysis of the gathered datasets than traditional methodologies. In addition, the performance of intelligent system can be increased by selecting appropriate algorithms, fine-tuning algorithm parameters, and integrating various types of data from different sensors [11, 12].

The rapid advancement in materials gives way to advanced flexible electronics [13–19], which enables human physiological signal detection, as shown in Figure 11.1. Flexible electronics that would well merge with our bodies are set to extend the way we perceive and interact with the world [24–26]. Flexible sensors can exist in various forms and be directly attached to the skin and clothes worn on the human body, such as glasses for eye motion detection, face masks for respiration monitoring, watches for healthcare recording, gloves for hand gesture recognition, socks and shoes for gait analysis. Such devices provide comprehensive monitoring of the user's health status and motions [20, 27, 28]. For instance, the flexible sensors attached to the skin or worn on the body can record body temperature, pulse, respiration rate, and blood pressure [29–31]. Smart glasses and smart watches are capable of monitoring blood pressure, pulse, blood, and oxygen level [32, 33]. Besides, the sensing masks would contribute an extended degree of freedom for healthcare assessment, by providing respiration diagnosis regarding the respiration rate and the exhaled gases [34–36]. The eventual realization of the exhaustive

Flexible Piezoelectric Energy Harvesters and Sensors, First Edition. Bin Yang, Zhiran Yi, and Chengkuo Lee.

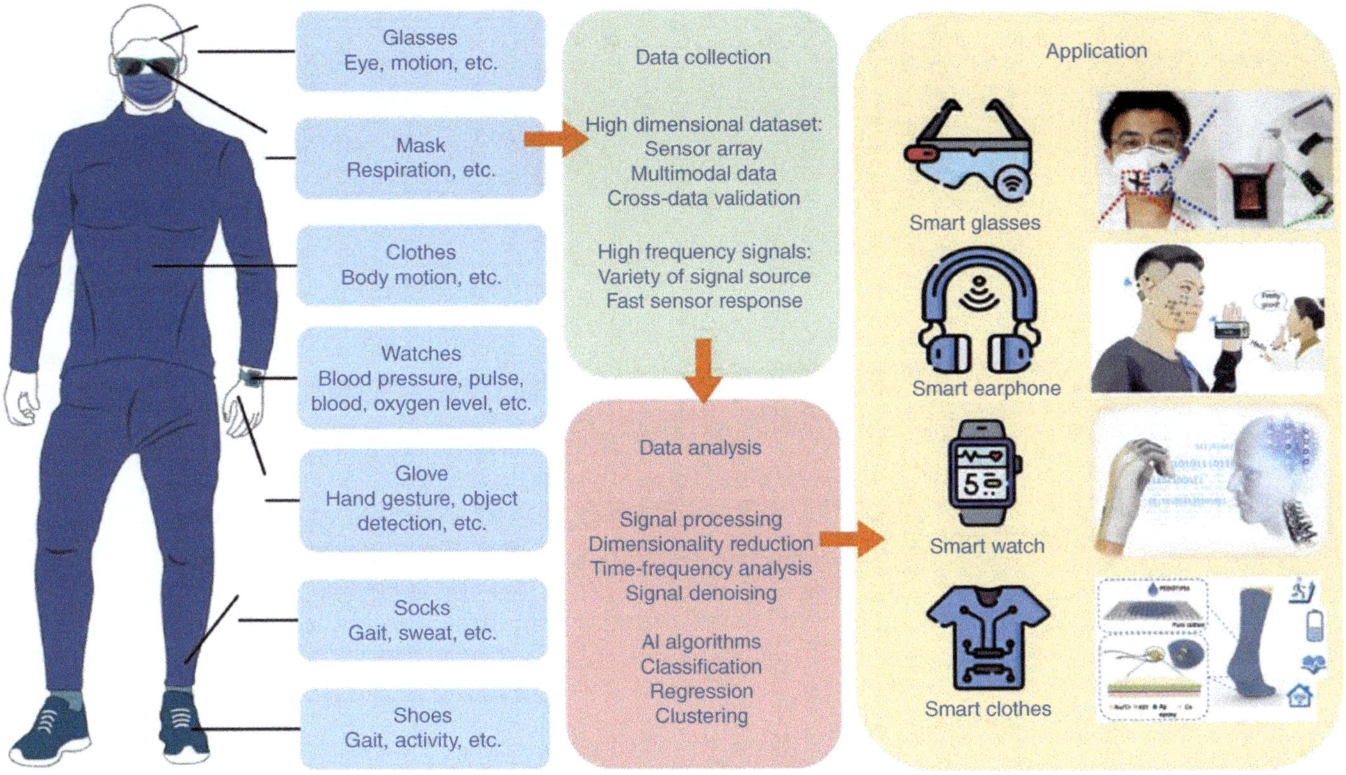

Figure 11.1 Schematic illustration of intelligent system based on wearable flexible sensor. PEDOT:PSS, poly(3,4-ethylenedioxythiophene):polystyrene sulfonate. Source: Reprinted with permission from Zhang et al. [20] / Springer Nature. Reproduced with permission. Wang et al. [21] / Springer Nature. Reproduced with permission Wen et al. [22] / Springer Nature. Reproduced with permission. Zhu et al. [23] / American Chemical Society.

wearable electronics will provide insightful information for disease diagnosis, preventive healthcare, rehabilitation, and even prompt treatment with the aid of surging drug-delivery systems [37–39]. Therefore, the flexible sensor developed toward more intelligent sensing applications enabling the potential large-scale use [40–42]. Therefore, the flexible sensors should be more sensitive, reliable, durable, and wearable for obtaining high-quality sensing data. The intelligent system based on flexible sensors needs to integrate power units toward self-powering, integrated battery, and wireless power [43–46]. Optimizing each component in the flexible sensor system in various aspects has witnessed many achievements, such as deliberately tailored materials, excellent permeability for wearing, long-term stability, and carefully designed soft circuits [47–50].

Overall, leveraging the 5G and IoT technologies, the approaches of data collection in flexible sensing systems can be improved to the high dimensional dataset, i.e. sensor array, multimodal data, cross-data validation, and high-frequency signals [51–53]. Such approach can get more useful information output from intelligent flexible sensor systems to meet the expectations of users even though a large volume of raw data [54, 55]. In addition, the collected data will be processed by various signal processing approaches, such as dimensionality reduction, time-frequency analysis, and signal denoising. Advanced data-processing strategies could narrow the gap between users' expectations and devices' performance with the fast-rising of AI [56–58].

In this chapter, we have introduced the approaches for sensing data from the flexible sensor to the fullest with the assistance of machine learning (ML) algorithms. Then the state of the art in AI-based systems is highlighted. Then we have focused on the intelligent glove sensor system based on various sensing mechanisms. Next, we have summarized the latest piezoelectric-based flexible sensor system in the different applications. Currently, triboelectric-based intelligent systems achieved more comprehensive practices, i.e. wearable electronics, robotic manipulators, smart homes, and self-sustainable AIoT systems. Therefore, in the end, conclusions on how to use AI algorithms to analyze various sensor systems, perspectives on the existing challenges of current intelligent systems, and future trends are provided, which give a glimpse of the further development in the 5G/IoT era.

11.2 Artificial Intelligence Algorithm

11.2.1 ML Algorithms and DL Algorithms

As an important subset of AI, ML has already been widely implemented in many fields of mathematics, physics, chemistry, engineering, and materials science [59–62]. Currently, AI algorithms are used to recognize image objects, transcribe speech into text, match news items, predict user interests, and select relevant results from web searches [21, 63, 64]. As a powerful tool for processing and analyzing raw data collected from wearable devices, ML algorithms facilitate the practicality of the flexible sensor system for a potential application as medical devices in many aspects.

Recently, deep learning (DL) has been widely used in various fields as a new type of AI algorithm. DL is a representation learning method with multiple representation levels. It is obtained by combining simple but nonlinear relations, each of which transforms the representation of the original input level into a higher and more abstract level. After combining enough such transformations, complex functions will be learned in the neural network structure. For classification tasks, higher-level representations amplify various aspects of the input information, which is important for distinguishing and suppressing irrelevant changes. Therefore, DL-based data analysis approaches have made great achievements in image recognition and speech recognition [65, 66]. In addition, DL has beaten other ML approaches, such as predicting the activity of epidemic disease, analyzing particle accelerator data, reconstructing the circuits of human brain, and predicting the mutations of noncoding DNA on gene expression and disease [67–70]. The most popular studies of DL-based systems include image information, tactile information, and voice information, which can be used to achieve the detection and control required in various applications, as shown in Figure 11.2.

The vision information is experiencing rapid development along with the computation power improvements of graphics processing unit (GPU). Correspondingly, the software part of computer vision has gone through significant advances in the past decades [74, 75]. Computer vision is a combination of concepts, technologies, and ideas in digital image processing, pattern recognition, AI, and computer graphics [76–78]. It has been most widely implemented in many applications, such as object detection, image recognition, and facial emotion decoding. Most of these tasks in computer vision are related to the process of obtaining information from digital images and feature extraction. As a way of emulating the human visual system, computer vision aims to develop an automated machine that can realize the tasks required for visual cognition [79–82]. First popular application of computer vision is object detection, which is a computer technology related to visual computing and image processing. The AI algorithms of object detection can deal with classifying

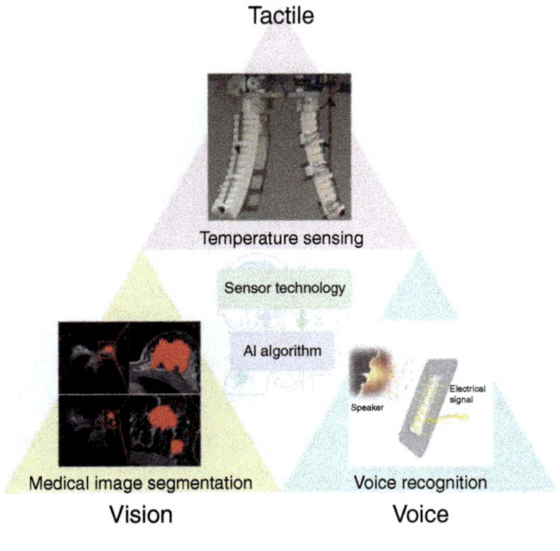

Figure 11.2 Schematic illustration of DL-based system includes image information, tactile information, and voice information. Source: Temperature sensing reprinted with permission from Sun et al. [71] / John Wiley & Sons. Medical image segmentation Wang et al. [72] / Springer Nature / CC BY 4.0. Voice recognition reprinted with permission from Han et al. [73] / Elsevier.

instances of semantic objects of a certain class, such as humans, buildings, or cars, in images and videos [83, 84]. Furthermore, well-researched domains of object detection include face detection and medical image segmentation. Therefore, object detection algorithm has already been applied in many areas of society, such as picture retrieval, security, observation, computerized vehicle systems, and machine investigation. For an instance, You Only Look Once (YOLO) is one of object detection algorithms that uses neural networks to provide real-time indications [82]. YOLO is much popular because of its high speed and accuracy of object detection. YOLO and its updated version have been successfully used in various applications to detect traffic signals, people, parking meters, and animals. In addition, the algorithm based on object detection also can be used to help soft robotics recognize the grasping object. Saxena et al. present a learning algorithm that can build a 3D model of the object from the process of grasping objects [81]. Given two or more images of an object as input, their algorithm can identify several good location points corresponding to the captured object in each image. Then triangulate the sparse point set to obtain the 3D position that you are trying to capture. In addition to object detection, image-to-image translation is another significant application of computer vision, which is a class of vision and graphics problems where the goal is to learn the mapping between an input image and an output image using a training set of aligned image pairs. Generative adversarial network (GAN) is one of the algorithms, which is a class of DL frameworks designed by Ian Goodfellow et al. in June 2014 [85]. GAN can learn to generate new data with the same statistics as the training set from the given dataset. For example, Zhu et al. captured the special features of an image collection and figured out how to transform these features into another image collection, all without any paired training examples [79]. With the amazing feats of AI and computer vision technology becoming more and more common in different industries, the future of computer vision-based intelligence seems to be full of unimaginable outcomes.

Voice information is also one of the most significant sensor information for intelligent systems. As a type of speech recognition, speaker recognition has attracted much attention in various applications, such as personalized voice assistants, smart home appliances, and biometric authentication [86, 87]. Traditional speaker recognition uses microelectromechanical system (MEMS)-based microphones to detect sound by measuring the capacitance between two conductive layers while providing continuous power [88, 89]. Therefore, recently, many other approaches show the potential to be complementary solutions to microphones [90–92]. For instance, Guo et al. proposed a self-powered triboelectric auditory sensor (TAS) as the component of electronic auditory system for an external hearing aid in intelligent robotic applications [93]. The proposed TAS has a broadband response from 100 Hz to 5 kHz by the annular and sectorial inner boundary architecture with systematic optimization. TAS shows high performance of music recording and voice recognition in the application of intelligent human–robot interaction. As a result, much more comprehensive voice information and optimal algorithms will help speaker recognition to achieve a high recognition rate [94–96]. For example, recognizing speakers in a scene involves complex processes, such as big data, massive training, speaker distance, and background noise. Concretely, these systems rely on running voice classification models to recognize the user's voice and execute user commands, such as automatic speech recognition (ASR) and speech generation based on DL models [97–99]. For instance, ASR system can recognize and detect who is speaking in a visual scene for one or more speakers, which is an essential frontend for a wide range of multi-modal applications, such as audio–visual speech recognition. Although complex DL-based algorithms are already utilized to solve these issues, fundamental research based on the sensor of sound collection is still required to enhance. In recent years, voice-based interaction is one of the most effective approaches and is widely used as personal assistants (such as Siri, Google Assistant) of smart phones [100, 101]. In the future, after the integration of the various sensor to collect abundant voice information, the voice-based system will be more intelligent to apply in healthcare monitoring, human–machine interface (HMI), robotic control, etc.

Tactile sensors measure information arising from physical interaction with its environment, which is generally modeled after the biological sense of cutaneous touch capable of detecting stimuli resulting from mechanical stimulation [94–96]. Tactile sensor is also an interaction between humans and external objects, which is suitable for detecting contact-based information, such as roughness, texture, and weight. There are different types of tactile sensors, including piezoresistive, piezoelectric, capacitive, and triboelectric sensors [94, 102–107]. In recent years, considerable advances in material science have enabled the rapid development of tactile sensors. Plenty of tactile sensors is capable of detecting multidimensional environmental information, in which the data acquired are processed by comprehensive algorithms [108, 109]. Sun et al. developed a TENG tactile sensor to realize the recognition of the grasped objects using a soft robotic manipulator. The final accuracy of 28 different shapes of objects is 97.143% with the assistance of ML for data processing [71]. In addition, Wang et al. proposed an ultrathin and flexible sensor (UFS) with high flexibility, good shape adaptability, and wide-range pressure. Sensitivity demonstrates it on a flexible phone, UFS can be used to unobtrusively monitor fingertip pulse waves, and the measured fingertip pulse can be shown on the phone's flexible screen [72]. Oren et al. proposed graphene-based nanomaterials onto tape for roll-to-roll production of tape-based wearable sensors [102]. The sensor can be integrated into smart glove for monitoring mechanical response of fingers when interacting with an object. Each finger was equipped with a pressure sensor at its fingertip and a strain sensor at the top of the finger joint, respectively. These sensors enabled real-time monitoring of changes in important mechanical parameters due to hand movements.

11.2.2 Application of ML Algorithms

Abundant raw data of different sensor signals can be collected from flexible electronics, which require further processing and analysis [89, 97, 110]. An accurate understanding of the information conveyed by these data from these sensors makes it possible for both users and clinicians to adjust treatments or further actions conveniently and efficiently. Conventional data analysis paradigms, including threshold limits, simple mathematical models, or manual selection, are insufficient for a large volume of raw data handling from many aspects, such as data structure complexity, high dimensionality, and multiple modalities [98, 99, 111]. Hence, AI algorithms have come to play a critical role in forging a new paradigm of data analysis that further facilitates the advances and practicality of smart and intelligent flexible electronics. Through implementing appropriate AI algorithms, useful information about various signal features can be extracted from raw data and utilized to the fullest, which, in turn, would make these wearable devices perform better in an intelligent manner. Selecting an appropriate algorithm for different types of raw data is important to establish the correct and robust correlation between the sensing signals and the physiological status. The trend of integrating AI techniques with various sensor systems has been promising for various application scenarios in the past few years. The seamless combination of these two technologies is expected to transform the conventional paradigms in many fields, such as healthcare (vital signs monitoring and tracking, disease diagnosis, and treatment), intelligent HMI, and multiple sensations-based virtual/augmented reality (VR/AR) applications [112–115]. From Section 11.2.1, we already introduced several intelligent systems, especially based on TENG sensing mechanism, to realize diversified applications in smart homes with the aid of AI analytics. In addition to the classification task as one of supervised learning, the AI applications can also be incorporated into the scheme with supervised regression, and unsupervised learning including clustering and dimensionality reduction, as shown in Figure 11.3. The main difference between supervised learning compared with unsupervised learning is whether the output data are given the various form of labeled data [125–129]. In this section, we will introduce some main applications of ML/DL analytics.

First of all, dimension reduction is used to compress a set of input data originating from a wide range of sources (e.g. sensors) with a consequently high degree of multidimensionality to a dataset of reduced dimensionalities. Common dimensionality reduction approaches are based on the principal component analysis (PCA), linear discriminant analysis (LDA), non-negative matrix factorization (NMF), and artificial neural network (ANN) approaches [130–132]. Then the resulting data can be used in subsequent data analysis. For instance, Kühner et al. demonstrated the reliable optical detection in the mid-infrared spectral range of pure glucose and fructose solutions as well as mixtures of both in aqueous solution by utilizing PCA for the extraction of vibrational data [116]. In addition, Zhu et al. reported a volatile organic compounds (VOCs) recognition system using a plasma-enhanced infrared (IR) absorption spectroscopy with fast response, accurate quantization, and good selectivity. PCA is utilized to visualize the relationship of different VOCs in the mixture, which demonstrates the feasibility of the VOC identification [117].

Secondly, clustering is the task of dividing the population or data points into a number of groups such that data points in the same groups are more similar to other data points in the same group than those in other groups. In contrast to classification tasks, the training data are not labeled, i.e. the output variables are unknown to the system. During training, the system learns to cluster the input data based on

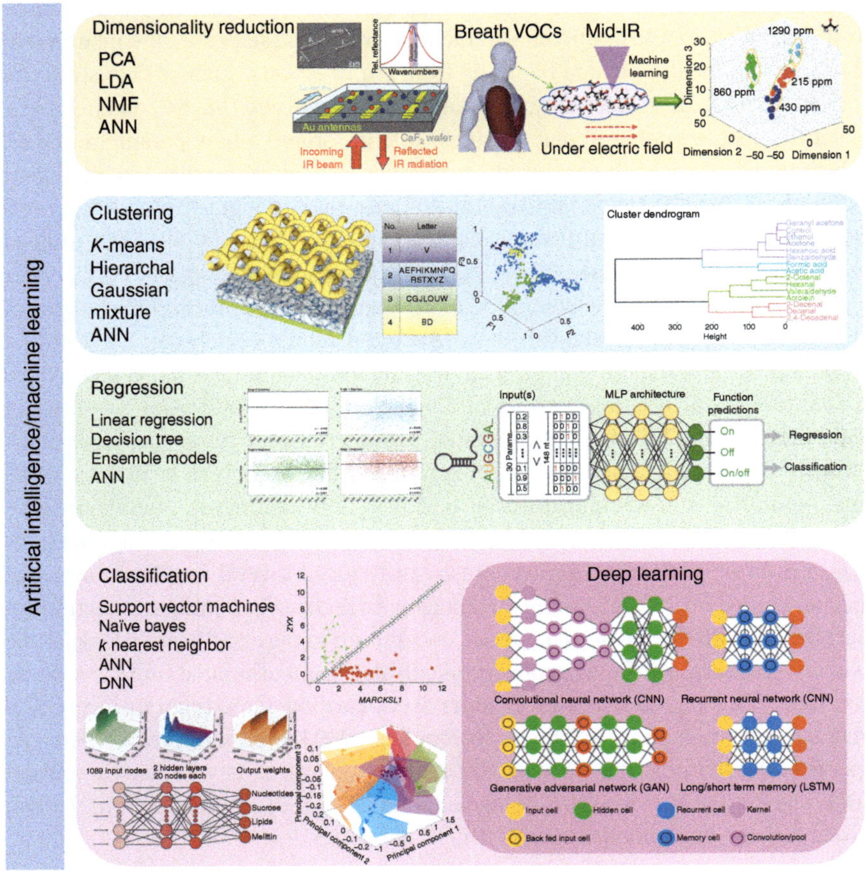

Figure 11.3 Main applications of AI for intelligent sensors and intelligent systems with exemplary algorithms. Source: Reprinted with permission from Kühner et al. [116] / American Chemical Society. Republished with permission from Zhu et al. [117] / American Chemical Society. Reprinted with permission from Ji et al. [109] / Wiley-VCH. Reprinted with permission from He et al. [118]. Copyright 2021 / Springer Nature. Wertheim et al. [119] / Springer Nature / CC BY-4.0. Angenent-Mari [120] / Springer Nature / CC BY-4.0. Reprinted with permission from Noble [121] / Springer Nature. Reprinted with permission from John-Herpin [122] / Wiley-VCH. Republished with permission from Meng et al. [123] / American Chemical Society. Reprinted with permission from Valeri et al. [124] / Springer Nature.

similar features and to make predictions on related features of new input data based on their association to a recognized group. The overall number of such groups can be prespecified or not. The identified clusters may serve as input in the form of features to further AI methods, or the user can assess the underlying meaning and consequences of the identified characteristics [133, 134]. Ji et al. proposed a TENG-based smart electronics with good sensitivity to the mechanical inputs during handwriting, and they used hierarchical cluster analysis to find the intrinsic relationship between 26 letters' handwriting [109]. Additionally, He et al. introduced an unsupervised and annotation-free framework, named as ClusterMap. This algorithm can incorporate the physical location and gene identity of RNAs, formulate the task as a point pattern analysis problem, and identify biologically meaningful structures by density peak clustering [118].

Thirdly, regression is a group of algorithms, which can predict the output values based on input features from the data fed into the system. The go-to methodology is the algorithm that builds a model on the features of training data and uses the model to predict the value of new data. Regression algorithm is a type of algorithm for estimating or predicting continuous output variables. It trains the input data and the assigned output data to the AI system to obtain a known response to each input variable. Regression problems range from simple statistical regression techniques to complex DL techniques. The goal is to use various AI methods to make predictions in highly complex systems, such as predicting machine failures, sensor service life, and gas concentration changes [135–137]. For instance, Wertheim et al. used the regression analysis to estimate the effect of viral load on clustering to provide an approximation of the impact of higher viral load on transmission fitness in the reconstructed HIV genetic transmission clusters [119]. Angenent-Mari et al. used multilayer perceptron (MLP) to build a comprehensive regression model to predict toehold switch function as a canonical riboswitch model in synthetic biology [120].

Lastly, the most wide-range application of AI is classification task. Common ML-based classification algorithms are support vector machine (SVM), naïve Bayes, *k*-nearest neighbor (*k*-NN), decision tree, random forest (RF), etc. [138–140]. Based on these algorithms, AI also can help to build many intelligent systems for various applications, such as disease diagnosis, chemical classifier, and speech recognition [121–123]. In addition to ML-based algorithms, DL-based algorithms, such as convolutional neural network (CNN), recurrent neural network (RNN), long/short-term memory (LSTM), GAN, and deep belief network (DBN), can deal with more complicated classification tasks because they are inspired by the way the human brain processes information [124, 141–144].

11.2.3 Intelligent Glove Platforms

Electronic skin (e-skin), as a significant branch of flexible electronics, is also widely adopted to decode sensory information between humans and robots and build human-centered intelligent systems, as shown in Figure 11.4 [10, 146]. In the past decade, e-skin has experienced rapid progress owing to the development of materials and manufacturing techniques, being more stretchable, integrated, and bionic [147–151]. To mimic laminated human skin with different mechanoreceptors, such as tactile, thermal, and feedback sensation, e-skin is generally multilayered with embedded pressure, temperature, and even feedback units [152–154]. While the soft characteristic of e-skin is endowed by with stretchable substrates, regarding applications, e-skin has been explored for human motion sensing and humanoid robotics (e.g. human–machine interaction, tactile sensing, and texture recognition) [155–159]. Upon human motion decoding, the glove HMI has unique advantages when compared with e-skin. It provides excellent mechanical strength, effectively avoiding the damage to electronic components during putting on and taking off.

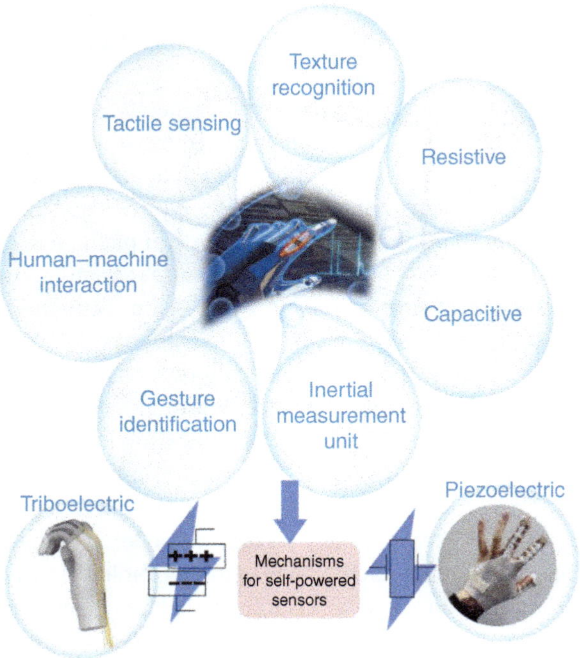

Figure 11.4 Intelligent glove platforms for human motion decoding. Further integrating with artificial intelligence toward more comprehensive applications. Source: Triboelectric glove reproduced with permission Wen et al. [22] / Springer Nature. Piezoelectric glove reproduced with permission Song et al. [145] / Springer Nature.

Besides, glove can lower the risk of skin inflammation due to good breathability of textile [13].

Recent wearable glove-based gesture recognition plays a significant role in the care of speech/hearing impaired. For instance, sign language recognition is an indispensable part of the signers' daily communication [160–162]. However, most of the population has less understanding of sign language without prior learning, which generates a communication barrier and is unfavorable for social participation of signers [29, 163, 164]. As an intuitive interface, glove can measure the hand motions in a comfortable and direct-contact manner, which is fitting into the sign language recognition. Generally, previous glove solutions mainly depend on amplitude, bandwidth, polarity, or peak number to achieve gesture translation. In this case, only a limited variety of gestures (i.e. numbers, letter, and words) can be recognized, far from the practical need of signers' daily communication that involves sentences. Thus, the integration of AI data analytics could be an optimal choice to realize more comprehensive gesture recognition. Classical mechanisms, including resistive, capacitive, and inertial measurement unit (IMU), are extensively used to fabricate smart gloves [165, 166]. However, the rigid component and the indispensable battery of IMU are limited for the detection of figure bending. In recent years, with the aid of the emerging flexible sensing technologies, based on resistive type, capacitive type, piezoelectric type, and triboelectric type, enable the glove system equipped with more functions, including the tactile sensing, texture recognition, human–machine interaction, and gesture identification [167–171].

For an instance, Figure 11.5 shows a scalable tactile glove (STAG) based on resistive sensor array of 548 sensors consisting of a piezoresistive film connected by a

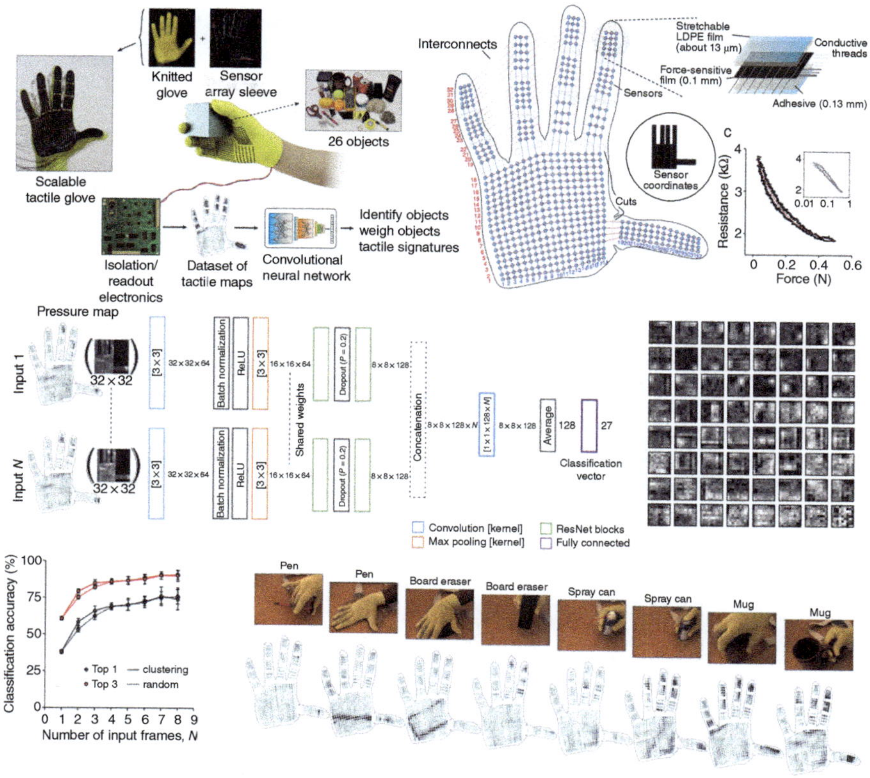

Figure 11.5 Scalable tactile glove based on resistive sensor with CNN analytics. LDPE, low-density polyethylene. Source: Reprinted with permission from Sundaram et al. [172] / Springer Nature.

network of conductive thread electrodes that are passively probed [172]. By analogizing the fundamental perception primitives between the visual and tactile domains, the STAG is worn on the palm for interacting with 26 different objects. After identifying the tactile map of the 32 × 32 arrays in the sensor coordinates, the STAG uses a ResNet-18-based architecture and reaches the maximal classification accuracy in seven random input frames. In addition to identifying objects, humans can easily estimate the weight of objects from tactile signals. The ability to estimate weights is of practical use in robotics and has been the focus of human perception experiments. To estimate the weights of objects from tactile interactions, a restricted dataset of multi-fingered grasps, where the object was picked up from above, was used. After an object is picked up, a single frame is used as an input to a CNN to predict its weight, which shows that the proposed network performs better than a naive linear model over the entire weight range. The above methods reveal that a much larger volume of information is accessible for studying interaction processes at a deeper level with the improvement of micro/nanosensors and the assistance of AI techniques, thereby aiding the future design and development of the next-generation wearable electronics and systems.

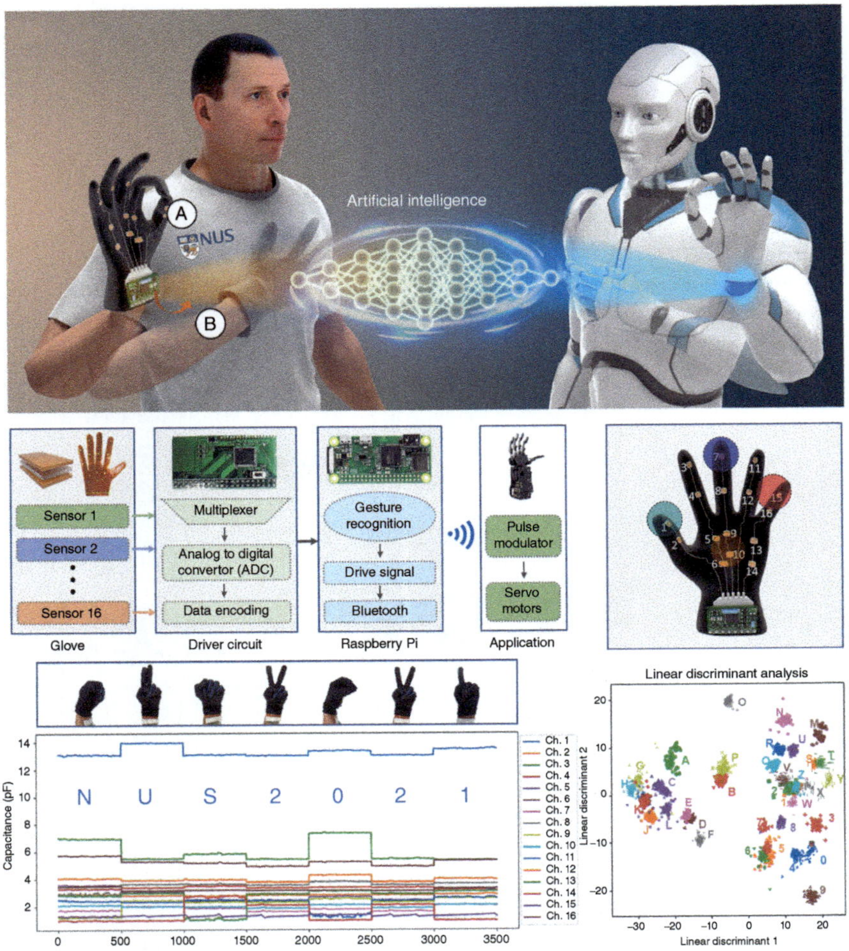

Figure 11.6 A hybrid-flexible capacitive sensing wearable glove. Source: Reprinted with permission from Pan et al. [163] / American Chemical Society.

In addition, as shown in Figure 11.6, Pan et al. developed a hybrid-flexible bimodal sensing wearable glove system that detects and recognizes hand gestures using an in-house low-power code-division multiple access (CDMA)-based multichannel-interfacing circuit in real time [173]. The proposed system uses aerosol printing of robust silver-multiwalled carbon nanotubes (MWCNTs) composite material on a low-cost flexible substrate to fabricate the capacitive sensor array. The proposed system is assembled by a simple bimodal capacitive sensor array to acquire both spatial and tactile data of a unique hand gesture concurrently, which are capable of providing a high-level interpretation of gestures between humans and robotics. Sixteen bimodal capacitance sensors can accurately recognize both 36 static sign gestures (99%) and 3 dynamic gesture phrases (91%), unveiling the potential for a more robust HMI wearable system. The proposed capacitive sensor can measure contact forces up to 30 N and detect a noncontact proximity distance as far as 16 cm. A CDMA-based interface chip with an extremely low-power consumption of 8.2 μW at a sampling rate of 2 kHz can drive our sensor array. This design enables high-resolution data acquisition, which is crucial for the development of gesture-recognition models. Then they develop a quadratic discriminant analysis (QDA) classifier and RNN for static and dynamic American sign language (ASL) recognition, respectively. Using only 16 capacitive sensors, our models have outperformed the majority of existing works in 36 static ASL gesture recognition and achieved an over 91% accuracy for three short dynamic ASL phrases. This is especially important for the ASL users who communicate mostly in phrases that rely on dynamic movement, the relative positioning of the hand, and directional gestures instead of the 36 static alphabets and numbers.

In recent years, the emerging self-powered technologies offer new potential for realizing large-scale batteryless IoT by reducing the power consumption of massive sensor nodes. In addition to being energy sources, self-powered sensors, such as triboelectric and piezoelectric nanogenerators, enable dynamic sensing that is complementary to static sensing of resistors and capacitors. Therefore, Song et al. explored a flexible glove including piezoelectric sensors for collecting the hand gesture signal and soft pneumatic actuator for providing tactile feedback, as shown in Figure 11.7 [145]. The pneumatic soft actuator is activated by an electrostatic force, which can be divided into a ring part where the electrostatic attractive force works, and the center part as the contact part. The pneumatic soft actuator can also be attached to the gloves to generate effective tactile feedback. In particular, when the

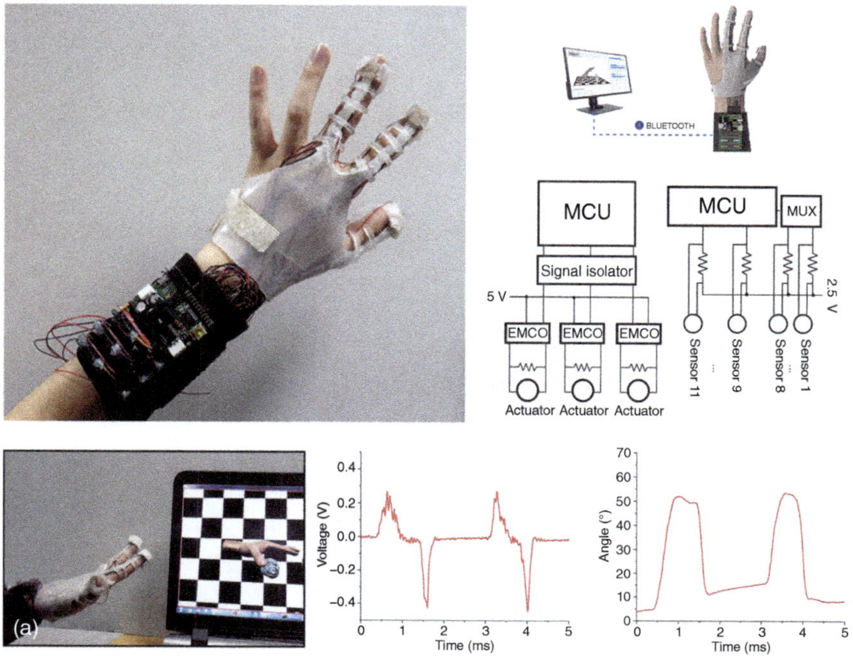

Figure 11.7 Flexible piezoelectric glove for collecting the hand gesture signal. Source: Adapted from Song et al. [145] / Springer Nature / CC BY-4.0.

user holds a virtual object, the actuator is well maintained in the on state, and when the virtual object is released, the actuator is switched to the off state. The actuator is actuated by electrostatic attraction. When the air space is reduced by the electrostatic attraction, the central part expands and is designed to generate haptic feedback. Especially, the actuator showed a larger movement as the period became faster and the applied voltage became larger. In addition, the designed silicone monolithic glove was able to detect movement of fingers with the polyvinylidene fluoride (PVDF) sensors and transmit data via Bluetooth. A voltage output by a piezoelectric sensor deformation provides finger motion information. To distinguish the moving state from the received information, the threshold value was specified, and the gain value was obtained through the initial calibration, which could be used in several ways by linking with various VR software.

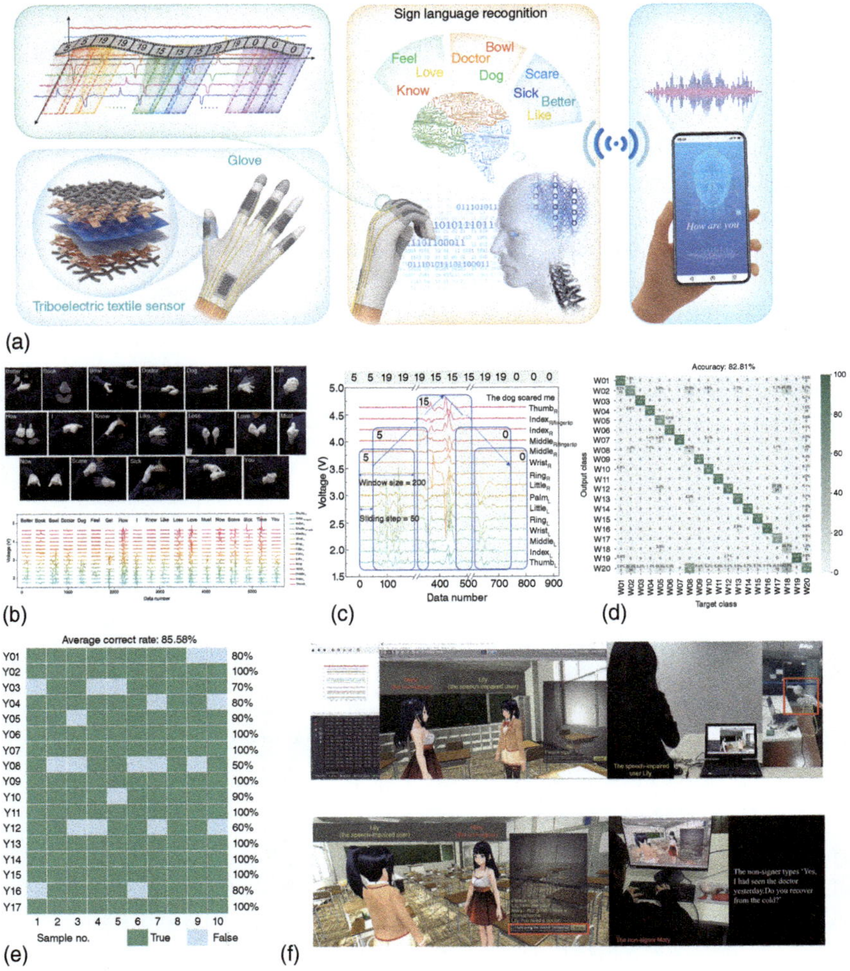

Figure 11.8 A sign language recognition and communication system based on triboelectric gloves, deep learning, and VR user interface. Source: Wen et al. [22] / Springer Nature / CC BY-4.0.

Different from piezoelectric nanogenerator (PENG)-based gloves, TENG has been attractive owing to less limitations on materials and uncomplicated fabrication. Combining triboelectric gloves with ML may give us inspiration for more intelligent applications. Correspondingly, as shown in Figure 11.8, Wen et al. showed a sign language recognition and communication system with the aid of gloves, DL, and VR user interface [22]. To make AI recognize sentences of sign language, the sentence samples (800 data points per sentence signal) in training dataset are split into word units by using a sliding window in the size of 200 data points and sliding step of 50 data points. Then the DL algorithm would identify these word elements first and then inversely reconstruct the original sentences, with the accuracy of 82.81% and 85.58%, respectively. Furthermore, this segmentation method offers new possibility for new/never-seen sentence recognition. Specifically, recognized word units can be arranged in a new/different order to create new sentences. In the meantime, the DL model will recognize all basic word elements in the new sentence and give a reasonable translation. In such way, the new sentences that are not included in the training dataset can be recognized. Finally, the recognition results of sentences are projected into virtual space in which the signer can use their familiar sign language to communicate, while the nonsigners directly type in their controlled VR interface. This advancement in recognizing existing sentences and new sentences improves the practicality of sign-language-recognition system, paving a way to lower the communication barrier between signers and nonsigners.

11.3 Applications for IoTs System

11.3.1 Application of Piezoelectric Sensor

As discussed in Section 11.2, voice recognition is the most intuitive user interface for bilateral communication between humans and smart devices. Speaker recognition has received spotlight as the next big thing in voice user interface (VUI), such as personalized voice-controlled assistants, smart home appliances, and biometric authentication based on AI and IoT infrastructure [174, 175]. Speaker recognition also suffers from a low-recognition rate, caused by limited voice information and optimal algorithms for a simple and accurate process. In addition, the conditions for speaker recognition involve complicated processes, such as DL of big data, massive training, speaker distance, and background noise. Although complex algorithms are utilized to solve these issues, fundamental research based on intrinsic sound information is required to enhance the voice recognition rate as the facile human mechanism of distinguished speakers. As shown in Figure 11.9a, ML-based speaker recognition is enabled by the multichannel flexible piezoelectric acoustic sensor (f-PAS) [126]. A flexible lead zirconate titanate (PZT) membrane providing piezoelectric responses was fabricated by utilizing a similar protocol of inorganic-based laser liftoff. A PZT sol–gel film was deposited on a sapphire substrate by spin coating and then crystallized using rapid thermal annealing at 700 °C for 75 minutes. The inorganic piezoelectric thin film was detached from the mother sapphire substrate

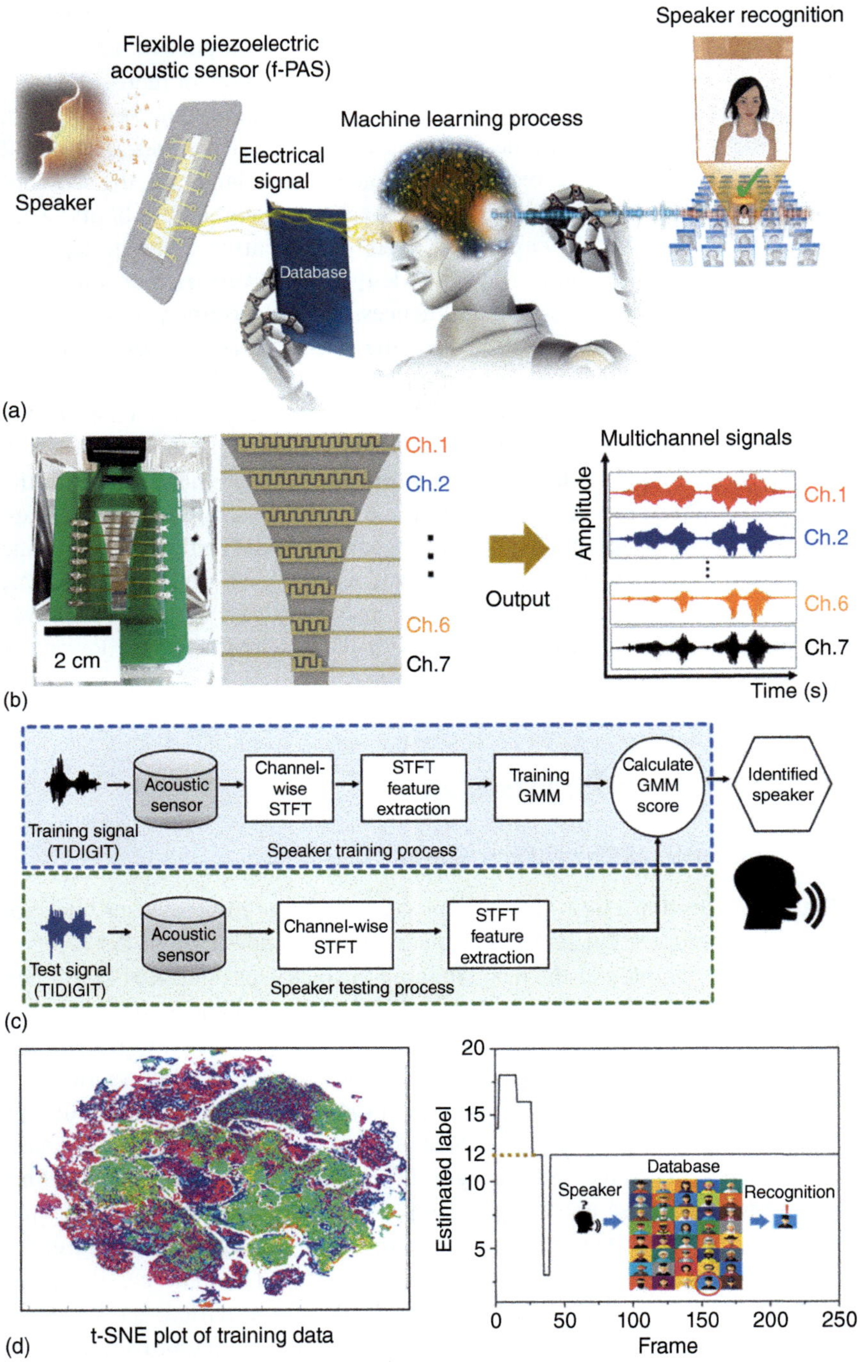

Figure 11.9 A ML-based speaker recognition enabled by the multichannel flexible piezoelectric acoustic sensor. (a) Schematic of the machine learning-based speaker recognition system. (b) Schematic of multi-channel flexible piezoelectric acoustic sensor consists of multiple signals. (c) GMM algorithm flow chart of speaker training process and test process using the standard TIDIGITS dataset. (d) The trained STFT features visualized in t-SNE and the majority voting method to test the speaker recognition. Source: Reprinted with permission from Han et al. [126] / Elsevier.

by irradiating an excimer laser of XeCl from the backside of the mother substrate and subsequently transferred onto a polymer membrane of 50 μm thickness. A concaved trapezoidal design was applied to the f-PAS to produce multiple piezoelectric signals. The multi-resonant band f-PAS was integrated into a small single chip that covered entire voice frequency range with high sensitivity. Seven channels of interdigitated electrodes were patterned on the PZT membrane using photolithography and an etching process. The self-powered flexible piezoelectric membrane was attached to a printed circuit board (PCB) to generate an electrical signal by the displacement vibration.

Human voice stimulation of the TIDIGITS dataset produced resonant oscillations in the flexible piezoelectric membrane, which is composed of acoustic signals of digit sequences to evaluate voice recognition. The multiple piezoelectric signals were generated by a resonant vibration of multichannel self-powered acoustic sensor, acquiring abundant voice information. ML-based training of TIDIGITS voice data including 2800 speeches from 40 people was modified for multi-signal processing by using a Gaussian Mixture Model (GMM) algorithm. Finally, the speeches of a random person were compared with the trained dataset for speaker recognition. Figure 11.9b shows a photograph and schematic illustration of the multichannel f-PAS and corresponding electrical output generated by sound input. The f-PAS output voltage as a function of time was generated from standard male speech of TIDIGITS. The self-powered sensing of sound waveforms is operated without an external power supply, which is a significant advantage for low-power IoT applications compared with commercialized condenser types of acoustic sensors, which require continuous external power sources to measure the capacitance between two conducting layers. Figure 11.9c shows the overall system of GMM-based speaker recognition from the multichannel voice data of the f-PAS. The GMM-based algorithm consists of training and testing using the standard TIDIGITS dataset, including 3080 voice data. The 40 speakers randomly selected from the TIDIGITS dataset were recorded by the multichannel f-PAS and then converted to short-time Fourier transform (STFT) features, which are calculated by weighted summation of STFT absolute value. Among the whole datasets of speaker-recognition algorithm, the 2800 data were used for training data and the remaining 280 data were utilized for testing data. A mixture was defined as the number of Gaussian distributions representing STFT feature. Figure 11.9d shows the trained STFT features visualized in t-distributed stochastic neighbor embedding (t-SNE) plot. The t-SNE plot embeds high-dimensional data of similar objects into a low-dimensional space, which is related to probability distribution of similar clusters among 2800 training data. The majority voting method was used to test the speaker-recognition process. When an input test signal is given, the STFT method generates multiple frames of fast Fourier transform (FFT) signals over time period. The likelihood of the test speaker was calculated according to each frame by comparing the training and testing results, and finally, majority voted speaker was determined as the frame increased. The significant improvement of the speaker-recognition rate indicates that the flexible piezoelectric sensor application can be further extended to voice-based biometric authentication and highly accurate speech recognition.

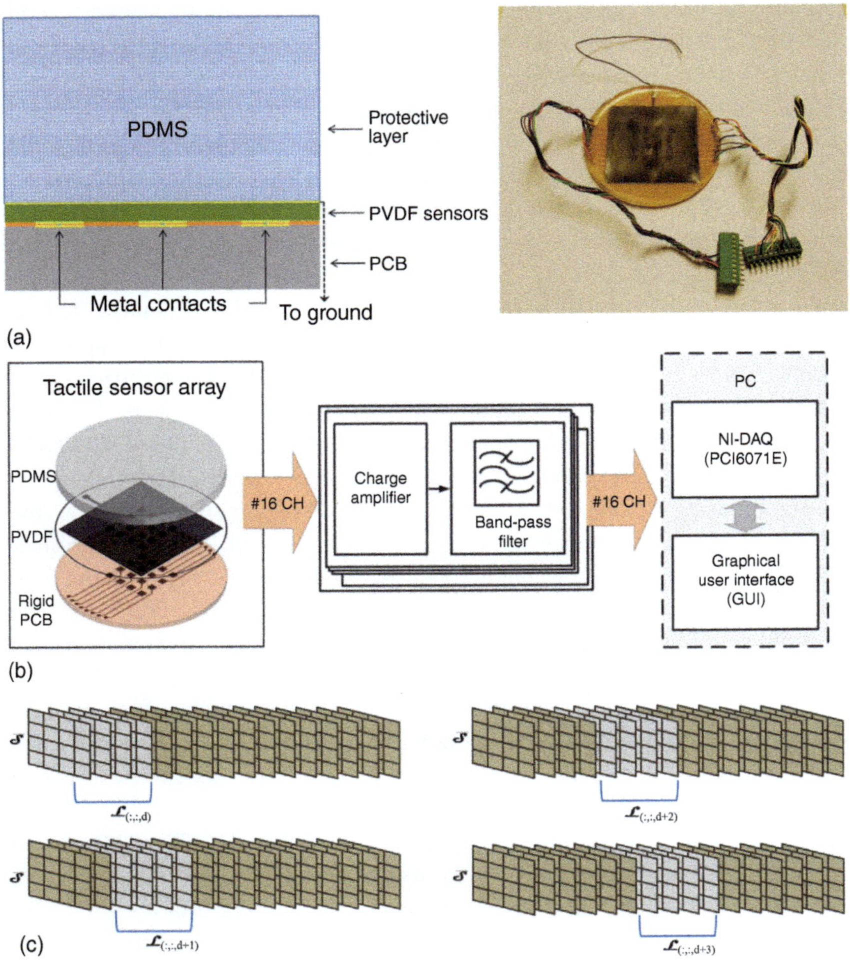

Figure 11.10 PVDF-based system for the measurement of mechanical stress and large bandwidth electromechanical transduction. (a) Multilayer skin structure, including the PCB substrate, the PVDF film and the PDMS (protective) layer. (b) Block diagram of the tactile sensing acquisition system. (c) A schematization of the pre-processing strategy based on the sliding window. DAQ, data acquisition. Source: Reprinted with permission from Gastaldo et al. [169] / Elsevier.

Another IoT application is using piezoelectric sensor arrays to measure contact forces for the connection of robotics. As shown in Figure 11.10a, PVDF has been used for direct measurement of mechanical stress and large bandwidth electromechanical transduction [176]. Additionally, a ML algorithm has been specifically designed to deal with the inherent tensor morphology of raw tactile data. The dataset involving 70 participants has been collected from the output signals under different modalities of touch. The proposed pattern-recognition system showed good performance in three gestures classification tasks for the application of human–robot interactions. Thin-film PVDF can build the e-skin due to its versatility – in principle it allows for the direct measurement of the mechanical stress, decoupling tensile, and shear stress components according to the way it is integrated into the skin. In addition, PVDF also has other advantages, including the large electro-mechanical transduction frequency bandwidth, pyroelectric behavior, bending-mode response, low cost, and low weight. Because PVDF must be integrated into structures that also include a substrate and a protective layer, the piezoelectric film is glued to a flexible PCB substrate, covered by an elastic layer to protect the sensor from physical damage or chemical contamination. To build a sensor array, the piezoelectric film features ad hoc metal contacts, which are deposited on both PVDF sides by means of inkjet printing. In particular, the film is provided with 16 patterned square electrodes (taxels) on the lower surface and with a continuous layer (ground) on the upper surface. The corresponding contact geometry is found on the underlying PCB substrate, which contains metal electrodes and tracks to extract the lower PVDF signals. The mechanical-to-electrical transduction operated by each PVDF taxel is measured as a generated charge and converted to voltage by a charge amplifier.

Then the ML technologies help tackle the classification of touch modalities. As tactile interactions are conveyed by stimuli that can only be fully characterized by exploiting tensor morphology, the research mainly addressed the deployment of a ML-based pattern-recognition system that was specifically aimed to deal with tensor signals, as shown in Figure 11.10b. The ability to process data represented as Zth order tensor is the key aspect of the proposed framework. As a major result, the present approach (i) does not require one to transform the tensor signal originated by the sensing layer into a n-dimensional vector, and (ii) can avoid the use of standard feature-extraction procedures. It proved that "tensor-SVM" and "tensor-RLS" could obtain consistent performances on a three-class classification task. Indeed, the availability of a tensorial kernel function may prove valuable when dealing with classification problems that admit a natural tensorial representation. The robustness and reliability of the proposed experiments are also confirmed by the fact that the dataset exploited in the simulations was collected from a wide spectrum of participants.

Alshurafa et al. performed classification of swallows using statistical features collected from spectrograms generated from [177]. This work shows promise in using spectrogram analysis in combination with piezoelectric sensors as opposed to audio sensors. The proposed necklace prototype has shown the ability to successfully distinguish between liquids and solids in two experiments using Random Forest Classifier with 100 trees, resulting in an F-measure above 90%. The system and framework are capable of distinguishing between hot and cold drinks with an F-measure of 90%. In addition, this system has the potential for distinguishing between solid food types with an F-measure of about 80%. A piezoelectric sensor, as a vibration sensor, produces a voltage when subjected to physical strain. By placing a piezoelectric sensor against the throat, the muscle contraction and motion of the skin during a swallow are represented in the output voltage of the sensor, when sampled at frequencies as low as 5 Hz. The necklace features a thin, lightweight piezoelectric vibration sensor attached to the inside of the necklace, along with a small microcontroller board capable of sampling the sensor and transmitting the data to a mobile phone via Bluetooth. The hardware is powered by a lightweight lithium-polymer battery. The conventional feature selection algorithms usually focus on specific metrics to quantify the relevance and redundancy to find the smallest subset of features that provides the maximum amount of useful information for prediction. Thus, the main goal of feature selection algorithms is to eliminate redundant or irrelevant features in each feature set. Applying an effective feature-selection algorithm not only decreases the computational complexity of the system by reducing the dimensionality and eliminating the redundancy but also increases the performance of the classifier by deleting irrelevant and confusing information. The two well-known feature selection categories are the filter and wrapper methods. Filter methods use a specific metric to score each individual feature (or a subset of features together) and are usually fast and much less computationally intensive. Wrapper methods usually utilize a classifier to evaluate feature subsets in an iterative manner according to their predictive power. They applied the wrapper method, testing on multiple combinations of feature subsets and classifiers, including k-NN, Bayesian Network, and Random Forest. Then the dimensionality of the features is reduced from $s = 360$ to l, where l depends on the feature selection algorithm used. The optimal feature extraction, feature selection, and classification combination are selected to run in real time. In addition, an optimal feature subset and classifier are trained to distinguish between food types.

11.3.2 Hybrid Self-Powered Flexible Sensor

In addition to the above applications based on piezoelectric sensors, the hybrid self-powered flexible sensor is also an important research direction for intelligent systems. Zhao et al. fabricated a wearable and flexible sensor via facile electrospinning, carbonization, and encapsulation process [178]. On the basis of the integration of piezoresistive, triboelectric, and piezoelectric effects, it had the advantages of multifunction, modularization, and high sensitivity. As shown in Figure 11.11a, the fabrication process of the dual-functional PAN-C/BTO-based sensors consists of two steps. The first step is that the PAN/BTO nanofibers were electrospun from the PAN solution mixed with BTO nanoparticles (NPs) and then carbonized under Ar atmosphere to endow them with good conductivity. Secondly, the sensors were obtained by polydimethylsiloxane (PDMS) encapsulation of the PAN-C/BTO nanofiber film connected with copper wires at two terminals. Herein, randomly aligned nanofibers were fabricated using a flat-plate collector, ensuring the generation of many mesoscopic joints with enhanced mechanical robustness, which was not as vulnerable as the oriented one that tended to be torn apart along the alignment direction.

This fabrication process provides the outstanding flexibility and versatility of PAN-C/BTO-based sensors, which promise various applications in flexible devices. Furthermore, the PAN-C/BTO-based sensor had high stability (>60 000 cycles) when working as either a flex sensor or a force sensor. According to the finite-element model, the high signal strength of the self-powered force

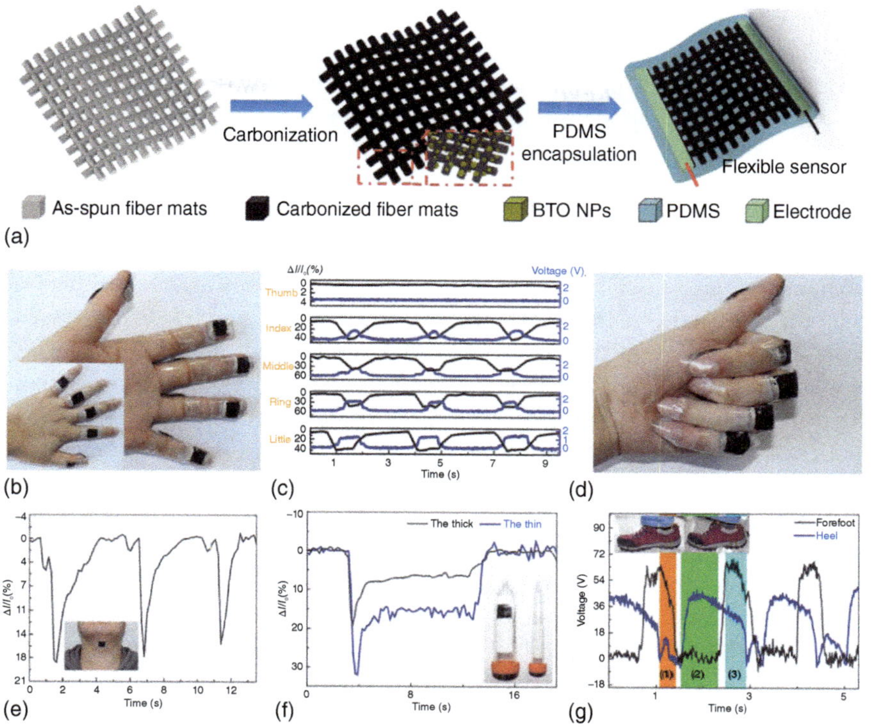

Figure 11.11 A wearable and flexible sensor via facile electrospinning, carbonization, and encapsulation process. (a) Illustration showing the fabrication process of the PAN-C/BTO-based sensor. (b) Smart sensing system was proposed with integrating two sensors on each finger. (c) Finger spray for (d) a "good" gesture. (e) Sensor can detect action of swallowing. (f) Flex sensor can measure the thickness of cylinders during the process of bending onto them because of different curvature radii. (g) When two PAN-C/BTO-based self-powered pressure sensors were attached under a shoe, they could detect human walking synergistically. Source: Reprinted with permission from Zhao et al. [170] / American Chemical Society.

senor was originated from the triboelectric output and piezoelectric output of the BTO NPs. When attached to the PAN-C/BTO-based flex sensor on the knuckle of a volunteer, the subtle motions of human fingers will be captured. Therefore, a 10-sensor hand-gesture-sensing system is shown in Figure 11.11b. In Figure 11.11c, the relationship between fingers' bending and stretching movements shows a "good" gesture language corresponding to Figure 11.11d. $\Delta I/I_0$ curves originated from the signals of the fingers bending (from sensors on the knuckle back), and voltage curves are signals of finger pulps contacting with palm (from sensors on the finger pads). There is no $\Delta I/I_0$ or voltage signal from the thumb because it was not moved during the course of action. Therefore, two sensors on different positions of one finger could detect the movements of the finger independently, and several modularized sensors could sense more complicated motions synergistically. The proposed flexible sensor can also monitor the swallowing actions during 13 seconds for human health-related applications, as shown in Figure 11.11e. When the larynx is raised and pressed forward during swallowing to prevent the food from entering the trachea, laryngeal prominence moves up and down. The PAN-C/BTO-based sensor on the laryngeal prominence of a male volunteer was stuck. The sensor was bent with the laryngeal prominence lifting when the volunteer is swallowing. Therefore, the sensor had the potential in health monitoring to detect the eating process of a patient. In addition, the flexible PAN-C/BTO-based sensor can distinguish the material shape and geometry when bending the sensor to capture the objective. By sticking the flex sensors onto the surface of two cylinders with different diameters, the flex sensor had different bending deformations to induce differentiated current signals, as shown in Figure 11.11f. Finally, the PAN-C/BTO-based sensors attached at the bottom of a shoe can also be used to detect the human's gait. Two sensors were attached under the forefoot and the heel, respectively. These two self-powered tactile sensors can detect the contact between the sole and the smooth marble floor. Generally, during human walking, the heel will touch the ground first with the forefoot falling and the process (1) is short in time. In process (2), the heel lifts and the forefoot remain in contact with the ground. Therefore, the sensor under the heel has a higher voltage signal, and the one under the forefoot maintains a low triboelectric potential. Process (2) takes longer time because the other leg is moving forward when the forefoot remains on the ground. In process (3), the foot leaves the floor and both sensors have high triboelectric outputs. In this way, the PAN-C/BTO-based sensors detected a complete process of one step from foot falling down to lifting the ground. If more sensors are integrated under soles, the sensing system would detect the gait pattern of different people due to the different plantar force distributions. Overall, on the basis of the numerous advantages, the PAN-C/BTO-based sensor has broad application prospects in future wearable electronics.

11.3.3 Application of Triboelectric Sensor

Benefiting from the advantages of various choices of materials, easy fabrication, self-power generation, and low cost, the TENG gradually becomes an optimal option based on textile format for sensors and energy harvesters. The flexible sensor in the form of textile exhibits outstanding ability of structural retention and fatigue resistance during washing [17, 179, 180]. In this regard, TENG has made a lot of research progress integrated with AI, including self-powered pressure sensors and HMI for IoT applications, and HMIs for robotic control and VR/AR interactions [181–184].

For instance, Zhu et al. developed a low-cost 3D-printed glove HMI with triboelectric finger bending sensors, palm sensor for sliding detection, and piezoelectric stimulator for feedback [19]. Projecting hand motions to cyberspace, the glove can achieve VR/AR control and human–machine collaboration. With further integration with AI, their proposed glove-based system realizes advanced object recognition. In detail, the glove HMI contains three main components for multifunctions. The triboelectric signals are generated when human skin contacts or separates with Ecoflex. Different gesture signals are defined as corresponding commands to control the virtual mouse for shopping. Similarly, the palm sensor is located in the center of four quarter-round electrodes. Varied sliding directions could induce output differences between four electrodes. Besides, the piezoelectric stimulator helps to provide feedback when touching virtual objects in unity by tuning applied voltage. As an alternative for optical/ultrasound and wearable tactile sensor-based 3D object recognition, the proposed glove has a minimalist design with fewer sensors. It achieves recognition of six objects facilitated by ML with an accuracy of 96.88%. To demonstrate the benefit and practicality of such recognition in medical field, the authors show VR surgical training based on object recognition. The left glove is responsible for operation mode and switching into recognition mode. By switching into the recognition mode using the left index finger, the left glove is then disabled, and the right glove enters the recognition mode to record the signals of grabbing gestures for corresponding recognition of selected tools. Overall, although there are only few sensors, the smart glove still verifies the potential applications in advanced multipurpose manipulation, in a simpler and more intuitive manner.

TENG also made a lot of progress in soft robot tactile/deformation perception [185–191]. To realize anthropomorphic robots with multimodal sensing capability, a highly integrated sensory system fusing sensors that respond to different kinds of stimuli is needed [192, 193]. Additionally, recent works on ML-enabled data analytics demonstrated the capability of AI to enrich the sensor functions by automatically extracting the features from the sensor signal [169, 172, 194]. By bringing this kind of technology into the robotic sensory system, more advanced interactive functions, e.g. object recognition, gait analysis, and pose estimation, could be realized to implement self-discriminable intelligent robots. Based on these considerations, Sun et al. report a smart soft robotic manipulator enabled by a multifunctional self-powered sensory system, as illustrated in Figure 11.12a, where two types of TENG sensors and a PVDF pyroelectric sensor are integrated. The gear structural TENG sensor is used to detect the continuous bending motion of the soft actuator. When the pneumatic finger is inflated and deformed, the strip mounted on the shaft of the gear will be stretched to drive the gear to rotate, resulting in the contact and separation movements between the gear teeth and the PTFE film, thus generating the triboelectric peaks. The tactile TENG sensor has distributed electrodes, with a silicone rubber thin-film covered on top serving as the tribo-layer. When the stimuli occur on the silicone rubber surface, the contact area and the position can be extracted from the generated output amplitude of the corresponding electrode and the voltage ratio between different electrodes, respectively. With a simple three-layer one-dimensional (1D) CNN for data fusion, the recognition accuracy of higher than 97% could be achieved for 28 grasped objects based on the features contributed by both two TENG sensors (Figure 11.12b), which is much higher than that just based on one sensor type, showing the effective complementary effect between multiple modalities. Moreover, self-powered temperature sensing function could also be realized by the PVDF sensor based on the pyroelectric mechanism, as shown in Figure 11.12c. With the temperature information further fused at the decision level, a more comprehensive perception of the grasped items could be achieved by this intelligent soft gripper, showing its great potential

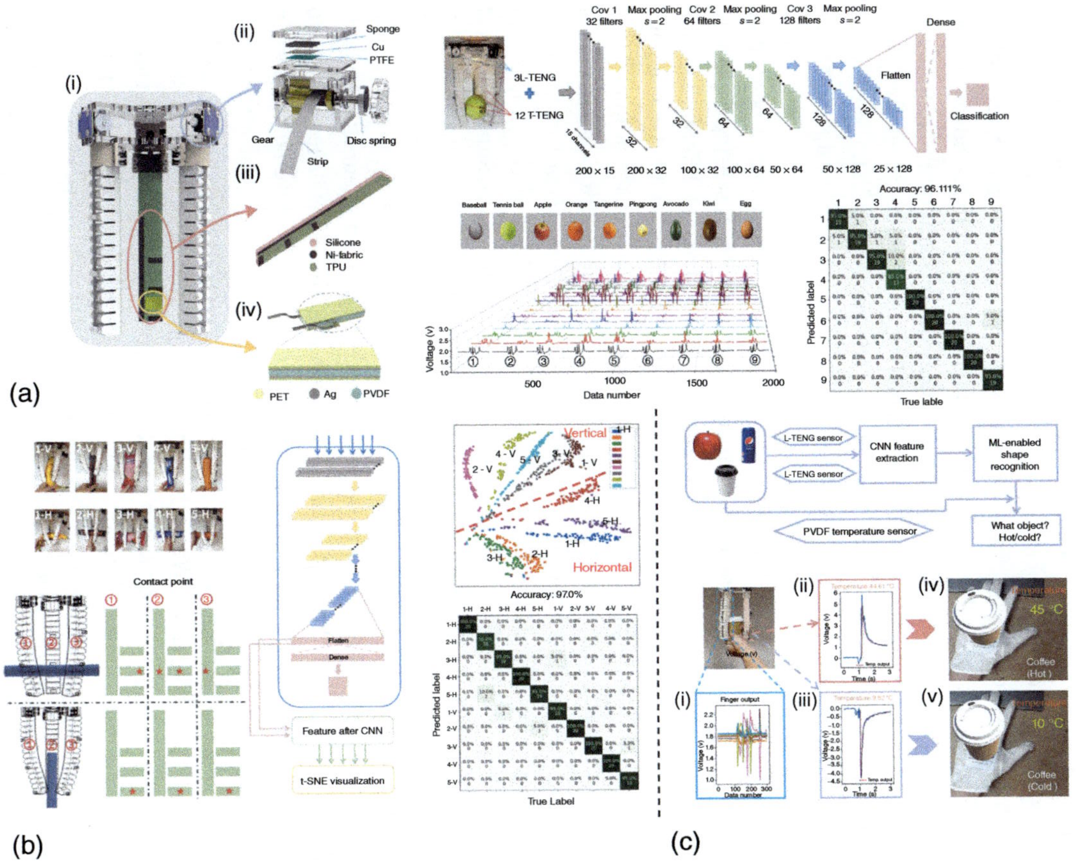

Figure 11.12 A TENG-enhanced smart soft robotic manipulator for AIoT virtual shop applications. L-TENG, length TENG sensor; TPUs, thermoplastic polyurethanes. (a) Overall structure of the system, (i) The configuration of the sensor-integrated smart manipulator. The structure and functionality of (ii) the L-TENG sensor, (iii) the TTENG sensor and (iv) the PVDF temperature sensor. (b) ML-enabled automatic grasped objects recognition system. (c) The flow chart for the shape-size-temperature fused sensory system. Source: Reproduced with permission from Sun et al. [71] / John Wiley & Sons.

to complete complex decision-making tasks in virtual shops and other unmanned working scenarios.

Walking is a fundamental trait that enables people to lead their daily lives and function as productive members of society. Walking gait is characterized by a repetitive sequence of limb motions that propels the body forward while maintaining stability. A typical walking gait is to have one foot forward and the other foot keeping the same distance from the first foot. Normal walking gait is a very efficient walking pattern in terms of strength and walking speed, allowing a human to walk comfortably for long periods of time. Furthermore, in addition, a normal gait allows people to easily switch between running and walking, as well as to easily go up and downstairs and change the walking direction, and even quickly avoid obstacles. However, some patients with neurological or muscular diseases have difficulty maintaining a normal gait, so even if they may have been severely damaged, they still need to work hard to recover and return to a normal gait. Therefore, it is significant to monitor the pattern of walking gait because of the useful information on healthcare and personal identification [195–197].

Meanwhile, the activity of footsteps is one of the major kinetic energy sources from the movement of human body, hence there are many studies on foot sensors and energy harvesters [198–203]. Therefore, TENG-based devices on foot can integrate the function of energy harvesting and gait analysis to realize the self-powered system [23, 204]. Currently, many studies have invested a lot of energy to develop insoles with the functionalities of energy harvesting and gait sensing, but in terms of flexibility and wearing comfort, socks are the best choice for gait analysis. In addition, socks can be more widely used in indoor scenes than the insoles, such as wearing socks at home is much more suitable than wearing shoes [37, 73, 205, 206]. As shown in Figure 11.13a, Zhang et al. developed low-cost triboelectric intelligent socks (LTISs) for the collection of gait patterns [207]. LTIS can also transmit wireless sensor signals using the waste energy harvested from low-frequency body motions. The whole system equips with self-powered functionality to deliver the users' information of identity, health status, and activity. As shown in Figure 11.13b, the sample triboelectric outputs from different participants show the differences in some important features, i.e. amplitude, frequency, interval of stages in gait cycle, etc. Such important information could represent the gait speed, contacting force between feet and shoes, sensors-triggering sequence, and the individual walking operation manners. After collecting gait signals from all participants in 500 seconds, the gait is stable and segmented by a 4-second sliding window to obtain two or three entire dynamic gait cycles. The whole data were divided into the training set, testing set, and validation set to evaluate the performance of AI. Then the training set will be fed into the training 1D CNN model directly without the complicated preprocessing steps. The 1D CNN is composed of four convolutional layers, four max-pooling layers, and one fully connected layer. As a result, the accuracy of training set and validation set for five participants can be up to 100% and 98.4%, and the testing set shows that 1D CNN method can assist the LTIS to achieve a high identity classification accuracy of 96%. When attaching the pieces of more sensor array on three optimized locations of sock, the LTIS is able to collect more comprehensive information from foot motions

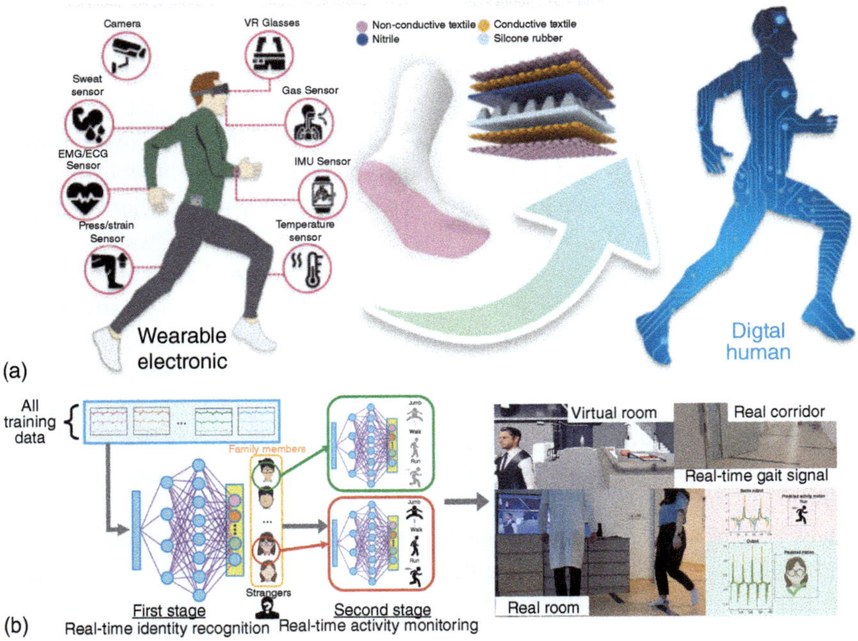

Figure 11.13 (a) Schematics of smart sock for the digital twin-based intelligent system. (b) The structure of the whole digital twin system, includes the first stage of identity recognition, second stage of activity monitoring, and real-time VR. ECG, electrocardiography. Source: Reprinted with permission from Zhang et al. [207] / Springer Nature.

for a larger group of participants and improve the accuracy performance. As a result, a 13-participant gait dataset achieves a high accuracy of 93.54%.

LTIS also gave two simple demonstrations of the future applications of digital twins in smart homes and smart classrooms. The LTIS-based smart building system realizes the automatic distinction between family members and strangers, and real-time monitoring of indoor activities of family members in a camera-free environment, by projecting the recognition results from real space into virtual space. Figure 11.13b summarizes the overall framework of the entire digital twin system, including the first stage of identification and the second stage of activity monitoring in real-time VR projection. Among them, LTIS has possibility to become a solution by using DL technology to provide a safer and smarter environment for smart buildings that do not require the support of cameras and microphones. Such wearable intelligent systems will promote the rapid development of digital twins with wearable formats in the future, in which real digital copies of humans can be created in virtual spaces.

In addition to using smart socks to achieve gait analysis, another common approach is using floor mat-based sensors [208–213]. Traditional approaches used for surveillance and identification are usually based on cameras in offices and home areas, but this method can cause serious privacy issues. For instance, laser beam scanning as one of optical methods can better protect people's privacy than camera approaches [214–216]. However, the above-reported floor sensors only present the possibility for the application of intelligent systems [217–219]. They do not use the AI algorithms to realize the software part of data analytics, and their demonstration seems to be difficult to realize the large-area floor sensing. Therefore, in the aspect of sensor design part, advanced electrode design is very promising to realize triboelectric floor mat sensor with simple configuration, cost-effectiveness, and high reliability. In the aspect of data analysis, it is easier to use advanced AI algorithms for practical large-area floor monitoring and interaction. Hence, Shi et al. proposed a deep-learning-enabled smart mat (DLES-mat) array integrated with DL analytics to realize a smart home [220]. The DLES-mat can realize not only position/activity sensing but also individual recognition. Since each person's step frequency and foot force are different, when a person walks on the triboelectric DLES-mat array, it can generate a unique output voltage pattern. After applying the part of data acquisition and data analysis, the overall structure of the smart floor monitoring system based on DLES-mat is shown in Figure 11.14a. Triboelectric output signals are generated on the two sensing electrodes due to the contact-separation motions between human steps and triboelectric surfaces. These generated signals are acquired by the signal acquisition module and built into a whole dataset from 10 different users. As a result, the CNN-based DL model provides a high recognition performance with an accuracy of 96.0%. The trained model can provide a real-time control interface for identifying similar gait patterns. In general, unlike traditional camera-based surveillance that usually involves video shooting, the intelligent floor monitoring system builds a virtual environment through a digital twin, which can better reflect human activities and their identity in the physical world. The digital twin-based intelligent system shows the potential for automation, control, healthcare, and security applications.

In addition, the schematic diagram of another 4×4-floor mat array is depicted in Figure 11.14b, where all the floor mats are fabricated with the same electrode pattern by screen printing [221]. The exploded view of one-floor mat clearly indicates the composition layers of polyvinyl chloride (PVC) substrate, silver (Ag) sensing electrode, and polyethylene terephthalate (PET) friction surface. The Ag electrode is designed with four individual electrodes – reference electrode, coding electrode 1, coding electrode 2, and sheet electrode, to overcome the limitation of humidity susceptibility. The large-area sheet electrode is designed to fully capture the different gait phases, while the other three electrodes are used for robust position monitoring, by taking output ratios from the two coding electrodes relative to the reference electrode. Since humidity has the same influence on all the electrodes, taking output ratios is an effective solution to achieve stable characteristics and minimize the adverse effect. As shown in Figure 11.14c, except for the regular walking trajectory detection of a single user, the smart floor monitoring system is also applied for multiuser detection, with additional small modifications on the electrode connections. With such electrode connections, user 1 first walks on trajectory 1 and trajectory 2, where alternative outputs are generated from electrode set 2 and electrode set 1, and electrode set 1 and electrode set 3, respectively. Afterward, user 1 and user 2 walks on the floor mats simultaneously following their respective trajectory 3, inducing the corresponding outputs on electrode set 3 and electrode set 4, and electrode set 1

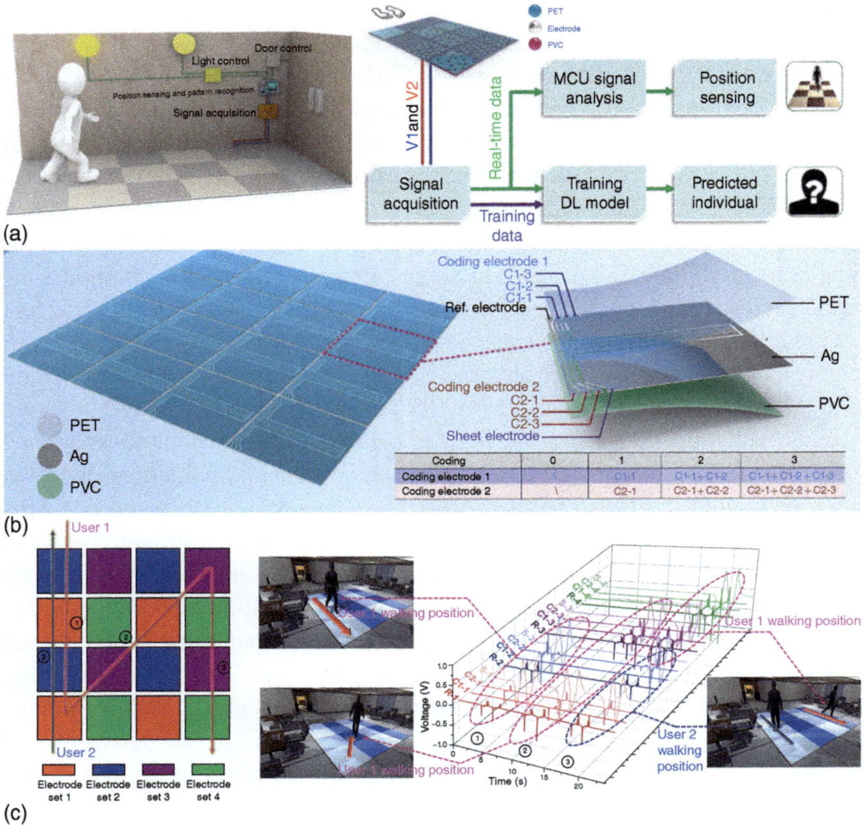

Figure 11.14 TENG-based floor mat for the application of smart home. (a) Schematics of smart floor mat for diversified applications. Source: Reprinted with permission from Shi et al. [220] / Springer Nature. (b) A 4 × 4 floor mat array with an enlarged view of the composition layers and the coding-electrode design. (c) Demonstration of a virtual smart office showing the detection capability of the system in normal/diagonal walking and multiuser walking. Source: Reprinted with permission for Shi et al. [221] / American Chemical Society.

and electrode set 2, respectively. Benefited by the modified electrode connection, the detection of diagonal walking and two-user walking can be realized with the proposed smart floor monitoring system, enabling more practical applications in smart homes, smart offices, and smart buildings. Overall, in these two demonstrations, real-time position and individual recognition of a person walking on the floor mat array are successfully achieved, showing the great potential of the smart floor monitoring system in smart buildings relative to automatic control and security access.

With the numerous self-powered sensors developed in the IoT system, the various monitoring functions with low power consumption can be also achieved in smart homes through equipping various temperature sensors, gas sensors, motion sensors, humidity sensors, etc. [222–225]. With the aid of AI, the different sensor information will be integrated into server via IoT nodes for further analysis. Zhang et al. developed a low-cost textile-based TENG (T-TENG) pressure sensor aiming for

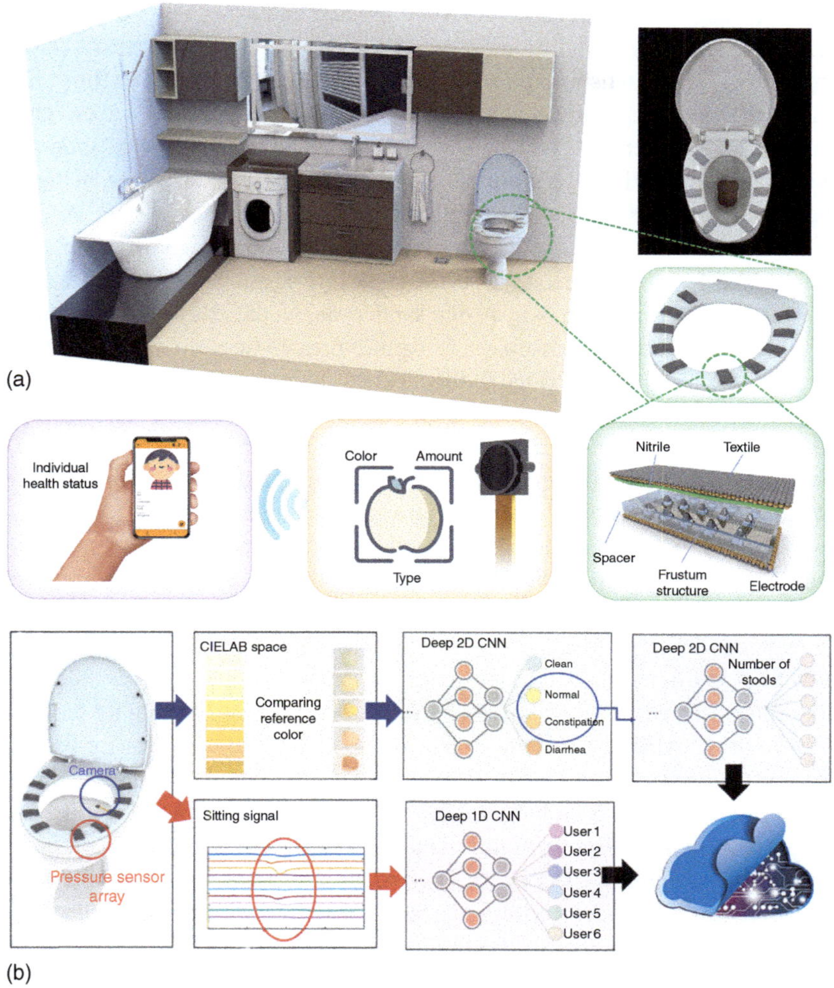

Figure 11.15 A low-cost T-TENG pressure sensor integrated with vision sensor aiming for an AI-toilet. (a) The schematics of AI-Toilet. (b) Schematic of the whole functions of AI-toilet, including urine detection, type and number recognition of stool state, seating time and defecation time recording, user identification and IoT application. Source: Reprinted with permission for Zhang et al. [226] / Elsevier.

an AI-toilet, as shown in Figure 11.15a [226]. The T-TENG contains four function layers – two conductive textile layers for charge collection and transfer, one nitrile thin film as the positive electrification layer, one silicone rubber film as the negative electrification layer, and the main sensing layer. The triboelectric layer with inner frustum structures and a spacer are made of silicone rubber material through 3D-printing molds, which can achieve different pressure sensing ranges by modifying the height. With 10 T-TENGs placed on a toilet seat, the detailed pressure distributions and characteristics of a user sitting on and standing up from the stool can be comprehensively recorded. In addition, as shown in Figure 11.15b, the AI-toilet integrates the image sensor to realize the AIoT application in smart home. AI-toilet will perform simple color recognition first and then compare it with the urine color chart to get the user's water shortage and preliminary health status. When the pressure sensor receives a trigger signal, the image recognition function is turned on. The neural network for excrement classification that has been burned on AIoT module will be activated first. When the stimulated defecation process starts by placing a small banana in the AI-toilet, the CNN will differentiate it as the stool state. At the same time, the CIELAB color space-based color recognition function is turned off. As the T-TENG-based pressure sensor detects the departure signal (tf), the total seating time can be obtained through $t_f - t_0$. At the same time, all data during this period, including urine color, excrement image recognition, excrement quantity, and the corresponding time, will be uploaded to the local PC. On the other hand, when the pressure sensor signal is detected at the beginning, the data of 10 pressure sensors will be also used to obtain the predicted user's identity information through the 1D CNN. In the end, all information will be uploaded to the cloud through the PC, and then transmitted to the user's mobile phone through the cloud. Therefore, it not only protects the privacy of users but also allows family members to better pay attention to the health of elder or children. Therefore, the proposed smart toilet will obtain various types of health information after interpreting the state and habits of the user, thereby realizing disease detection, whose intelligence and versatility will become one of the final solutions of smart toilets in smart homes.

Further integrated with AI technology, such sensors can achieve a higher convenience for multiple AIoT smart home applications, acting as a caregiver for users' healthcare monitoring. Aiming for the caregiving of elderly and motion-impaired people, the population of which has already exceeded one billion in the world and is still continuously increasing, a multifunctional walking stick has also been proposed, as shown in Figure 11.16a [227]. The proposed walking stick contains two main functional units (Figure 11.16b), i.e. the hybridized unit and the rotational unit. The hybridized unit consists of a top press TENG (P-TENG), a middle electromagnetic harvester (EMG), and a bottom rotational TENG (R-TENG), while the rotational unit only consists of the EMG part. The P-TENG contains two aluminum layers, one nitrile layer, and one Ecoflex layer, which can generate various output voltages regarding the applied pressure. The bottom aluminum has been divided into five electrodes, which are able to record the entire process of a walking stick when contacting and separating the ground, including the contact point, contact force, contact time, and contact sequence. With the DL with 1D-CNN structure to analyze the output from P-TENG, the walking stick can distinguish 5 different movements (stand up, sit down, walk, upstairs, and downstairs), evaluate 3 different statuses, and identify 10 different users. At the same time, the R-TENG can also detect the gait abnormity, such as the fall of the user through irregular patterns of output. Figure 11.16c shows the real-time motion status of the user where a virtual environment has also been built to mimic real life. The output signals of P-TENG and R-TENG are collected by an microcontroller unit (MCU) module and further sent to the computer wirelessly for analysis. Through the DL model, the real-time motion status of the user in the home can be easily obtained and reflected in the virtual

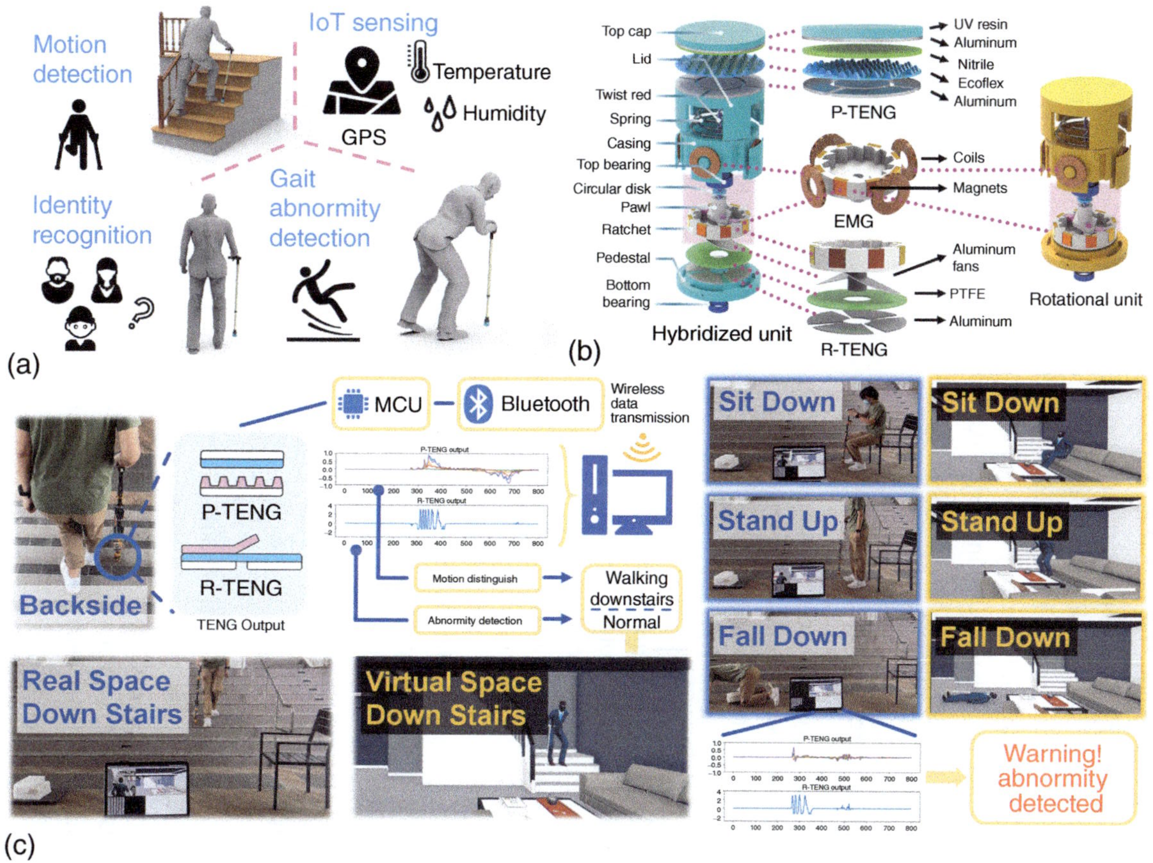

Figure 11.16 (a) A multifunctional walking stick aiming for the caregiving of elderly and motion-impaired people. (b) The hybridized unit consists of a top press TENG (PTENG), a middle EMG, and a bottom rotational TENG (R-TENG). (c) Demonstration for the application of indoor monitoring with the caregiving walking stick. Source: Reprinted with permission for Guo et al. [227] / American Chemical Society.

environment. And when the user falls in the home, that gait abnormity will also be acquired instantaneously and able to call for immediate help. This caregiving walking stick only monitors the user's motion status as an important well-being indicator to avoid privacy concerns from traditional camera-based in-door healthcare monitoring. In the meantime, through the linear-to-rotary structure, which can transfer the low-frequency linear motion to high-speed rotation, the two units can efficiently harvest the ultralow frequency of the motion-impaired people. A maximum average power density of 0.595 mW/cm^3 under 1 Hz driven frequency and the capability to charge a 4 mF capacitor to 5 V in eight seconds have been successfully achieved. The harvested biomechanical energy can serve as the power supply for a self-sustainable IoT system with global positioning system (GPS) location tracking and environmental temperature and humidity sensing, achieving an Omni-monitoring for users.

11.3.4 Future Directions

Due to recent breakthroughs in computing power, AI technology, and software architecture, intelligent systems are undergoing rapid development. Flexible sensors will play more and more important roles in IoT applications. In particular, intelligent systems based on TENG and PENG, due to their low-power consumption, provide a new possibility for realizing self-sustainable system. In addition, the feature direction of intelligent system relies on the sensor fusion with multi-modality data. With the evolving AI technologies, more advanced sensors could be realized for intelligent systems ranging from wearable sensors, e.g. glove, sock, touchpad, exoskeleton, e-skin, to the comprehensive applications of the healthcare, robot perception, and smart home. Intelligent systems will play an indispensable role on various occasions. The application of AI will bring process automation and process optimization to a new level, due to its ability in accurate feature classification and prediction of sensor data, efficient processing of massive and dimensional datasets, and multi-modal physiological signal analysis.

References

1 Li, S., Ni, Q., Sun, Y. et al. (2018). Energy-efficient resource allocation for industrial cyber-physical IoT systems in 5G era. *IEEE Transactions on Industrial Informatics* 14 (6): 2618–2628. https://doi.org/10.1109/TII.2018.2799177.

2 Dong, B., Shi, Q., Yang, Y. et al. (2021). Technology evolution from self-powered sensors to AIoT enabled smart homes. *Nano Energy* 79: 105414. https://doi.org/10.1016/j.nanoen.2020.105414.

3 Ren, Z., Xu, J., Le, X., and Lee, C. (2021). Heterogeneous wafer bonding technology and thin-film transfer technology-enabling platform for the next generation applications beyond 5G. *Micromachines* 12 (8): https://doi.org/10.3390/mi12080946.

4 Cai, X., Geng, S., Wu, D. et al. (2020). A multi-cloud model based many-objective intelligent algorithm for efficient task scheduling in internet of things.

IEEE Internet of Things Journal 14 (8): 9645–9653. https://doi.org/10.1109/JIOT.2020.3040019.

5 Stergiou, C., Psannis, K.E., Kim, B.G., and Gupta, B. (2018). Secure integration of IoT and cloud computing. *Future Generation Computer Systems* 78: 964–975. https://doi.org/10.1016/j.future.2016.11.031.

6 Kim, J., Campbell, A.S., de Ávila, B.E.F., and Wang, J. (2019). Wearable biosensors for healthcare monitoring. *Nature Biotechnology* 37 (4): 389–406. https://doi.org/10.1038/s41587-019-0045-y.

7 Bariya, M., Nyein, H.Y.Y., and Javey, A. (2018). Wearable sweat sensors. *Nature Electronics* 1 (3): 160–171. https://doi.org/10.1038/s41928-018-0043-y.

8 Ray, T.R. et al. (2019). Bio-integrated wearable systems: a comprehensive review. *Chemical Reviews* 119 (8): 5461–5533. https://doi.org/10.1021/acs.chemrev.8b00573.

9 Yang, Y. and Gao, W. (2019). Wearable and flexible electronics for continuous molecular monitoring. *Chemical Society Reviews* 48 (6): 1465–1491. https://doi.org/10.1039/c7cs00730b.

10 Yang, J.C., Mun, J., Kwon, S.Y. et al. (2019). Electronic skin: recent progress and future prospects for skin-attachable devices for health monitoring, robotics, and prosthetics. *Advanced Materials* 31 (48): 1904765, 1–50. https://doi.org/10.1002/adma.201904765.

11 Niu, S. et al. (2019). A wireless body area sensor network based on stretchable passive tags. *Nature Electronics* 2 (8): 361–368. https://doi.org/10.1038/s41928-019-0286-2.

12 Tian, X. et al. (2019). Wireless body sensor networks based on metamaterial textiles. *Nature Electronics* 2 (6): 243–251. https://doi.org/10.1038/s41928-019-0257-7.

13 Wen, F. et al. (2020). Machine learning glove using self-powered conductive superhydrophobic triboelectric textile for gesture recognition in VR/AR applications. *Advanced Science* 7 (14): 1–15. https://doi.org/10.1002/advs.202000261.

14 Liu, K. et al. (2021). Augmented reality navigation method for recontouring surgery of craniofacial fibrous dysplasia. *Scientific Reports* 11 (1): 1–7. https://doi.org/10.1038/s41598-021-88860-x.

15 Miehlbradt, J., Cuturi, L.F., Zanchi, S. et al. (2021). Immersive virtual reality interferes with default head–trunk coordination strategies in young children. *Scientific Reports* 11 (1): 1–13. https://doi.org/10.1038/s41598-021-96866-8.

16 Kwon, K. et al. (2019). Miniaturized, light-adaptive, wireless dosimeters autonomously monitor exposure to electromagnetic radiation. *Science Advances* 5 (12): 1–10. https://doi.org/10.1126/sciadv.aay2462.

17 Fan, W. et al. (2020). Machine-knitted washable sensor array textile for precise epidermal physiological signal monitoring. *Science Advances* 6 (11): 1–11. https://doi.org/10.1126/sciadv.aay2840.

18 Pu, X. et al. (2017). Eye motion triggered self-powered mechnosensational communication system using triboelectric nanogenerator. *Science Advances* 3 (7): 1–8. https://doi.org/10.1126/sciadv.1700694.

19 Zhu, M. et al. (2020). Haptic-feedback smart glove as a creative human–machine interface (HMI) for virtual/augmented reality applications. *Science Advances* 6 (19): 1–15. https://doi.org/10.1126/sciadv.aaz8693.

20 Zhang, G.H. et al. (2020). High-performance particulate matter including nanoscale particle removal by a self-powered air filter. *Nature Communications* 11 (1): https://doi.org/10.1038/s41467-020-15502-7.

21 Wang, Y. et al. (2021). All-weather, natural silent speech recognition via machine-learning-assisted tattoo-like electronics. *npj Flexible Electronics* 5 (1): 1–9. https://doi.org/10.1038/s41528-021-00119-7.

22 Wen, F., Zhang, Z., He, T., and Lee, C. (2021). AI enabled sign language recognition and VR space bidirectional communication using triboelectric smart glove. *Nature Communications* 12 (1): 1–13. https://doi.org/10.1038/s41467-021-25637-w.

23 Zhu, M. et al. (2019). Self-powered and self-functional cotton sock using piezoelectric and triboelectric hybrid mechanism for healthcare and sports monitoring. *ACS Nano* 13: 1940–1952. https://doi.org/10.1021/acsnano.8b08329.

24 Wang, S. et al. (2013). Motion charged battery as sustainable flexible-power-unit. *ACS Nano* 7 (12): 11263–11271. https://doi.org/10.1021/nn4050408.

25 Todaro, M.T. et al. (2018). Biocompatible, flexible, and compliant energy harvesters based on piezoelectric thin films. *IEEE Transactions on Nanotechnology* 17 (2): 220–230. https://doi.org/10.1109/TNANO.2017.2789300.

26 Dong, K., Peng, X., and Wang, Z.L. (2020). Fiber/fabric-based piezoelectric and triboelectric nanogenerators for flexible/stretchable and wearable electronics and artificial intelligence. *Advanced Materials* 32 (5): 1902549. https://doi.org/10.1002/adma.201902549.

27 Charroux, B., Daian, F., and Royet, J. (2020). *Drosophila* aversive behavior toward *Erwinia carotovora carotovora* is mediated by bitter neurons and leukokinin. *iScience* 23 (6): 101152. https://doi.org/10.1016/j.isci.

28 Kalavakonda, R.R., Masna, N.V.R., Mandal, S., and Bhunia, S. (2021). A smart mask for active defense against airborne pathogens. *Scientific Reports* 11 (1): https://doi.org/10.1038/s41598-021-99150-x.

29 Jayathilaka, W.A.D.M. et al. (2019). Significance of nanomaterials in wearables: a review on wearable actuators and sensors. *Advanced Materials* 31 (7): 1–21. https://doi.org/10.1002/adma.201805921.

30 Zhang, F., Zang, Y., Huang, D. et al. (2015). Flexible and self-powered temperature-pressure dual-parameter sensors using microstructure-frame-supported organic thermoelectric materials. *Nature Communications* 6: 1–10. https://doi.org/10.1038/ncomms9356.

31 Wu, R. et al. (2019). Silk composite electronic textile sensor for high space precision 2D combo temperature–pressure sensing. *Small* 15 (31): 1–11. https://doi.org/10.1002/smll.201901558.

32 Narayanaswami, C. and Raghunath, M.T. (2000). Application design for a smart watch with a high resolution display. *International Symposium on Wearable Computers, Digest of Papers*, pp. 7–14. https://doi.org/10.1109/iswc.2000.888452.

33 Ouyang, H. et al. (2017). Self-powered pulse sensor for antidiastole of cardiovascular disease. *Advanced Materials* 29 (40): 1703456. https://doi.org/10.1002/adma.201703456.

34 Su, Y. et al. (2020). Alveolus-inspired active membrane sensors for self-powered wearable chemical sensing and breath analysis. *ACS Nano* 14 (5): 6067–6075. https://doi.org/10.1021/acsnano.0c01804.

35 Su, Y. et al. (2020). A wireless energy transmission enabled wearable active acetone biosensor for non-invasive prediabetes diagnosis. *Nano Energy* 74: https://doi.org/10.1016/j.nanoen.2020.104941.

36 Su, Y. et al. (2019). Improving sensitivity of self-powered room temperature NO_2 sensor by triboelectric-photoelectric coupling effect. *Applied Physics Letters* 115 (7): https://doi.org/10.1063/1.5110334.

37 Tamm, T., Pärlin, K., Tiimus, T. et al. (2014). Smart insole sensors for sports and rehabilitation. *Nanosensors, Biosensors, and Info-Tech Sensors and Systems 2014*, Volume 9060, April 2014, p. 90600L. https://doi.org/10.1117/12.2045062.

38 Wang, J., Wang, H., He, T. et al. (2019). Investigation of low-current direct stimulation for rehabilitation treatment related to muscle function loss using self-powered TENG system. *Advanced Science* 6 (14): 1900149. https://doi.org/10.1002/advs.201900149.

39 Wang, J., He, T., and Lee, C. (2019). Development of neural interfaces and energy harvesters towards self-powered implantable systems for healthcare monitoring and rehabilitation purposes. *Nano Energy* 65: 104039. https://doi.org/10.1016/j.nanoen.2019.104039.

40 Simonyan, K. and Zisserman, A. (2015). Very deep convolutional networks for large-scale image recognition. *3rd International Conference on Learning Representations, ICLR 2015 - Conference Track Proceedings*, pp. 1–14.

41 Wang, A.C., Wu, C., Pisignano, D. et al. (2018). Polymer nanogenerators: opportunities and challenges for large-scale applications. *Journal of Applied Polymer Science* 135 (24): 1–17. https://doi.org/10.1002/app.45674.

42 Liu, M. et al. (2017). Large-area all-textile pressure sensors for monitoring human motion and physiological signals. *Advanced Materials* 29 (41): 1703700. https://doi.org/10.1002/adma.201703700.

43 Zhu, J. et al. (2020). Development trends and perspectives of future sensors and MEMS/NEMS. *Micromachines* 11 (1): 1–29. https://doi.org/10.3390/mi11010007.

44 Wang, T. et al. (2018). Investigation of the temperature fluctuation of single-phase fluid based microchannel heat sink. *Sensors (Switzerland)* 18 (5): 1–11. https://doi.org/10.3390/s18051498.

45 Zhou, H., Kropelnicki, P., and Lee, C. (2015). CMOS compatible midinfrared wavelength-selective thermopile for high temperature applications. *Journal of Microelectromechanical Systems* 24 (1): 144–154. https://doi.org/10.1109/JMEMS.2014.2322675.

46 Hasan, D., Pitchappa, P., Pei Ho, C., and Lee, C. (2017). High temperature coupling of IR inactive C=C mode in complementary metal oxide semiconductor metamaterial structure. *Advanced Optical Materials* 5 (3): 1600778. https://doi.org/10.1002/adom.201600778.

47 Wicaksono, I. et al. (2020). A tailored, electronic textile conformable suit for large-scale spatiotemporal physiological sensing in vivo. *npj Flexible Electronics* 4 (1): 5. https://doi.org/10.1038/s41528-020-0068-y.

48 Qiao, Q., Peng, C., Xia, J. et al. (2019). Ultra-small photonic crystal (PhC)-based test tool for gas permeability of polymers. *Optics Express* 27 (24): 35600. https://doi.org/10.1364/oe.27.035600.

49 Hasan, D. and Lee, C. (2018). Hybrid metamaterial absorber platform for sensing of CO_2 gas at mid-IR. *Advanced Science* 5 (5): https://doi.org/10.1002/advs.201700581.

50 Hassani, F.A. and Lee, C. (2015). A triboelectric energy harvester using low-cost, flexible, and biocompatible ethylene vinyl acetate (EVA). *Journal of Microelectromechanical Systems* 24 (5): 1338–1345. https://doi.org/10.1109/JMEMS.2015.2403256.

51 Dhakar, L., Pitchappa, P., Tay, F.E.H., and Lee, C. (2016). An intelligent skin based self-powered finger motion sensor integrated with triboelectric nanogenerator. *Nano Energy* 19: 532–540. https://doi.org/10.1016/j.nanoen.2015.04.020.

52 Bai, Z., Xu, Y., Lee, C., and Guo, J. (2021). Autonomously adhesive, stretchable, and transparent solid-state polyionic triboelectric patch for wearable power source and tactile sensor. *Advanced Functional Materials* 31 (37): 1–11. https://doi.org/10.1002/adfm.202104365.

53 Dhakar, L., Tay, F.E.H., and Lee, C. (2015). Development of a broadband triboelectric energy harvester with SU-8 micropillars. *Journal of Microelectromechanical Systems* 24 (1): 91–99. https://doi.org/10.1109/JMEMS.2014.2317718.

54 Wang, W. et al. (2021). Neuromorphic motion detection and orientation selectivity by volatile resistive switching memories. *Advanced Intelligent Systems* 3 (4): 2000224. https://doi.org/10.1002/aisy.202000224.

55 Liang, F.-X., Wang, I.-T., and Hou, T.-H. (2021). Progress and benchmark of spiking neuron devices and circuits. *Advanced Intelligent Systems* 3 (8): 2100007. https://doi.org/10.1002/aisy.202100007.

56 Sohn, K.S. et al. (2017). An extremely simple macroscale electronic skin realized by deep machine learning. *Scientific Reports* 7 (1): 1–10. https://doi.org/10.1038/s41598-017-11663-6.

57 Lee, J.W. et al. (2018). Deep-learning technique to convert a crude piezoresistive carbon nanotube-ecoflex composite sheet into a smart, portable, disposable, and extremely flexible keypad. *ACS Applied Materials & Interfaces* 10 (24): 20862–20868. https://doi.org/10.1021/acsami.8b04914.

58 Cho, M.Y., Lee, J.H., Kim, S.H. et al. (2019). An extremely inexpensive, simple, and flexible carbon fiber electrode for tunable elastomeric piezo-resistive sensors and devices realized by LSTM RNN. *ACS Applied Materials & Interfaces* 11 (12): 11910–11919. https://doi.org/10.1021/acsami.9b00464.

59 Tkatchenko, A. (2020). Machine learning for chemical discovery. *Nature Communications* 11 (1): 8–11. https://doi.org/10.1038/s41467-020-17844-8.

60 Moosavi, S.M., Jablonka, K.M., and Smit, B. (2020). The role of machine learning in the understanding and design of materials. *Journal of the American Chemical Society* 142 (48): 20273–20287. https://doi.org/10.1021/jacs.0c09105.

61 Lei, Y., Yang, B., Jiang, X. et al. (2020). Applications of machine learning to machine fault diagnosis: a review and roadmap. *Mechanical Systems and Signal Processing* 138: 106587. https://doi.org/10.1016/j.ymssp.2019.106587.

62 Fathi, S., Srinivasan, R., Fenner, A., and Fathi, S. (2020). Machine learning applications in urban building energy performance forecasting: a systematic review. *Renewable and Sustainable Energy Reviews* 133: 110287. https://doi.org/10.1016/j.rser.2020.110287.

63 Kim, J.Y., Park, G., Lee, S.A., and Nam, Y. (2020). Analysis of machine learning-based assessment for elbow spasticity using inertial sensors. *Sensors (Switzerland)* 20 (6): 1–15. https://doi.org/10.3390/s20061622.

64 Zhu, J. et al. (2021). Machine learning-enabled textile-based graphene gas sensing with energy harvesting-assisted IoT application. *Nano Energy* 86: 106035. https://doi.org/10.1016/j.nanoen.2021.106035.

65 Xie, C., Tan, M., Gong, B. et al. (2020). Adversarial examples improve image recognition. *Proceedings of the IEEE Computer Society Conference on Computer Vision and Pattern Recognition*, pp. 816–825. https://doi.org/10.1109/CVPR42600.2020.00090.

66 Dosovitskiy, A. et al. (2020). An image is worth 16×16 words: transformers for image recognition at scale, October 2020. http://arxiv.org/abs/2010.11929.

67 Ciodaro, T., Deva, D., de Seixas, J.M., and Damazio, D. (2012). Online particle detection with neural networks based on topological calorimetry information. *Journal of Physics: Conference Series* 368 (1): 012030. https://doi.org/10.1088/1742-6596/368/1/012030.

68 Hussain, A.A., Bouachir, O., Al-Turjman, F., and Aloqaily, M. (2020). AI techniques for COVID-19. *IEEE Access* 8: 128776–128795. https://doi.org/10.1109/ACCESS.2020.3007939.

69 Helmstaedter, M., Briggman, K.L., Turaga, S.C. et al. (2013). Connectomic reconstruction of the inner plexiform layer in the mouse retina. *Nature* 500 (7461): 168–174. https://doi.org/10.1038/nature12346.

70 Xiong, H.Y. et al. (2015). The human splicing code reveals new insights into the genetic determinants of disease. *Science* 347 (6218): https://doi.org/10.1126/science.1254806.

71 Sun, Z. et al. (2021). Artificial intelligence of things (AIoT) enabled virtual shop applications using self-powered sensor enhanced soft robotic manipulator. *Advanced Science* 8: 2100230. https://doi.org/10.1002/advs.202100230.

72 Wang, X. et al. (2021). Enabling the unconstrained epidermal pulse wave monitoring via finger-touching. *Advanced Functional Materials* 31 (32): 1–11. https://doi.org/10.1002/adfm.202102378.

73 Han, Y. et al. (2019). Self-powered gait pattern-based identity recognition by a soft and stretchable triboelectric band. *Nano Energy* 56: 516–523. https://doi.org/10.1016/j.nanoen.2018.11.078.

74 Cosido, O. et al. (2014). Hybridization of convergent photogrammetry, computer vision, and artificial intelligence for digital documentation of cultural heritage-a case study: the magdalena palace. *Proceedings - 2014 International Conference on Cyberworlds, CW 2014*, pp. 369–376. https://doi.org/10.1109/CW.2014.58.

75 Panwar, H., Gupta, P.K., Siddiqui, M.K. et al. (2020). A deep learning and grad-CAM based color visualization approach for fast detection of COVID-19 cases using chest X-ray and CT-scan images. *Chaos, Solitons & Fractals* 140: 110190. https://doi.org/10.1016/j.chaos.2020.110190.

76 Wiley, V. and Lucas, T. (2018). Computer vision and image processing: a paper review. *International Journal of Artificial Intelligence Research* 2 (1): 22. https://doi.org/10.29099/ijair.v2i1.42.

77 Kortylewski, A., Egger, B., Schneider, A. et al. (2018). Empirically analyzing the effect of dataset biases on deep face recognition systems. *IEEE Computer Society Conference on Computer Vision and Pattern Recognition Workshops*, Volume 2018-June, pp. 2174–2183. https://doi.org/10.1109/CVPRW.2018.00283.

78 Li, R., Cheong, L.F., and Tan, R.T. (2019). Heavy rain image restoration: integrating physics model and conditional adversarial learning. *Proceedings of the IEEE Computer Society Conference on Computer Vision and Pattern Recognition*, Volume 2019-June, no. 1, pp. 1633–1642. https://doi.org/10.1109/CVPR.2019.00173.

79 Zhu, J.Y., Park, T., Isola, P., and Efros, A.A. (2017). Unpaired image-to-image translation using cycle-consistent adversarial networks. *Proceedings of the IEEE International Conference on Computer Vision*, Volume 2017-October, pp. 2242–2251. https://doi.org/10.1109/ICCV.2017.244.

80 Wang, S. et al. (2021). Annotation-efficient deep learning for automatic medical image segmentation. *Nature Communications* 12 (1): 1–13. https://doi.org/10.1038/s41467-021-26216-9.

81 Saxena, A., Driemeyer, J., and Ng, A.Y. (2008). Robotic grasping of novel objects using vision. *International Journal of Robotics Research* 27 (2): 157–173. https://doi.org/10.1177/0278364907087172.

82 Redmon, J. and Farhadi, A. (2018). YOLOv3: an incremental improvement, April 2018. http://arxiv.org/abs/1804.02767.

83 Luo, S., Mou, W., Althoefer, K., and Liu, H. (2015). Novel tactile-SIFT descriptor for object shape recognition. *IEEE Sensors Journal* 15 (9): 5001–5009. https://doi.org/10.1109/JSEN.2015.2432127.

84 Jiang, Y., Moseson, S., and Saxena, A. (2011). Efficient grasping from RGBD images: learning using a new rectangle representation. *Proceedings - IEEE International Conference on Robotics and Automation*, pp. 3304–3311. https://doi.org/10.1109/ICRA.2011.5980145.

85 Goodfellow, I.J. et al. (2014). Generative adversarial networks. In: *Advances in Neural Information Processing Systems*, vol. 2017, 4089–4099. https://arxiv.org/abs/1406.2661.

86 Gide, A. (1967). 'Who' is saying 'what'? Brain-based decoding of human voice and speech. *Angewandte Chemie International Edition* 6 (11): 951–952, vol. 322, 5–24.

87 Perrachione, T.K., del Tufo, S.N., and Gabrieli, J.D.E. (2011). Human voice recognition depends on language ability. *Science* 333 (6042): 595. https://doi.org/10.1126/science.1207327.

88 van den Oord, A. et al. (2016). WaveNet: a generative model for raw audio, pp. 1–15, September 2016. http://arxiv.org/abs/1609.03499.

89 Graves, A., Mohamed, A.R., and Hinton, G. (2013). Speech recognition with deep recurrent neural networks. *ICASSP, IEEE International Conference on Acoustics, Speech and Signal Processing - Proceedings*, no. 6, pp. 6645–6649. https://doi.org/10.1109/ICASSP.2013.6638947.

90 Han, J.H. et al. (2018). Basilar membrane-inspired self-powered acoustic sensor enabled by highly sensitive multi tunable frequency band. *Nano Energy* 53: 198–205. https://doi.org/10.1016/j.nanoen.2018.08.053.

91 Wang, C. et al. (2017). Detection of non-joint areas tiny strain and anti-interference voice recognition by micro-cracked metal thin film. *Nano Energy* 34: 578–585. https://doi.org/10.1016/j.nanoen.2017.02.050.

92 Liu, H. et al. (2020). An epidermal sEMG tattoo-like patch as a new human–machine interface for patients with loss of voice. *Microsystems & Nanoengineering* 1–13. https://doi.org/10.1038/s41378-019-0127-5.

93 Guo, H. et al. (2018). A highly sensitive, self-powered triboelectric auditory sensor for social robotics and hearing aids. *Science Robotics* 3 (20): eaat2516. https://doi.org/10.1126/scirobotics.aat2516.

94 Hua, Q. et al. (2018). Skin-inspired highly stretchable and conformable matrix networks for multifunctional sensing. *Nature Communications* 9 (1): 1–11. https://doi.org/10.1038/s41467-017-02685-9.

95 Li, J. et al. (2020). Visually aided tactile enhancement system based on ultra-thin highly sensitive crack-based strain sensors. *Applied Physics Reviews* 7 (1): https://doi.org/10.1063/1.5129468.

96 He, J. et al. (2020). Recent advances of wearable and flexible piezoresistivity pressure sensor devices and its future prospects. *Journal of Materiomics* 6 (1): 86–101. https://doi.org/10.1016/j.jmat.2020.01.009.

97 Zhu, P., Wang, Y., Sheng, M. et al. (2019). A flexible active dual-parameter sensor for sensitive temperature and physiological signal monitoring: via integrating thermoelectric and piezoelectric conversion. *Journal of Materials Chemistry A* 7 (14): 8258–8267. https://doi.org/10.1039/c9ta00682f.

98 Manap, H.H., Tahir, N.Md., and Yassin, A.I.M. (2011). Statistical analysis of Parkinson disease gait classification using artificial neural network. *IEEE International Symposium on Signal Processing and Information Technology, ISSPIT 2011*, pp. 060–065. https://doi.org/10.1109/ISSPIT.2011.6151536.

99 Lee, G. et al. (2020). Parallel signal processing of a wireless pressure-sensing platform combined with machine-learning-based cognition, inspired by the human somatosensory system. *Advanced Materials* 32 (8): 1906269. https://doi.org/10.1002/adma.201906269.

100 Ravanelli, M., Parcollet, T., and Bengio, Y. (2019). The Pytorch-kaldi speech recognition toolkit. *ICASSP, IEEE International Conference on Acoustics, Speech* *and Signal Processing - Proceedings*, Volume 2019-May, pp. 6465–6469. https://doi.org/10.1109/ICASSP.2019.8683713.

101 Alcázar, J.L. et al. (2020). Active speakers in context. *Proceedings of the IEEE Computer Society Conference on Computer Vision and Pattern Recognition*, pp. 12462–12471. https://doi.org/10.1109/CVPR42600.2020.01248.

102 Oren, S., Ceylan, H., Schnable, P.S., and Dong, L. (2017). High-resolution patterning and transferring of graphene-based nanomaterials onto tape toward roll-to-roll production of tape-based wearable sensors. *Advanced Materials Technologies* 2 (12): 1–14. https://doi.org/10.1002/admt.201700223.

103 Rogers, J.A., Someya, T., and Huang, Y. (2010). Materials and mechanics for stretchable electronics. *Science* 327 (5973): 1603–1607. https://doi.org/10.1126/science.1182383.

104 Larson, C. et al. (2016). Highly stretchable electroluminescent skin for optical signaling and tactile sensing. *Science* 351 (6277): 1071–1074. https://doi.org/10.1126/science.aac5082.

105 Kim, D.C., Shim, H.J., Lee, W. et al. (2020). Material-based approaches for the fabrication of stretchable electronics. *Advanced Materials* 32 (15): 1–29. https://doi.org/10.1002/adma.201902743.

106 Tien, H.C., Huang, Y.W., Chiu, Y.C. et al. (2021). Intrinsically stretchable polymer semiconductors: molecular design, processing and device applications. *Journal of Materials Chemistry C* 9 (8): 2660–2684. https://doi.org/10.1039/d0tc06059c.

107 Chortos, A., Liu, J., and Bao, Z. (2016). Pursuing prosthetic electronic skin. *Nature Materials* 15 (9): 937–950. https://doi.org/10.1038/nmat4671.

108 Tcho, I.W., Kim, W.G., and Choi, Y.K. (2020). A self-powered character recognition device based on a triboelectric nanogenerator. *Nano Energy* 70: 104534. https://doi.org/10.1016/j.nanoen.2020.104534.

109 Ji, X., Zhao, T., Zhao, X. et al. (2020). Triboelectric nanogenerator based smart electronics via machine learning. *Advanced Materials Technologies* 5 (2): 1–11. https://doi.org/10.1002/admt.201900921.

110 Yu, J. et al. (2020). Ultra-flexible and high-sensitive triboelectric nanogenerator as electronic skin for self-powered human physiological signal monitoring. *Nano Energy* 69: 104437. https://doi.org/10.1016/j.nanoen.2019.104437.

111 Malawski, F. and Kwolek, B. (2016). Classification of basic footwork in fencing using accelerometer. *2016 Signal Processing: Algorithms, Architectures, Arrangements, and Applications (SPA)*, September 2016, Volume 1, pp. 51–55. https://doi.org/10.1109/SPA.2016.7763586.

112 Hougne, P., Imani, M.F., Diebold, A.v. et al. (2020). Learned integrated sensing pipeline: reconfigurable metasurface transceivers as trainable physical layer in an artificial neural network. *Advanced Science* 7 (3): 1901913. https://doi.org/10.1002/advs.201901913.

113 Lecun, Y., Bengio, Y., and Hinton, G. (2015). Deep learning. *Nature* 521 (7553): 436–444. https://doi.org/10.1038/nature14539.

114 Wang, J., Chen, Y., Hao, S. et al. (2019). Deep learning for sensor-based activity recognition: a survey. *Pattern Recognition Letters* 119: 3–11. https://doi.org/10.1016/j.patrec.2018.02.010.

115 Syafrudin, M., Alfian, G., Fitriyani, N., and Rhee, J. (2018). Performance analysis of IoT-based sensor, big data processing, and machine learning model for real-time monitoring system in automotive manufacturing. *Sensors* 18 (9): 2946. https://doi.org/10.3390/s18092946.

116 Kühner, L., Semenyshyn, R., Hentschel, M. et al. (2019). Vibrational sensing using infrared nanoantennas: toward the noninvasive quantitation of physiological levels of glucose and fructose. *ACS Sensors* 4 (8): 1973–1979. https://doi.org/10.1021/acssensors.9b00488.

117 Zhu, J., Ren, Z., and Lee, C. (2021). Toward healthcare diagnoses by machine-learning-enabled volatile organic compound identification. *ACS Nano* 15 (1): 894–903. https://doi.org/10.1021/acsnano.0c07464.

118 He, Y. et al. (2021). ClusterMap for multi-scale clustering analysis of spatial gene expression. *Nature Communications* 12 (1): https://doi.org/10.1038/s41467-021-26044-x.

119 Wertheim, J.O. et al. (2019). Natural selection favoring more transmissible HIV detected in United States molecular transmission network. *Nature Communications* 10 (1): 1–10. https://doi.org/10.1038/s41467-019-13723-z.

120 Angenent-Mari, N.M., Garruss, A.S., Soenksen, L.R. et al. (2020). A deep learning approach to programmable RNA switches. *Nature Communications* 11 (1): 1–12. https://doi.org/10.1038/s41467-020-18677-1.

121 Noble, W.S. (2006). What is a support vector machine? *Nature Biotechnology* 24 (12): 1565–1567. https://doi.org/10.1038/nbt1206-1565.

122 John-Herpin, A., Kavungal, D., von Mücke, L., and Altug, H. (2021). Infrared metasurface augmented by deep learning for monitoring dynamics between all major classes of biomolecules. *Advanced Materials* 33 (14): https://doi.org/10.1002/adma.202006054.

123 Meng, J., Cadusch, J.J., and Crozier, K.B. (2021). Plasmonic mid-infrared filter array-detector array chemical classifier based on machine learning. *ACS Photonics* 8 (2): 648–657. https://doi.org/10.1021/acsphotonics.0c01786.

124 Valeri, J.A. et al. (2020). Sequence-to-function deep learning frameworks for engineered riboregulators. *Nature Communications* 11 (1): 1–14. https://doi.org/10.1038/s41467-020-18676-2.

125 Huang, H., Li, X., Liu, S. et al. (2018). TriboMotion: a self-powered triboelectric motion sensor in wearable internet of things for human activity recognition and energy harvesting. *IEEE Internet of Things Journal* 5 (6): 4441–4453. https://doi.org/10.1109/JIOT.2018.2817841.

126 Han, J.H. et al. (2018). Machine learning-based self-powered acoustic sensor for speaker recognition. *Nano Energy* 53: 658–665. https://doi.org/10.1016/j.nanoen.2018.09.030.

127 Zhang, W., Wang, P., Sun, K. et al. (2019). Intelligently detecting and identifying liquids leakage combining triboelectric nanogenerator based self-powered sensor with machine learning. *Nano Energy* 56: 277–285. https://doi.org/10.1016/j.nanoen.2018.11.058.

128 Ordóñez, F.J. and Roggen, D. (2016). Deep convolutional and LSTM recurrent neural networks for multimodal wearable activity recognition. *Sensors (Switzerland)* 16 (1): https://doi.org/10.3390/s16010115.

129 Kiaghadi, A., Baima, M., Gummeson, J. et al. (2018). Fabric as a sensor: towards unobtrusive sensing of human behavior with triboelectric textiles. *SenSys 2018 - Proceedings of the 16th Conference on Embedded Networked Sensor Systems*, pp. 199–210. https://doi.org/10.1145/3274783.3274845.

130 Wen, J. et al. (2019). Robust sparse linear discriminant analysis. *IEEE Transactions on Circuits and Systems for Video Technology* 29 (2): 390–403. https://doi.org/10.1109/TCSVT.2018.2799214.

131 Becht, E. et al. (2019). Dimensionality reduction for visualizing single-cell data using UMAP. *Nature Biotechnology* 37 (1): 38–47. https://doi.org/10.1038/nbt.4314.

132 Ren, B., Pueyo, L., Ben Zhu, G. et al. (2018). Non-negative matrix factorization: robust extraction of extended structures. *The Astrophysical Journal* 852 (2): 104. https://doi.org/10.3847/1538-4357/aaa1f2.

133 Lyu, B., Wu, W., and Hu, Z. (2021). A novel bidirectional clustering algorithm based on local density. *Scientific Reports* 11 (1): 1–10. https://doi.org/10.1038/s41598-021-93244-2.

134 Ranjan, B. et al. (2021). DUBStepR is a scalable correlation-based feature selection method for accurately clustering single-cell data. *Nature Communications* 12 (1): 5849. https://doi.org/10.1038/s41467-021-26085-2.

135 Holloway, J. and Mengersen, K. (2018). Statistical machine learning methods and remote sensing for sustainable development goals: a review. *Remote Sensing* 10 (9): https://doi.org/10.3390/rs10091365.

136 Song, S., Li, S., Zhang, T. et al. (2021). Research on time series characteristics of the gas drainage evaluation index based on lasso regression. *Scientific Reports* 11 (1): 20593. https://doi.org/10.1038/s41598-021-00210-z.

137 Miki, T., Higuchi, D., Takebayashi, T., and Samukawa, M. (2021). Factors associating with disability of non-specific low back pain in different subgroups: a hierarchical linear regression analysis. *Scientific Reports* 11 (1): 1–7. https://doi.org/10.1038/s41598-021-97569-w.

138 Tao, P., Sun, Z., and Sun, Z. (2018). An improved intrusion detection algorithm based on GA and SVM. *IEEE Access* 6: 13624–13631. https://doi.org/10.1109/ACCESS.2018.2810198.

139 Ali, M., Jung, L.T., Abdel-Aty, A.H. et al. (2020). Semantic-k-NN algorithm: an enhanced version of traditional k-NN algorithm. *Expert Systems with Applications* 151: https://doi.org/10.1016/j.eswa.2020.113374.

140 Speiser, J.L., Miller, M.E., Tooze, J., and Ip, E. (2019). A comparison of random forest variable selection methods for classification prediction modeling. *Expert Systems with Applications* 134: 93–101. https://doi.org/10.1016/j.eswa.2019.05.028.

141 Leijnen, S. and van Veen, F. (2020). The neural network zoo. *Proceedings* 47 (1): 9. https://doi.org/10.3390/proceedings2020047009.

142 He, K., Zhang, X., Ren, S., and Sun, J. (2016). Deep residual learning for image recognition. *Proceedings of the IEEE Computer Society Conference on Computer Vision and Pattern Recognition*, Volume 2016-December, pp. 770–778. https://doi.org/10.1109/CVPR.2016.90.

143 Syu, M.H., Guan, Y.J., Lo, W.C., and Fuh, Y.K. (2020). Biomimetic and porous nanofiber-based hybrid sensor for multifunctional pressure sensing and human gesture identification via deep learning method. *Nano Energy* 76 (300): 105029. https://doi.org/10.1016/j.nanoen.2020.105029.

144 Ma, S., Fu, J., Chen, C.W., and Mei, T. (2018). DA-GAN: instance-level image translation by deep attention generative adversarial networks (with supplementary materials), February 2018. http://arxiv.org/abs/1802.06454.

145 Song, K. et al. (2019). Pneumatic actuator and flexible piezoelectric sensor for soft virtual reality glove system. *Scientific Reports* 9 (1): https://doi.org/10.1038/s41598-019-45422-6.

146 Kim, K.K. et al. (2020). A deep-learned skin sensor decoding the epicentral human motions. *Nature Communications* 11 (1): 1–8. https://doi.org/10.1038/s41467-020-16040-y.

147 Zhao, S. and Zhu, R. (2017). Electronic skin with multifunction sensors based on thermosensation. *Advanced Materials* 29 (15): https://doi.org/10.1002/adma.201606151.

148 Lee, W.W. et al. (2019). A neuro-inspired artificial peripheral nervous system for scalable electronic skins. *Science Robotics* 4 (32): eaax2198. https://doi.org/10.1126/scirobotics.aax2198.

149 Park, Y. et al. (2020). Wireless, skin-interfaced sensors for compression therapy. *Science Advances* 6 (49): 1–10. https://doi.org/10.1126/sciadv.abe1655.

150 You, I. et al. (2020). Artificial multimodal receptors based on ion relaxation dynamics. *Science* 370 (6519): 961–965. https://doi.org/10.1126/science.aba5132.

151 Son, D. et al. (2018). An integrated self-healable electronic skin system fabricated via dynamic reconstruction of a nanostructured conducting network. *Nature Nanotechnology* 13 (11): 1057–1065. https://doi.org/10.1038/s41565-018-0244-6.

152 Osborn, L.E. et al. (2018). Prosthesis with neuromorphic multilayered e-dermis perceives touch and pain. *Science Robotics* 3 (19): https://doi.org/10.1126/scirobotics.aat3818.

153 Boutry, C.M. et al. (2018). A hierarchically patterned, bioinspired e-skin able to detect the direction of applied pressure for robotics. *Science Robotics* 3 (24): eaau6914. https://doi.org/10.1126/scirobotics.aau6914.

154 Zou, Z., Zhu, C., Li, Y. et al. (2018). Rehealable, fully recyclable, and malleable electronic skin enabled by dynamic covalent thermoset nanocomposite. *Science Advances* 4 (2): 1–9. https://doi.org/10.1126/sciadv.aaq0508.

155 Lee, G. et al. (2021). Fingerpad-inspired multimodal electronic skin for material discrimination and texture recognition. *Advanced Science* 8 (9): 1–8. https://doi.org/10.1002/advs.202002606.

156 Chossat, J.B., Tao, Y., Duchaine, V., and Park, Y.L. (2015). Wearable soft artificial skin for hand motion detection with embedded microfluidic strain sensing. *Proceedings - IEEE International Conference on Robotics and Automation*, Volume 2015-June, pp. 2568–2573. https://doi.org/10.1109/ICRA.2015.7139544.

157 Hammond, F.L., Menguc, Y., and Wood, R.J. (2014). Toward a modular soft sensor-embedded glove for human hand motion and tactile pressure measurement. *IEEE International Conference on Intelligent Robots and Systems*, IROS, pp. 4000–4007. https://doi.org/10.1109/IROS.2014.6943125.

158 Khatib, M., Zohar, O., Saliba, W., and Haick, H. (2020). A multifunctional electronic skin empowered with damage mapping and autonomic acceleration of self-healing in designated locations. *Advanced Materials* 32 (17): 1–7. https://doi.org/10.1002/adma.202000246.

159 Kim, J. et al. (2014). Stretchable silicon nanoribbon electronics for skin prosthesis. *Nature Communications* 5: https://doi.org/10.1038/ncomms6747.

160 Liu, X., Sacks, J., Zhang, M. et al. (2017). The virtual trackpad: an electromyography-based, wireless, real-time, low-power, embedded hand-gesture-recognition system using an event-driven artificial neural network. *IEEE Transactions on Circuits and Systems II: Express Briefs* 64 (11): 1257–1261. https://doi.org/10.1109/TCSII.2016.2635674.

161 Kim, J.H., Thang, N.D., and Kim, T.S. (2009). 3-D hand motion tracking and gesture recognition using a data glove. *IEEE International Symposium on Industrial Electronics*, no. ISlE, pp. 1013–1018. https://doi.org/10.1109/ISIE.2009.5221998.

162 Zhang, S.L. et al. (2019). Energy harvesting-storage bracelet incorporating electrochemical microsupercapacitors self-charged from a single hand gesture. *Advanced Energy Materials* 9 (18): 1–7. https://doi.org/10.1002/aenm.201900152.

163 Zhou, Z. et al. (2020). Sign-to-speech translation using machine-learning-assisted stretchable sensor arrays. *Nature Electronics* 3 (9): 571–578. https://doi.org/10.1038/s41928-020-0428-6.

164 Camgoz, N.C., Koller, O., Hadfield, S., and Bowden, R. (2020). Multi-channel transformers for multi-articulatory sign language translation. *Lecture Notes in Computer Science (including subseries Lecture Notes in Artificial Intelligence and Lecture Notes in Bioinformatics)*, Vol. 12538 LNCS, pp. 301–319. https://doi.org/10.1007/978-3-030-66823-5_18.

165 Weber, P., Rueckert, E., Calandra, R. et al. (2016). A low-cost sensor glove with vibrotactile feedback and multiple finger joint and hand motion sensing for human–robot interaction. *25th IEEE International Symposium on Robot and Human Interactive Communication, RO-MAN 2016*, pp. 99–104. https://doi.org/10.1109/ROMAN.2016.7745096.

166 Fang, B., Sun, F., Liu, H., and Liu, C. (2018). 3D human gesture capturing and recognition by the IMMU-based data glove. *Neurocomputing* 277: 198–207. https://doi.org/10.1016/j.neucom.2017.02.101.

167 Suzuki, K. et al. (2016). Rapid-response, widely stretchable sensor of aligned MWCNT/elastomer composites for human motion detection. *ACS Sensors* 1 (6): 817–825. https://doi.org/10.1021/acssensors.6b00145.

168 Pan, J., Luo, Y., Li, Y. et al. (2020). A wireless multi-channel capacitive sensor system for efficient glove-based gesture recognition with AI at the edge. *IEEE Transactions on Circuits and Systems II: Express Briefs* 67 (9): 1624–1628. https://doi.org/10.1109/TCSII.2020.3010318.

169 Wang, M. et al. (2020). Gesture recognition using a bioinspired learning architecture that integrates visual data with somatosensory data from stretchable sensors. *Nature Electronics* 3 (9): 563–570. https://doi.org/10.1038/s41928-020-0422-z.

170 He, T. et al. (2019). Self-powered glove-based intuitive interface for diversified control applications in real/cyber space. *Nano Energy* 58: 641–651. https://doi.org/10.1016/j.nanoen.2019.01.091.

171 Mishra, R.K. et al. (2017). Wearable flexible and stretchable glove biosensor for on-site detection of organophosphorus chemical threats. *ACS Sensors* 2 (4): 553–561. https://doi.org/10.1021/acssensors.7b00051.

172 Sundaram, S., Kellnhofer, P., Li, Y. et al. (2019). Learning the signatures of the human grasp using a scalable tactile glove. *Nature* 569 (7758): 698–702. https://doi.org/10.1038/s41586-019-1234-z.

173 Pan, J. et al. (2021). Hybrid-flexible bimodal sensing wearable glove system for complex hand gesture recognition. *ACS Sensors* 6 (11): 4156–4166. https://doi.org/10.1021/acssensors.1c01698.

174 Myers, C.M., Furqan, A., and Zhu, J. (2019). The impact of user characteristics and preferences on performance with an unfamiliar voice user interface, May 2019. https://doi.org/10.1145/3290605.3300277.

175 Pyae, A. and Joelsson, T.N. (2018). Investigating the usability and user experiences of voice user interface: a case of Google home smart speaker. in *MobileHCI 2018 - Beyond Mobile: The Next 20 Years - 20th International Conference on Human–Computer Interaction with Mobile Devices and Services, Conference Proceedings Adjunct*, September 2018, pp. 127–131. https://doi.org/10.1145/3236112.3236130.

176 Gastaldo, P., Pinna, L., Seminara, L. et al. (2015). A tensor-based approach to touch modality classification by using machine learning. *Robotics and Autonomous Systems* 63 (P3): 268–278. https://doi.org/10.1016/j.robot.2014.09.022.

177 Alshurafa, N. et al. (2015). Recognition of nutrition intake using time-frequency decomposition in a wearable necklace using a piezoelectric sensor. *IEEE Sensors Journal* 15 (7): 3909–3916. https://doi.org/10.1109/JSEN.2015.2402652.

178 Zhao, G. et al. (2018). Piezoelectric polyacrylonitrile nanofiber film-based dual-function self-powered flexible sensor. *ACS Applied Materials & Interfaces* 10 (18): 15855–15863. https://doi.org/10.1021/acsami.8b02564.

179 Huang, Q. et al. (2020). Additive functionalization and embroidery for manufacturing wearable and washable textile supercapacitors. *Advanced Functional Materials* 30 (27): 1910541. https://doi.org/10.1002/adfm.201910541.

180 Zhao, Z. et al. (2020). Machine-washable and breathable pressure sensors based on triboelectric nanogenerators enabled by textile technologies. *Nano Energy* 70: https://doi.org/10.1016/j.nanoen.2020.104528.

181 Li, G., Liu, S., Wang, L., and Zhu, R. (2020). Skin-inspired quadruple tactile sensors integrated on a robot hand enable object recognition. *Science Robotics* 5 (49): eabc8134. https://doi.org/10.1126/scirobotics.abc8134.

182 Shih, B. et al. (2020). Electronic skins and machine learning for intelligent soft robots. *Science Robotics* 5 (41): eaaz9239. https://doi.org/10.1126/scirobotics.aaz9239.

183 Chen, S., Tan, M.W.M., Gong, X., and Lee, P.S. (2021). Low-voltage soft actuators for interactive human–machine interfaces. *Advanced Intelligent Systems* 2100075: 2100075. https://doi.org/10.1002/aisy.202100075.

184 Parida, K., Xiong, J., Zhou, X., and Lee, P.S. (2018, 2019). Progress on triboelectric nanogenerator with stretchability, self-healability and bio-compatibility. *Nano Energy* 59: 237–257. https://doi.org/10.1016/j.nanoen.2019.01.077.

185 Mitchell, S.K. et al. (2019). An easy-to-implement toolkit to create versatile and high-performance HASEL actuators for untethered soft robots. *Advanced Science* 6 (14): 1900178. https://doi.org/10.1002/advs.201900178.

186 Mishra, A.K. et al. (2020). Autonomic perspiration in 3D-printed hydrogel actuators. *Science robotics* 5 (38): eaaz3918. https://doi.org/10.1126/scirobotics.aaz3918.

187 Acome, E. et al. (2018). Hydraulically amplified self-healing electrostatic actuators with muscle-like performance. *Science* 359 (6371): 61–65. https://doi.org/10.1126/science.aao6139.

188 Cacucciolo, V., Shintake, J., Kuwajima, Y. et al. (2019). Stretchable pumps for soft machines. *Nature* 572 (7770): 516–519. https://doi.org/10.1038/s41586-019-1479-6.

189 Usevitch, N.S., Hammond, Z.M., Schwager, M. et al. (2020). An untethered isoperimetric soft robot. *Science Robotics* 5 (40): eaaz0492. https://doi.org/10.1126/scirobotics.aaz0492.

190 Terryn, S., Brancart, J., Lefeber, D. et al. (2017). Self-healing soft pneumatic robots. *Science Robotics* 2 (9): eaan4268. https://doi.org/10.1126/scirobotics.aan4268.

191 Lee, H. et al. (2020). 3D-printed programmable tensegrity for soft robotics. *Science Robotics* 5 (45): eaay9024. https://doi.org/10.1126/scirobotics.aay9024.

192 Zhao, H., O'Brien, K., Li, S., and Shepherd, R.F. (2016). Optoelectronically innervated soft prosthetic hand via stretchable optical waveguides. *Science Robotics* 1 (1): eaai7529. https://doi.org/10.1126/scirobotics.aai7529.

193 Bai, H., Li, S., Barreiros, J. et al. (2020). Stretchable distributed fiber-optic sensors. *Science* 370 (6518): 848–852. https://doi.org/10.1126/science.aba5504.

194 Zhu, M., He, T., and Lee, C. (2020). Technologies toward next generation human machine interfaces: from machine learning enhanced tactile sensing to neuromorphic sensory systems. *Applied Physics Reviews* 7 (3): 031305. https://doi.org/10.1063/5.0016485.

195 Zhang, R. et al. (2019). Sensing body motions based on charges generated on the body. *Nano Energy* 63: 103842. https://doi.org/10.1016/j.nanoen.2019.06.038.

196 Li, G., Deng, C., Wu, J. et al. (2019). Sensor data-driven bearing fault diagnosis based on deep convolutional neural networks and S-transform. *Sensors* 19 (12): 2750. https://doi.org/10.3390/s19122750.

197 Begg, R. and Kamruzzaman, J. (2005). A machine learning approach for automated recognition of movement patterns using basic, kinetic and kinematic gait data. *Journal of Biomechanics* 38 (3): 401–408. https://doi.org/10.1016/j.jbiomech.2004.05.002.

198 Choi, S.-Il., Moon, J., Park, H.C., and Choi, S.T. (2019). User identification from gait analysis using multi-modal sensors in smart insole. *Sensors (Switzerland)* 19 (17): 1–13. https://doi.org/10.3390/s19173785.

199 Ricciardi, C. et al. (2019). Using gait analysis' parameters to classify parkinsonism: a data mining approach. *Computer Methods and Programs in Biomedicine* 180: 105033. https://doi.org/10.1016/j.cmpb.2019.105033.

200 Hssayeni, M.D., Jimenez-Shahed, J., Burack, M.A., and Ghoraani, B. (2019). Wearable sensors for estimation of parkinsonian tremor severity during free body movements. *Sensors (Switzerland)* 19 (19): 1–17. https://doi.org/10.3390/s19194215.

201 Bamberg, S., Benbasat, A.Y., Scarborough, D.M. et al. (2008). Gait analysis using a shoe-integrated wireless sensor system. *IEEE Transactions on Information Technology in Biomedicine* 12 (4): 413–423. https://doi.org/10.1109/TITB.2007.899493.

202 Jiang, W. and Yin, Z. (2015). Human activity recognition using wearable sensors by deep convolutional neural networks. *MM 2015 - Proceedings of the 2015 ACM Multimedia Conference*, pp. 1307–1310. https://doi.org/10.1145/2733373.2806333.

203 Dehzangi, O., Taherisadr, M., and ChangalVala, R. (2017). IMU-based gait recognition using convolutional neural networks and multi-sensor fusion. *Sensors* 17 (12): 2735. https://doi.org/10.3390/s17122735.

204 Ishida, K. et al. (2013). Insole pedometer with piezoelectric energy harvester and 2 V organic circuits. *IEEE Journal of Solid-State Circuits* 48 (1): 255–264. https://doi.org/10.1109/JSSC.2012.2221253.

205 Lin, Z. et al. (2019). A triboelectric nanogenerator-based smart insole for multifunctional gait monitoring. *Advanced Materials Technologies* 4 (2): 1800360. https://doi.org/10.1002/admt.201800360.

206 Turner, A. and Hayes, S. (2019). The classification of minor gait alterations using wearable sensors and deep learning. *IEEE Transactions on Biomedical Engineering* 66 (11): 3136–3145. https://doi.org/10.1109/TBME.2019.2900863.

207 Zhang, Z. et al. (2020). Deep learning-enabled triboelectric smart socks for IoT-based gait analysis and VR applications. *npj Flexible Electronics* 4 (1): https://doi.org/10.1038/s41528-020-00092-7.

208 Pan, S., Wang, N., Qian, Y. et al. (2015). Indoor person identification through footstep induced structural vibration. *HotMobile 2015 - 16th International Workshop on Mobile Computing Systems and Applications*, pp. 81–86. https://doi.org/10.1145/2699343.2699364.

209 Alhamoud, A., Nair, A.A., Gottron, C. et al. (2014). Presence detection, identification and tracking in smart homes utilizing bluetooth enabled smartphones. *Proceedings - Conference on Local Computer Networks, LCN*, Volume 2014-November, pp. 784–789. https://doi.org/10.1109/LCNW.2014.6927735.

210 Zhang, J., Wei, B., Hu, W., and Kanhere, S.S. (2016). WiFi-ID: Human identification using WiFi signal. *Proceedings - 12th Annual International Conference on Distributed Computing in Sensor Systems, DCOSS 2016*, pp. 75–82. https://doi.org/10.1109/DCOSS.2016.30.

211 Xiong, J., Li, F., Zhao, N., and Jiang, N. (2014). Tracking and recognition of multiple human targets moving in a wireless pyroelectric infrared sensor network. *Sensors (Switzerland)* 14 (4): 7209–7228. https://doi.org/10.3390/s140407209.

212 Zeng, Y., Pathak, P.H., and Mohapatra, P. (2016). WiWho: WiFi-based person identification in smart spaces. *2016 15th ACM/IEEE International Conference on Information Processing in Sensor Networks, IPSN 2016 – Proceedings*. https://doi.org/10.1109/IPSN.2016.7460727.

213 Helfenbein, T., Király, R., Töröcsik, M. et al. (2017). Extension of RFID based indoor localization systems with smart tags. *Infocommunications Journal* 9 (3): 25–31.

214 Zou, Z., Baghaei-Nejad, M., Tenhunen, H., and Zheng, L.R. (2007). An efficient passive RFID system for ubiquitous identification and sensing using impulse UWB radio. *Elektrotechnik und Informationstechnik* 124 (11): 397–403. https://doi.org/10.1007/s00502-007-0483-y.

215 Sousa, M., Techmer, A., Steinhage, A. et al. (2013). Human tracking and identification using a sensitive floor and wearable accelerometers. *2013 IEEE International Conference on Pervasive Computing and Communications, PerCom 2013*, pp. 166–171. https://doi.org/10.1109/PerCom.2013.6526728.

216 Mokhtari, G., Zhang, Q., Hargrave, C., and Ralston, J.C. (2017). Non-wearable UWB sensor for human identification in smart home. *IEEE Sensors Journal* 17 (11): 3332–3340. https://doi.org/10.1109/JSEN.2017.2694555.

217 Ma, L. et al. (2020). A machine-fabricated 3D honeycomb-structured flame-retardant triboelectric fabric for fire escape and rescue. *Advanced Materials* 32 (38): 1–10. https://doi.org/10.1002/adma.202003897.

218 Jung, S. et al. (2020). 3D cu ball-based hybrid triboelectric nanogenerator with non-fullerene organic photovoltaic cells for self-powering indoor electronics. *Nano Energy* 77: 105271. https://doi.org/10.1016/j.nanoen.2020.105271.

219 Islam, E. et al. (2020). Electromagnetic-triboelectric-hybrid energy tile for biomechanical green energy harvesting. *Nano Energy* 77: 105250. https://doi.org/10.1016/j.nanoen.2020.105250.

220 Shi, Q. et al. (2020). Deep learning enabled smart mats as a scalable floor monitoring system. *Nature Communications* 11 (1): 4609. https://doi.org/10.1038/s41467-020-18471-z.

221 Shi, Q., Zhang, Z., Yang, Y. et al. (2021). Artificial intelligence of things (AIoT) enabled floor monitoring system for smart home applications. *ACS Nano* 15 (11): 18312–18326. https://doi.org/10.1021/acsnano.1c07579.

222 Li, Z., Zheng, Q., Wang, Z.L., and Li, Z. (2020). Nanogenerator-based self-powered sensors for wearable and implantable electronics. *Research* 2020: 1–25. https://doi.org/10.34133/2020/8710686.

223 Wen, F., He, T., Liu, H. et al. (2020). Advances in chemical sensing technology for enabling the next-generation self-sustainable integrated wearable system in the IoT era. *Nano Energy* 78: 105155. https://doi.org/10.1016/j.nanoen.2020.105155.

224 Chu, Y. et al. (2018). Human pulse diagnosis for medical assessments using a wearable piezoelectret sensing system. *Advanced Functional Materials* 28 (40): 1803413. https://doi.org/10.1002/adfm.201803413.

225 Zhou, H. et al. (2020). Stretchable piezoelectric energy harvesters and self-powered sensors for wearable and implantable devices. *Biosensors and Bioelectronics* 168: 112569. https://doi.org/10.1016/j.bios.2020.112569.

226 Zhang, Z. et al. (2021). Artificial intelligence of toilet (AI-toilet) for an integrated health monitoring system (IHMS) using smart triboelectric pressure sensors and image sensor. *Nano Energy* 90 (PA): 106517. https://doi.org/10.1016/j.nanoen.2021.106517.

227 Guo, X. et al. (2021). Artificial intelligence-enabled caregiving walking stick powered by ultra-low-frequency human motion. *ACS Nano* https://doi.org/10.1021/acsnano.1c04464.

Index

Flexible Piezoelectric Energy Harvesters and Sensors, First Edition. Bin Yang, Zhiran Yi, and Chengkuo Lee.